The NASA STI Program Office ... in Profile

Since its founding, NASA has been dedicated to the advancement of aeronautics and space science. The NASA Scientific and Technical Information (STI) Program Office plays a key part in helping NASA maintain this important role.

The NASA STI Program Office is operated by Langley Research Center, the lead center for NASA's scientific and technical information. The NASA STI Program Office provides access to the NASA STI Database, the largest collection of aeronautical and space science STI in the world. The Program Office is also NASA's institutional mechanism for disseminating the results of its research and development activities. These results are published by NASA in the NASA STI Report Series, which includes the following report types:

- TECHNICAL PUBLICATION. Reports of completed research or a major significant phase of research that present the results of NASA programs and include extensive data or theoretical analysis. Includes compilations of significant scientific and technical data and information deemed to be of continuing reference value. NASA counterpart of peer-reviewed formal professional papers, but having less stringent limitations on manuscript length and extent of graphic presentations.

- TECHNICAL MEMORANDUM. Scientific and technical findings that are preliminary or of specialized interest, e.g., quick release reports, working papers, and bibliographies that contain minimal annotation. Does not contain extensive analysis.

- CONTRACTOR REPORT. Scientific and technical findings by NASA-sponsored contractors and grantees.

- CONFERENCE PUBLICATION. Collected papers from scientific and technical conferences, symposia, seminars, or other meetings sponsored or co-sponsored by NASA.

- SPECIAL PUBLICATION. Scientific, technical, or historical information from NASA programs, projects, and missions, often concerned with subjects having substantial public interest.

- TECHNICAL TRANSLATION. English-language translations of foreign scientific and technical material pertinent to NASA's mission.

Specialized services that complement the STI Program Office's diverse offerings include creating custom thesauri, building customized databases, organizing and publishing research results ... even providing videos.

For more information about the NASA STI Program Office, see the following:

- Access the NASA STI Program Home Page at *http://www.sti.nasa.gov*

- E-mail your question via the Internet to help@sti.nasa.gov

- Fax your question to the NASA STI Help Desk at (301) 621-0134

- Phone the NASA STI Help Desk at (301) 621-0390

- Write to:
  NASA STI Help Desk
  NASA Center for AeroSpace Information
  7121 Standard Drive
  Hanover, MD 21076-1320

NASA/TM-2005-213758

Comet/Asteroid Protection System (CAPS): Preliminary Space-Based System Concept and Study Results

Daniel D. Mazanek, Carlos M. Roithmayr, and Jeffrey Antol
Langley Research Center, Hampton, Virginia

Sang-Young Park, Robert H. Koons, and James C. Bremer
Swales Aerospace, Inc., Hampton, Virginia

Douglas G. Murphy, James A. Hoffman, Renjith R. Kumar, and Hans Seywald
Analytical Mechanics Associates, Inc., Hampton, Virginia

Linda Kay-Bunnell and Martin R. Werner
Joint Institute for Advancement of Flight Sciences (JIAFS)
The George Washington University, Hampton, Virginia

Matthew A. Hausman
Colorado Center for Astrodynamics Research
The University of Colorado, Boulder, Colorado

Jana L. Stockum
San Diego State University, San Diego, California

National Aeronautics and
Space Administration

Langley Research Center
Hampton, Virginia 23681-2199

May 2005

The use of trademarks or names of manufacturers in the report is for accurate reporting and does not constitute an official endorsement, either expressed or implied, of such products or manufacturers by the National Aeronautics and Space Administration.

Available from:

NASA Center for AeroSpace Information (CASI)
7121 Standard Drive
Hanover, MD 21076-1320
(301) 621-0390

National Technical Information Service (NTIS)
5285 Port Royal Road
Springfield, VA 22161-2171
(703) 605-6000

Contents

Acknowledgments v

Foreword vii

Comet/Asteroid Protection System: Concept Study Executive Summary
Daniel D. Mazanek 1

Near-Earth Object (NEO) Hazard Background
Daniel D. Mazanek 11

Accurate Determination of Comet and Asteroid Orbits Leading to Collision With Earth
*Carlos M. Roithmayr, Linda Kay-Bunnell, Daniel D. Mazanek, Renjith R. Kumar,
Hans Seywald, and Matthew A. Hausman* 26

Detection Element Concepts—Initial Design
Robert H. Koons and James C. Bremer 79

Near-Earth Object Astrometric Interferometry
Martin R. Werner 102

Mission Functionality for Deflecting Earth-Crossing Asteroids/Comets
Sang-Young Park and Daniel D. Mazanek 121

Orbit Modification of Earth-Crossing Asteroids/Comets Using Rendezvous Spacecraft and
Laser Ablation
Sang-Young Park and Daniel D. Mazanek 148

CAPS Simulation Environment Development
Douglas G. Murphy and James A. Hoffman 163

Comparison of Detector Technologies for CAPS
Jana L. Stockum 193

Survey of Enabling Technologies for CAPS
Jeffrey Antol, Daniel D. Mazanek, and Robert H. Koons 202

Chapter Notes: Nomenclature 217

Acknowledgments

The Comet/Asteroid Protection System (CAPS) conceptual study described herein was conducted as part of the NASA Revolutionary Aerospace System Concepts Program managed at Langley Research Center. The CAPS study was performed largely between January 2001 and January 2003. The study benefited from the contributions and advice of many individuals who were participants or consulted on various topics during this period. The contributing authors would like to recognize the support and involvement of the following individuals:

Langley Research Center
- Brian J. Boland
- George G. Ganoe
- Dr. Stephen J. Katzberg
- Victor F. Lucas
- Patrick A. Troutman

Marshall Spaceflight Center
- Dr. Jonathan W. Campbell

Analytical Mechanics Associates, Inc.
- Dr. Scott R. Angster
- W. Joel Derrick
- George R. Kline
- Shawn A. Krizan
- Frank M. McQuarry
- Jeffery G. Murch
- Joshua E. Sams

Swales Aerospace, Inc.
- Christopher V. Strickland

Joint Institute for Advancement of Flight Sciences (JIAFS), The George Washington University
- Dr. Robert H. Tolson

Colorado Center for Astrodynamics Research, The University of Colorado
- Dr. George H. Born
- Dr. R. Steven Nerem

Study Leader and Compiler: Daniel D. Mazanek, Langley Research Center

Editing and illustration preparation: Denise M. Stefula, Sr. Editor, CLASIC Contract

Foreword

In recent years, impacts from Earth-crossing asteroids and comets have been identified as a threat that is real and credible. Although infrequent in relation to a human lifetime, devastating impacts have occurred countless times during Earth's history and will occur again in the future. In 1990, the United States House of Representatives directed the National Aeronautics and Space Administration (NASA) to study the impact problem. NASA organized an international conference and conducted two workshop studies. The first workshop, comprising three formal meetings in 1991, defined a program for dramatically increasing the detection rate of large Earth-orbit-crossing asteroids, and established the requirements for determining the orbits of such bodies. The second workshop, in January of 1992, focused on defining systems and technologies to alter the orbits of Earth threatening asteroids, or obliterate them. The major result of these efforts was the establishment of a coordinated ground-based effort, known as the Spaceguard Survey, whose primary goal is to search for large asteroids capable of global devastation. This logical first step has significantly reduced the chances that Earth will be impacted by a large near-Earth asteroid without significant warning time. Just over 2 years later, from July 16 to 22, 1994, the inhabitants of the Earth witnessed the colossal impacts on Jupiter caused by more than 20 fragments of the Shoemaker-Levy 9 comet. Each impact zone was approximately the size of the entire Earth and provided a timely and vivid illustration of the impact hazard.

Although the current efforts make sense under the constraint of present funding levels, concerns have been raised that they are not sufficient. For example, in the American Institute of Aeronautics and Astronautics "6th International Space Cooperation Workshop Report" (March 2001), the Working Group on Earth-Threatening Asteroids and Comets concluded that current efforts are "quite inadequate for smaller near-Earth asteroids, such as a few hundred meters in diameter, and other short-period bodies. Moreover, the present capability for detecting long-period comets is almost nonexistent." Additionally, the working group emphasized the need for a long-term planetary defense system that "should deflect rather than fragment an incoming near-Earth object, and this should be done by nonnuclear means if possible." The impactor size capable of causing significant disruption to terrestrial activities is not well understood. An impact by an object significantly smaller than the 1- to 2-km global threat threshold could have devastating physical and economic consequences that could affect civilization on a global scale. Protecting against objects that are an immediate threat, particularly long-period impactors, is an extremely difficult problem, technically and financially, and the impact probability is presently believed to be significantly lower than that of asteroids of a similar size. However, the consequences from a long-period impactor could be more catastrophic because these objects, despite likely having lower densities, can impact at velocities several times greater than an asteroid. Due to their extremely long orbital periods, it is likely that a collision with this type of object will not be known decades, or even years, in advance with our current detection efforts. Simply taking the attitude of passive acceptance of this aspect of the impact hazard, or waiting for technology to "solve" the problem, is not consistent with the goal of protecting Earth and its inhabitants from a globally devastating impact event.

Comet/Asteroid Protection System: Concept Study Executive Summary[1]

DANIEL D. MAZANEK
NASA Langley Research Center

RASC Charter and CAPS Vision

The charter of the Revolutionary Aerospace Systems Concepts (RASC) activity is to develop and assess revolutionary aerospace system concepts and architectures and to identify the requirements for critical technologies to enable the realization of these concepts 25 or more years from now. These concepts and architectures serve to identify technology investment areas that the National Aeronautics and Space Administration (NASA) needs to support presently in order to elevate the progress of these technologies to be of benefit in the future.

When the RASC program was soliciting study proposals, it was suggested to take the initial steps in conceptualizing a future protection system that could provide a revolutionary advancement in the detection and precision orbit determination of impacting comets and asteroids coupled with a pre-established, robust method for altering their orbits in a controlled and rapid manner. This document provides a summary of the preliminary concept for the space-based Comet/Asteroid Protection System (CAPS), which was one of the studies performed under RASC in 2001 and 2002. It should be emphasized that at this time CAPS is only a conceptual study, and no direct plans are currently being made to implement CAPS. The vision for this future system architecture is primarily to provide planetary defense. However, the system should also provide productive science, resource utilization, and technology development when it is not needed for the extremely infrequent diversion of impacting comets and asteroids. A future for mankind is envisioned wherein asteroids and comets are routinely moved to processing facilities with a permanent infrastructure that is capable of and prepared to divert those objects posing a hazard.

Background

An enormous number of asteroids and comets orbit the Sun, ranging in size from pebbles to mountains. Earth approaching asteroids and comets are collectively termed near-Earth objects (NEOs). Fortunately, only a tiny number of these objects cross the Earth's orbit, and our atmosphere protects us from small and structurally weak objects; however, smaller NEOs vastly outnumber the larger, thus resulting in a higher impact frequency. The goal of current search efforts is to catalog and characterize by 2008 the orbits of 90 percent of near-Earth asteroids (NEAs) larger than 1 km in diameter, currently estimated to number between 900 and 1300. Devastating impacts can also occur from smaller NEAs, short-period comets (SPCs) in asteroid-like orbits, and long-period comets (LPCs) that do not regularly enter near-Earth space because their orbital periods range from 200 years to millions of years.

Impacts are extremely infrequent events relative to a human lifetime, but have the potential for massive loss of life and property. Impacts have occurred in the past and will occur in the future. The energy released from the impact of even a small NEO, like the one that exploded in the sky over the Tunguska region of Siberia in 1908, is equivalent to the explosive energy of a thermonuclear device (without the radioactivity). The energy released from an impactor capable of causing surface damage ranges from ≈10 megatons (Mt) of TNT to billions of megatons (1 Mt = 4.184×10^{15} joules). A 10-Mt impact can result from an object approximately 50 m in diameter and is roughly equal to 700 Hiroshima-size explosions. This class of impact is estimated to occur every several hundred years (or possibly less) and

---

[1]Chapter nomenclature available in chapter notes, p. 217.

can cause regional destruction. An impact with a 1-km diameter object, capable of releasing more than 100 000 Mt and resulting in a global catastrophe, can be expected to occur every several hundred thousand years to a million years. An impact from a 10-km object, like the one believed to have caused the great dinosaur extinction 65 million years ago, can be expected on an interval of 10 million years or greater. Because the last significant NEO impact with the Earth occurred nearly a century ago, the ability for mankind to internalize the reality of this threat is understandably difficult.

Despite enormous destructive potential, there are many positive aspects of asteroids and comets. They represent a significant resource for commercial exploitation, space exploration, and scientific research. They may even provide clues regarding the origin of life on Earth. It is worthwhile to identify and understand these planetary neighbors both for what they can provide us and for what they are capable of taking away.

Motivation for CAPS

While terrestrial-based telescopes can address many aspects of NEO detection, the ability to discover and track faint and/or small comets and asteroids is tremendously enhanced, if not enabled, from space. It is recognized, and appreciated, that the currently funded terrestrial-based detection efforts are a vital and logical first step. Focusing on the detection of large asteroids capable of global destruction is the best expenditure of limited resources; however, various aspects of the impact threat are largely unaddressed by these efforts. Currently there is no specific search for LPCs, small NEAs, or small SPCs. Additionally, coordinated follow-up observations are critical to limit the likelihood of losing a newly discovered NEO and to precisely determine the object's orbit. One shortcoming of current ground-based efforts is the difficulty in providing these follow-up measurements, which are provided in part by amateur astronomers. Looking for much smaller and fainter targets is likely to exceed the current capabilities of many asteroid and comet "hunters."

Just as the Hubble Space Telescope has expanded our ability to see the universe without the limitations imposed by Earth's atmosphere, a space-based NEO detection system would allow us to expand the range of observable comets and asteroids and to provide coordinated follow-up observations. A space-based detection system is capable of making observations on a continuous basis without the various constraints (daylight, weather, etc.) imposed on Earth-based systems, and NEO searches need not be focused on the solar opposition point. If detection systems can be designed to observe faint NEOs that appear to be near the Sun, which is impossible from the ground because the atmosphere scatters sunlight during the daytime, it would be possible to see objects close to the Sun and on the solar far side where solar illumination conditions are favorable. Additionally, it is critical to ascertain, to the greatest extent possible, the composition and physical characteristics of these objects. A space-based approach can solve this aspect of the problem through both remote observations and rendezvous missions with the NEO.

It is likely that the next object to impact Earth will be a small near-Earth asteroid or comet. The most significant danger from smaller NEOs (several hundred meters in diameter) could come from ocean impacts, which may generate tsunamis capable of massive destruction on distant shorelines. Even if this type of impact is not preventable, a limited amount of warning time could permit the evacuation of coastal populations, which could save millions of lives, prevent untold suffering, and mitigate loss of property. Finally, it is likely that a globally devastating impact with a 1-km class LPC will not be known decades or even years in advance with our current detection efforts. Searching for and protecting ourselves against these types of impactors is a worthwhile endeavor. A space-based detection system, despite being more costly and complex than Earth-based initiatives, is the most promising way of expanding the range of detectable objects and surveying the entire celestial sky on a regular basis. Current ground-based efforts should certainly be expanded in the near term, and a coordinated space-based system should be defined and implemented in the future. CAPS is an attempt to begin the definition of that future space-based system and identify the technology development areas that are needed to enable its implementation.

Finally, any attempt to deflect an impacting NEO with reasonable lead time is likely to be accomplished only by using a space-based deflection system. Many deflection approaches are possible given sufficient warning time, particularly decades of advance notice. Immediate threats are extremely difficult to defend against, and likely require highly capable spacecraft that can quickly engage the target and provide large, rapid changes in the object's orbital velocity to avert a collision. These requirements, plus the desire to modify the orbit in a controlled manner, considerably restrict the potential methods that could be used to alter an impactor's trajectory. It is essential to understand that the issues associated with detection and deflection of an impactor are intimately connected, particularly if we are not afforded decades of warning time that would be necessary for large NEAs. The requirements for the detection system could be significantly reduced given an extremely robust deflection capability. However, due to the enormous amounts of energy required to move these massive bodies, any additional warning time is an extremely valuable asset.

Concept Overview

CAPS is a future (25 or more years from now) space-based system concept designed to detect and protect against the entire range of threatening comets and asteroids. The initial focus is to determine the feasibility of protecting against 1-km class long-period comets, including inactive nuclei. The system is designed also to protect against smaller LPCs, as well as NEAs and SPCs capable of regional destruction. Although the primary motivation for CAPS is to provide protection against impacting comets and asteroids, it is anticipated that the system and technologies developed would have many additional benefits extending to governments (U.S. and international), the commercial sector, the scientific community, and academia. The CAPS detection system would provide an astronomical asset that could observe extremely faint or small targets (both planetary bodies and extrasolar objects), allowing an unprecedented level of scientific observations while surveying the entire celestial sky on a regular basis. The CAPS orbit modification system could enable exploitation of the vast economic resources available from NEOs and promote synergistic technologies for other future space missions. Technologies that will permit the future exploration and colonization of the solar system (e.g., high power and thermal management systems, high thrust and specific impulse propulsion, and power beaming) are also applicable to the deflection of Earth impacting comets and asteroids. Additionally, there is tremendous benefit in "practicing" how to move these objects from a threat mitigation standpoint; developing the capability to alter the orbits of comets and asteroids routinely for nondefensive purposes could greatly increase the probability that we can successfully divert a future impactor.

Combining the words "comet" and "asteroid" in the CAPS acronym intends to convey the idea of utilizing a combined approach for protection against both types of cosmic projectiles. Conventional ground-based telescopes may provide little or no advanced warning of collisions with small NEAs or LPCs, and developing and maintaining separate space-based systems may be impractical. Precision orbit determination of most NEAs and SPCs can be expected to be obtained several orbital periods prior to a collision, provided that we actually have the ability to observe them. This would not be the case for impacting LPCs, whose orbits need to be characterized very accurately over a small observation arc on their first observed perihelial passages through the solar system. If the situation occurs wherein a small NEA is first detected on its final approach, preventing an impact may prove problematic. In this case, CAPS would at least provide an accurate assessment of where the object would impact and enough warning time to allow some appropriate civil defense effort to be carried out successfully.

The timely detection of LPCs, even those of significant size, presents many intractable problems. LPCs can be extremely faint (albedos of ≈ 0.02) until the sublimation of their volatile frozen gases begins.

Moreover, comets can remain in a dormant state during perihelial passage, or they can exhaust their volatiles and become extinct nuclei. Observing LPCs at significant distances from the Sun is a formidable task. The ability to predict their orbits accurately, and hence determine whether or not they represent a threat, is dependent upon the number, resolution, and spacing of observations of these objects. Finally, the comet's trajectory can be significantly altered by nongravitational forces if it becomes active, affecting our ability to predict its path and properly alter it. The ability to observe faint LPCs and rapidly determine their orbits is consistent with protection against small, previously undiscovered NEAs. A system capable of protecting against LPCs, placed properly in heliocentric space, should also be capable of protecting against small NEAs and small SPCs.

The baseline detection concept advocates the use of high-resolution telescopes with advanced detector arrays, coordinated telescope control for NEO surveying and tracking, rapid spectral imaging for NEO identification and characterization, and interferometric techniques to obtain precision orbit determination when required. Detection telescopes would be orbiting and/or lunar surface-based, providing surveys of nearly the entire celestial sky approximately every 30 days. Orbiting telescopes could be placed in heliocentric orbits, including Earth-Sun libration points, or around a planetary body or moon. The CAPS detection system would provide a high probability that impacting NEOs are detected, and their orbits accurately characterized with significant warning time, even upon a first observed near-Earth approach.

The primary orbit modification approach uses a spacecraft that combines a multimegawatt or gigawatt-class electrical power system, a high thrust and specific impulse propulsion system for rapid rendezvous, and a pulsed laser ablation payload for changing the target's orbit. This combination of technologies may offer a future orbit modification system that could deflect impactors of various compositions and provide an effective method for altering the orbits of NEOs for resource utilization. If laser power levels required for a single spacecraft are prohibitive, multiple spacecraft with more modest laser payloads could be deployed to the target.

Detection

It is worthwhile to describe what is meant by "detection" in relation to the CAPS detection system. Detection includes initial NEO discovery, follow-up observations, precision orbit determination, and some level of physical characterization. Although all aspects of the detection problem are critical, for objects with a very limited observational period the accurate assessment of their trajectories is vital.

The initial benchmark for the CAPS detection system is to be able to identify an impactor with a diameter of 1 km or greater at a distance of at least 5 astronomical units (au) from Earth, and to identify objects as small as 50 m in diameter at a distance of 0.2 au from Earth. In general, these distance limits would provide warning times of approximately 1 year for a 1-km LPC and a few weeks to approximately a month for a small NEA that has not been previously cataloged. A system possessing the sensitivity to observe 1-km objects at 5 au would be capable of detecting many 50-m class asteroids significantly farther away than 0.2 au, so the warning times for uncataloged NEAs could be significantly longer. Ultimately, the ability to identify an LPC on an impact course at a distance of 5 au from Earth may not provide sufficient warning time, and observing further out may be valuable. Conversely, due to the unpredictable nature of comets, both with respect to their orbits and structural integrities, it may not be prudent to take any defensive action until the object is much closer. The threat of impact may change significantly if the comet becomes active, or if it fragments into a number of sizable objects. The extremely short warning times for LPCs, the large changes in orbital velocity required for averting an impact, and the orbital and compositional uncertainties make this aspect of the impact hazard particularly difficult to solve.

Figure 1. Depiction of detection system using a lunar-based approach.

The envisioned CAPS detection system would feature large aperture (3- to 6-m diameter), high-resolution telescopes capable of imaging in the ultraviolet, optical, and infrared wavelengths. Coordinated telescope control for NEO surveying and tracking would be incorporated to maximize follow-up observations, and baffling and/or shading would be employed to permit observations close to the Sun. Figure 1 depicts a lunar-based option with a detection node consisting of a wide field-of-view (FOV) survey telescope located in the center and three narrow FOV tracking telescopes (telescope enclosures and/or baffling are not shown). Two detection nodes, located in the northern and southern lunar hemispheres, could provide nearly complete sky coverage every month. Each telescope would have large area mosaic detector arrays (approximately 36000×36000 pixels), with the survey telescopes having a 1.0×1.0-degree FOV and the tracking telescopes having a 0.1×0.1-degree FOV. Spectral imaging would be implemented as early as possible in the detection process. Advanced detectors capable of rapid identification of NEOs and their spectral signals could greatly simplify operations and minimize requirements on the tracking telescopes. If NEOs could be uniquely identified in multiple survey images, a preliminary orbit could be determined with minimal risk of "losing" the object. The tracking telescopes would be used in an interferometric mode when higher precision astrometric observations are needed to confirm an object is on an impacting trajectory. Finally, active laser ranging could be used to provide range and range-rate data to augment precision orbit determination. Active laser ranging is preferable to radar systems due to the potentially large distances between the target and the detection system. The tracking telescopes could be used as receivers for the laser ranging system, or the return signal of faint NEOs could be enhanced through active illumination to aid in interferometry measurements.

Deflection/Orbit Modification

Altering the trajectory of a confirmed impactor as early as possible minimizes the required change in velocity (ΔV). This is particularly true for comets and asteroids that are an immediate threat because the required ΔV can increase by several orders of magnitude during the final months before impact. Besides rapid, controlled trajectory modification, one goal of CAPS orbit modification is to be effective against

NEOs of various compositions. Asteroids range from primarily stony to mostly metallic, with various proportions of each type of material, and may contain deep, powdery regolith that can affect deflection efforts, particularly landing and attaching to the object. Comets contain a mixture of nonvolatile materials and large amounts of frozen volatiles. When a comet becomes active, these volatiles create a diffuse cloud surrounding the nucleus called the coma. This variety of compositions and environments makes the issue of deflection difficult and suggests orbit modification methods that can move the NEO without landing on it may be highly advantageous. Because time may be critical, this approach could also diminish the need for detailed physical characteristic observations to be made before dispatching a deflection effort.

Many methods for altering the trajectory of a comet or asteroid have been proposed (ref. 1), but the most feasible approaches require a spacecraft to intercept or rendezvous with the target. Given a spacecraft with an advanced propulsion system (such as plasma or nuclear) capable of rapid rendezvous with the target, one deflection approach would be to physically attach to the object and thrust in the proper direction to change the object's orbit. There are many technical issues associated with this approach, but the one that is fundamentally limiting is the propellant required. Providing large quantities of propellant to permit a rapid rendezvous or intercept is difficult, but delivering enough propellant to alter the orbit of a massive asteroid or comet nucleus is impractical for an immediate threat. This situation is also possible for large NEAs found many years before impact; although the required ΔV is small, they can be extremely massive. One approach that can circumvent this problem and alter the trajectory of the object in a highly controlled manner is to use pulsed laser ablative propulsion. A sufficiently intense laser pulse ablates the surface of the NEO by causing plasma blow off. The spacecraft would station-keep with the object at a "small" standoff distance while the laser ablation is performed. The momentum change from a single laser pulse is very small; however, the cumulative result is very effective because the laser can interact with the object over significant periods of time. The laser ablation technique can overcome the mass penalties associated with other nondisruptive approaches because no propellant is required to generate the ΔV (the material of the celestial object is the propellant source). Additionally, laser ablation is effective against a wide range of surface materials and does not require any landing or physical attachment to the object.

For diverting asteroids and comets at significant heliocentric distances, the power and optical requirements of a laser ablation system located on or near Earth may be too extreme to contemplate in the next few decades. The CAPS hybrid solution, depicted in figure 2, can minimize these requirements by utilizing the spacecraft to deliver the laser as a payload to a particular celestial body and by making dual use of the power system. The laser ablation system would require an extremely powerful electrical generator, which is likely needed for the propulsion system to provide the rapid rendezvous phase of the mission. Even this approach requires the development of highly capable space-based power, propulsion, and laser systems. An alternative approach would be to send multiple spacecraft with laser payloads to the target, where they would work cooperatively to provide the required change in orbital velocity. This approach would be highly desirable from redundancy and mission risk standpoints.

Ultimately, a spacecraft capable of rapid interception of an incoming impactor is extremely beneficial, and several approaches for modifying the NEO's orbit could be incorporated into the deflection system. One of the most commonly cited methods for deflecting or pulverizing a threatening NEO on its final approach is the use of a nuclear detonation (ref. 1); however, there are many issues associated with this technique (such as fragmentation or radiation), and it is unlikely that the CAPS goal of controlled orbit modification can be achieved in such a fashion. Moreover, there is a great deal of uncertainty as to how effective a nuclear explosion would be against a porous or nonmonolithic object that is effectively a

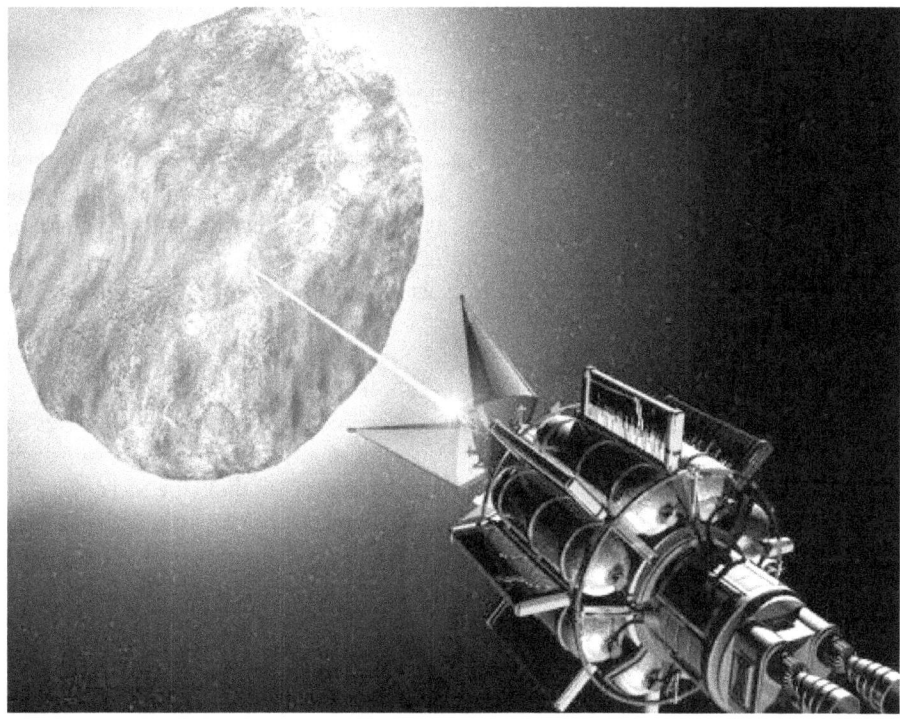

Figure 2. Depiction of rendezvous spacecraft with laser ablation payload.

gravitationally bound "rubble pile" (ref. 2). Compatible secondary payloads could be carried as an additional level of redundancy for a deflection mission, or a phased approach using rendezvous and intercept trajectories with various payloads could provide a robust defense. Regardless of the deflection method used, rapid engagement of an object is critical for preventing an impact from a newly discovered LPC or asteroid.

Study Focus Areas

The focus of the CAPS study has been on understanding the various aspects of this extremely complex problem and identifying a spaced-based detection system concept that maximizes the range of detectable objects and provides a high probability that the objects will be detected with significant warning time, *even upon a first observed near-Earth approach.*

The CAPS study efforts were focused on several key areas: precision orbit determination, preliminary detection element design to allow astrometric interferometry, simulation environment development, and orbit modification mission functionality using a rapid rendezvous spacecraft with a laser ablation payload. Resource limitations prevented many aspects of the system from being addressed during this conceptual study. Although extremely important for implementation, the following are example areas that were not addressed in-depth during the CAPS study: launch, cost, assembly and deployment, and system reliability, maintainability, and availability (RMA).

Technologies Identified

Finally, the CAPS study was tasked with identifying revolutionary technologies and techniques that permit the detection and orbit modification of potentially hazardous near-Earth objects, as well as nonimpacting asteroids and comets that could be used as a resource. The definition of "revolutionary" can vary significantly from individual to individual; it may be seen by one as radical new technologies or by another as a novel combination of existing technologies. It is hoped that the study will identify technologies that result in a fundamental change in approaching the impact hazard. A more realistic outcome, however, is that a combination of options and techniques requiring varying levels of advancement will be combined into a viable system solution.

There are many possible technologies that can be applied to both the detection and deflection of NEOs. By understanding the requirements for CAPS, the applicability of new technologies can be identified and their performances can be evaluated. It is also important to point out that one strongly desired outcome for the RASC program is to identify synergistic technologies that can be applied across a wide range of future space missions. For example, future technologies that permit human missions to traverse the solar system rapidly could be highly compatible with the rapid interception of an impactor. Likewise, laser power beaming (such as visible, ultraviolet, or microwave) may be applicable for space-based energy transfer for remote power applications, as well as NEO orbit modification. The following paragraph provides some of the key technologies that have been identified as having possible application for CAPS.

The enabling technologies required for the development of a viable protection system can be divided into the two areas of detection and deflection/orbit modification. Many of the detection technologies are currently in development for advanced in-space telescope systems such as the James Webb Telescope (formerly known as the Next Generation Space Telescope). Orbit modification technologies are also currently being studied as part of advanced power and propulsion research. With the proper funding levels, many of the technologies needed to support a CAPS architecture could be achievable within the next 15 to 20 years. The high power, propulsion, heat rejection, and directed energy systems would likely be farther term than some of the detection technologies.

The following are key technologies for CAPS detection capability:

- Large aperture, high-resolution advanced telescopes (ultraviolet, optical, and infrared) will be required for detection as well as tracking tasks.

- Advanced lightweight mirrors could be used to reduce the launch weight of CAPS detection assets and thus reduce the overall system cost. Examples of these technologies include low-mass membrane mirror optics and liquid surface mirrors. It is likely that active control will also be required to maintain precise mirror shape.

- Large area mosaic charge-injected device (CID) sensor arrays (approximately $36\,000 \times 36\,000$ pixels) are needed for rapid surveying and tracking (precise determination of a target object's angular position). CIDs exhibit less light bloom from pixel to pixel when subjected to high intensity light as compared with charge-coupled device (CCD) arrays, and they are also less sensitive to radiation.

- Advanced detectors capable of rapid NEO identification would be extremely valuable. The S-Cam, currently under development by the European Space Agency (ESA), uses superconducting tunneling junctions (STJs), which can count individual photons and provide associated spectral information. These data could be used to detect and "tag" asteroids and comets for simplified follow-up observations, cataloging, and future identification.

- Active cooling is required to achieve optimal performance from the sensors, whether they are CCDs, CIDs, or STJs, with temperature requirements being only a fraction of a degree Kelvin in the case of STJs.

- Baffling and/or shading technologies would permit observations close to the Sun (possibly within 15° of the Sun line), increasing the area of sky that can be sampled. The shading could be in the form of an attached sunshade, a large deployable shade flying in formation with the telescope, or an internal occulting disk such as that used in particular coronagraphs.

- Optical interferometric techniques and/or laser ranging systems would allow for precision orbit determination.

- Precision spacecraft and detector pointing will be needed to provide star field accuracy for guide stars to better than 0.001 arcsec. Accurate position and time knowledge is also needed so that the CAPS detection system can precisely acquire targets. If interferometric techniques are employed, the linear distance between two telescopes must be known to within approximately 1 nmi, and precise position determination and control with nano-/picometer knowledge must be available.

- Advanced data management systems and rapid communications will be needed for processing observation data and cataloging NEOs. Significant image data will be generated by multiple large CCD/CID arrays from multiple telescopes potentially at remote locations. These data will have to be processed and downlinked, the resulting image data stored, and an object database created. Ultra-high data rates for downlink may be achievable using optical communications technology. Potential high bandwidth intersatellite communications may also be needed for interferometry or database synchronization.

The following are key technologies for CAPS deflection/orbit modification capability:

- High thrust, high specific impulse propulsion systems (such as plasma or nuclear) would allow delivery of orbit modification systems to target NEOs.

- Multimegawatt to gigawatt-class electrical power systems are required for propulsion and laser applications.

- Advanced thermal management systems are critical to reject large amounts of waste heat.

- Reliable, high-power pulsed laser ablation systems with adaptive laser optics, precision beam-width focusing, and closed-loop control system would provide continuous orbit modification capability. Systems could also potentially be used as an active ranging system for precision orbit determination.

- Advanced autonomous or semi-autonomous rendezvous and station-keeping capability provide for engaging the NEO at close distances. Formation flying capability and precise attitude control may also be needed for interferometry using orbiting detectors.

Summary

Many of the major issues have been identified for a futuristic capability to protect against impacting comets and asteroids, and a preliminary space-based concept has been envisioned. Some of the basic concept elements, approaches, methodologies, and features have been identified. When contemplating the ability to monitor comets and asteroids continuously, there are many trade-offs between orbiting observatories and detection systems on planetary bodies without an atmosphere. Future orbit modification techniques have the potential for rapid and controlled alteration of NEO orbits, provided that high-power, compatible thermal management systems are developed. Much additional work and analysis are required to identify a final system concept, and many trade studies need to be performed to select the best mix of system capability, reliability, maintainability, and cost. Finally, it is fully appreciated that at the present time space systems are much more costly than terrestrial-based systems. Hopefully, this will change in the future. Regardless, understanding what it would take to defend against a much wider range of the impact threat will foster ideas, innovations, and technologies that could one day enable the development of such a system. This understanding is vital to provide ways of reducing the costs and quantifying the benefits that are achievable with a system like CAPS.

References

1. Canavan, G. H.; Solem, J. C.; and Rather, J. D. G., eds.: *Proceedings of the Near-Earth-Object Interception Workshop*, Los Alamos National Laboratory, Los Alamos, NM, Feb. 1993.

2. Holsapple, K. A.: Geology of Asteroids: Implication of Spin States Regarding Internal Structure and Some Implications of That Structure on Mitigation Methods. Presentation at the 2002 NASA Workshop on Scientific Requirements for Mitigation of Hazardous Comets and Asteroids, Arlington, VA, Sept. 3–6, 2002.

Near-Earth Object (NEO) Hazard Background[2]

DANIEL D. MAZANEK
NASA Langley Research Center

Introduction

The fundamental problem regarding NEO hazards is that the Earth and other planets, as well as their moons, share the solar system with a vast number of small planetary bodies and orbiting debris. Objects of substantial size are typically classified as either comets or asteroids. Although the solar system is quite expansive, the planets and moons (as well as the Sun) are occasionally impacted by these objects. We live in a cosmic shooting gallery where collisions with Earth occur on a regular basis. Because the number of smaller comets and asteroids is believed to be much greater than larger objects, the frequency of impacts is significantly higher. Fortunately, the smaller objects, which are much more numerous, are usually neutralized by the Earth's protective atmosphere. It is estimated that between 1000 and 10 000 tons of debris fall to Earth each year, most of it in the form of dust particles and extremely small meteorites (ref. 1). With no atmosphere, the Moon's surface is continuously impacted with dust and small debris. On November 17 and 18, 1999, during the annual Leonid meteor shower, several lunar surface impacts were observed by amateur astronomers in North America (ref. 2). The Leonids result from the Earth's passage each year through the debris ejected from Comet Tempel-Tuttle. These annual showers provide a periodic reminder of the possibility of a much more consequential cosmic collision, and the heavily cratered lunar surface acts a constant testimony to the impact threat. The impact problem and those planetary bodies that are a threat have been discussed in great depth in a wide range of publications and books, such as "The Spaceguard Survey" (ref. 3), *Hazards Due to Comets and Asteroids* (ref. 4), and *Cosmic Catastrophes* (ref. 5). The following sections give a brief overview on the background of this problem and address some limitations of ground-based surveys for detection of small and/or faint near-Earth objects.

Range of Threat

Threatening near-Earth objects (NEOs) are typically divided among three classifications based on their orbital characteristics and telescopic appearances: near-Earth asteroids (NEAs), short-period comets (SPCs), and long-period comets (LPCs). Many publications also use the terms Earth-crossing asteroids (ECAs) and Earth-crossing comets (ECCs) to specifically identify objects whose orbits can intersect the Earth. Although the primordial population of NEAs has long been cleaned out from the solar system by collisions and gravitational ejections, it is important to recognize that the population of NEOs is constantly being replenished through a variety of mechanisms. A description of each classification and its source is provided subsequently. A great many publications have been written, such as *Asteroids II* (ref. 6), that deal with these planetary bodies in great depth and address the composition and characteristics of the objects. The following sections are designed to provide a brief overview of each object type and describe the differing characteristics that directly relate to the Comet/Asteroid Protection System (CAPS) study effort.

Although comets and asteroids have become the typical classifications applied to NEOs, there are asteroids that exhibit some amount of comet-like behavior, and some extinct comet nuclei may be classified currently as asteroids. The impact threat can also be further classified as short-period objects and

---

[2]Chapter nomenclature available in chapter notes, p. 217.

long-period objects, which more aptly distinguishes between the difficulties in protecting against these impactors. Precise orbit knowledge is the paramount factor for determining an impactor. More rapidly determining an impacting object's orbit allows more warning time and opportunity to divert the object or mitigate against impact effects. The following four categories define the impact threat from a warning time standpoint assuming current detection methods:

1. Well-defined Orbits
   Detected ECAs
   Warning time = Decades

2. Uncertain Orbits
   Newly discovered ECAs and SPCs
   Warning time = Years

3. Immediate Threat
   LPCs; small ECAs
   Warning time = Months

4. No Warning
   LPCs; unknown ECAs
   Warning time = Days to seconds

Near-Earth Asteroids

NEAs are organized into three groups based on their orbits in relation to the Earth's orbit—Atens, Apollos, and Amors. Atens and Apollos are asteroids whose orbits cross that of the Earth. Atens have orbital periods of less than 1 year and aphelial distances greater than 0.983 astronomical units (au), while Apollos have orbital periods greater than 1 year and perihelial distances less than 1.017 au. Amors have orbits that lie completely outside Earth's orbit (perihelial distance between 1.017 and 1.3 au) but have the potential to be perturbed into Earth-crossing trajectories. There is another group of NEAs that is likely to exist, but it has not been observed to date. These are asteroids with orbits that lie completely within the Earth's orbit (aphelial distances less than 0.983 au), and like the Amors could be perturbed into becoming Earth crossers. These objects have been referred to as interior-Earth asteroids (IEAs) and await an official designation after their existence has been confirmed.

There are believed to be two main sources of NEAs. The first is the main asteroid belt, which is believed to replenish the NEAs through collisions and chaotic orbital dynamics. The main asteroid belt is a vast toroidal region between the orbits of Mars and Jupiter (approximately between 2 and 4 au from the Sun) containing most of the asteroids that orbit the Sun. The main asteroid belt is estimated to contain millions of asteroids ranging from the size of a pebble to approximately 1000 km and is believed to be the origin of most NEAs. The observed composition of NEAs is very similar to main-belt asteroids, and it is believed that collisions and orbital resonances with Jupiter can result in changes in their orbits that can put them into Earth-crossing orbits. The recent Sloan Digital Sky Survey provides an estimate of 700 000 asteroids in the main belt that are 1 km in diameter or greater, a number significantly lower than previous estimates of approximately 2 million (ref. 7). This lower number, together with the millions of smaller asteroids, still provides an ample supply of debris that could eventually find itself on a collision course with Earth. The second source of asteroids is believed to be extinct comet nuclei. Several asteroids have orbits very similar to short-period comets, and at least one cataloged asteroid is associated with a significant meteor shower on Earth. Meteor showers typically result when the Earth passes through the debris trail of a comet. The Geminid meteor shower is associated with asteroid 3200 Phaethon (ref. 8). Additionally, nongravitational forces appear to alter the orbits of some asteroids, indicating that some cometary activity is present in these bodies (ref. 3).

NEAs are classified into three major categories based on their reflectance spectra (primarily in the visible through infrared wavelengths) and their geometric albedos (visible). Although there is significant heterogeneity in the major categories, and a number of additional classes, the main asteroid classifications are identified as S, C, and M types. S-type asteroids are reddish in appearance (similar to stony-iron meteorites) and have moderate albedos (0.07 to 0.23). C-type asteroids are dark with some having extremely low albedos (0.02 to 0.07) and are similar in appearance to carbonaceous chondrites. M-type asteroids are believed to be mainly metallic in composition (primarily nickel-iron), have moderate albedos (0.10 to 0.22), and exhibit high radar reflectivity (refs. 9 and 10). Approximately 55 NEAs have been spectrally characterized to some extent and the results published. Unpublished data exist on approximately 40 other NEAs. From this limited population of imperfectly characterized NEAs, approximately 50 percent are S-type, 24 percent are C-type, and 6 percent are M-type. Several other classes exist (each containing a single object or a few objects) and provide the remainder of the classified NEA population. Because the photometric and spectroscopic techniques required to classify NEAs favor the relatively bright objects, the actual proportion of very dark C-type NEAs is likely to be closer to 50 percent (ref. 11).

The density of asteroids is assumed to correspond to the range of meteorites that have been collected. The density of these meteorites can vary widely due to differences in composition as well as porosity of the material. The densities of stony meteorites range from approximately 2300 to 4000 kg/m^3, while iron rich meteorites can possess a maximum density of approximately 7900 kg/m^3 (ref. 10).

Estimates of the NEA population differ somewhat and are constantly being refined. Recent data gathered by NASA's Near-Earth Asteroid Tracking (NEAT) system indicate that the population of large NEAs (diameter > 1 km) may be only half the previously estimated range of 1000 to 2000. The population estimated from the NEAT data estimates 700 ± 230 NEAs with an absolute magnitude (H) < 18, and assuming an albedo of 0.10 (ref. 12). As additional data are acquired these estimates will be refined further; however, it is certain that the number of smaller NEAs is much greater than the larger objects, as the curve demonstrates in figure 1 (adapted in ref. 13 and originally from ref. 14). This plot, with the error band in gray, shows the much larger population of asteroids less than 1 km in diameter that is believed to exist, and the area under this curve highlights the vast number of potentially threatening ECAs.

Binary Asteroids

Asteroids are not necessarily lone wanderers moving through the solar system. A significant fraction of the asteroid population, including ECAs, have companions. Of approximately 28 known terrestrial impact craters larger than 20 km, at least 3 are confirmed to be double craters. Researchers estimate that approximately 16 percent of NEAs larger than 200 km in diameter are likely to be binary systems (ref. 15). In binary systems, the main asteroid is significantly larger than the smaller body, and rotates much faster than most solitary NEAs. Significantly more information can be obtained from observing binary asteroid pairs, particularly if radar measurements can be made. Whether a potential impactor consists of one or two bodies is important information when formulating a deflection or mitigation strategy.

Comets

While NEAs are a relatively recent discovery (the first NEA, Eros, was not discovered until 1898), comets have been observed by humans for thousands of years. Many comets can be easily observed by the naked eye as they approach the Sun, and the characteristic coma and bright tail are visible. Comets are made up of five main parts: the nucleus, the coma, the dust tail, the ion tail, and the hydrogen cloud

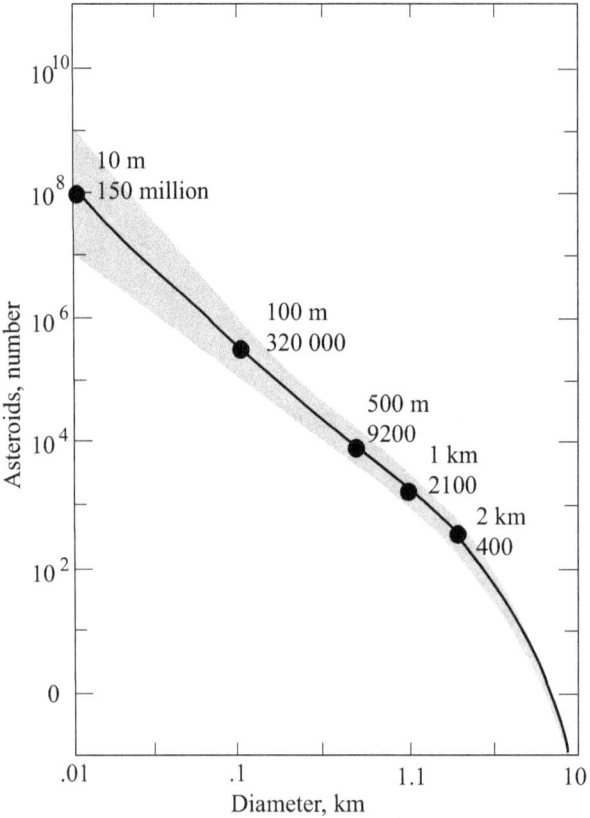

Figure 1. Estimated number of Earth-crossing asteroids.

(sometimes called the corona). Comet nuclei, typically described as a "dirty iceberg" or "dirty snowball," are a mixture of dust and hydrocarbons and volatile ices (predominately water ice and carbon dioxide ice). There is a great deal of uncertainty in the densities of comet nuclei, but they are thought to range from 200 to 1000 kg/m^3 (ref. 11). If some NEAs are actually extinct comet nuclei, the density of some nuclei could be higher. Comets can be extremely difficult to observe before becoming active because their nuclei can be extremely dark. A cometary nucleus was previously assumed to have a fairly high albedo due to its icy composition. However, spacecraft rendezvous missions in 1986 showed that the nucleus of Comet Halley is coated with dark hydrocarbons and has an albedo of only 0.03. This makes Comet Halley's nucleus blacker than coal and one of the darkest objects in the solar system (ref. 16). As the comet nucleus approaches the Sun, the increasing temperature causes volatile ices to sublimate and then release gas and dust from the comet's coma, which can be thousands of kilometers across. As the dust in the coma reflects more sunlight, the coma absorbs ultraviolet radiation and begins to fluoresce. A cloud of ionized hydrogen atoms, much larger than the coma, also develops as the comet approaches the Sun. This hydrogen cloud is visible in the extreme ultraviolet region of the spectrum and can only be observed from space. Two tails are formed when the comet becomes active. The Sun's radiation pressure and solar wind accelerate the gas and dust away from the coma at different rates. The ion tail extends in a nearly straight line in the antisolar direction. The dust tail tends to bend toward the orbital path of the comet because the dust particles are more massive than the ionized gases.

Short-Period Comets

SPCs have orbital periods of less than 200 years, although some researchers classify comets with orbital periods from 20 to 200 years as intermediate-period comets. SPCs are often referred to as periodic comets because their relatively short orbital periods allow observations during multiple perihelial passages. It is estimated that the population of SPCs on Earth-crossing orbits comprises 30 ± 10 larger than 1 km, 125 ± 30 larger than 0.5 km, and 3000 ± 1000 larger than 0.1 km (ref. 3). Many of these objects are thought to originate from the Kuiper belt, a vast population of small bodies orbiting the Sun in a thick ring beyond Neptune and extending 30 to 1000 au from the Sun. The Kuiper belt is estimated to contain 10^8 to 10^9 cometary bodies, with between 35 000 to 70 000 objects larger than 100 km residing in the region between 30 and 50 au (ref. 17). SPCs may also originate as LPCs with planetary interactions (primarily with Jupiter) perturbing them into short-period orbits.

Long-Period Comets

In 1950, Jan Oort noticed three main points when analyzing the trajectories of known LPCs with orbital periods greater than 200 years (up to 14 million years). First, no comet had been observed whose orbit conclusively indicated it originated from interstellar space. Second, in general the aphelial distance of LPCs was greater than 20 000 au. Finally, the orbits of these comets had no preferential incoming direction (approximately 50 percent of currently known LPC orbits are retrograde). These observations led him to theorize that a massive cloud of comets surrounds the solar system. This vast reservoir of comets is known today as the Oort cloud (ref. 18). There is no direct evidence of the Oort cloud because no comet has ever been observed at this great distance. Although the number and mass estimates are not precisely known, the cloud may contain 10^{12} to 10^{13} comets with a total mass of approximately 30 Earth masses (ref. 17). It is believed that the inner Oort cloud begins approximately 1000 au from the Sun and may extend out to 100 000 au (almost halfway to the Sun's nearest stellar neighbor). Figure 2 shows a three-dimensional depiction of the Oort cloud. At such large distances, these objects are only loosely bound to the Sun, and various perturbations can eject them into interstellar space or into the inner solar where they can become impact hazards. The flux of LPCs in the inner solar system is difficult to accurately characterize. Based on the fact that approximately 700 LPCs have been observed during recorded history, it is currently believed that LPCs are only 5 to 10 percent of the ECA population. However, their much higher relative velocities, compared with NEAs, contribute disproportionately to impact threat. It is believed that LPCs could contribute about 25 percent of the total NEO hazards (ref. 3). Because LPCs are not likely to have been previously observed, they represent an impact threat with potentially very little warning time.

In general, LPCs do not become active until they are within approximately 5 au of the Sun, which is just inside the orbit of Jupiter (ref. 3). Before developing a coma and characteristic dust and ion tails, the nucleus may be quite dark with a geometric albedo approaching a limit of 0.02 (ref. 19). This value is approximately the same as the minimum albedo for C-type asteroids. Because the CAPS detection goal is to be able to confirm that an LPC is on impact trajectory when the object reaches a distance within approximately 5 au of the Earth (not just observing the object), it is likely that initial discovery and many follow-up observations would need to be made while the nucleus is inactive. As a worst case, an albedo of 0.02 was assumed for both LPCs and NEAs during this study.

The "Missing" Comet Problem

Our current understanding of the population and the evolution of cometary bodies is still far from complete. A deficiency exists in the number of observed LPCs and evolved SPCs when compared with

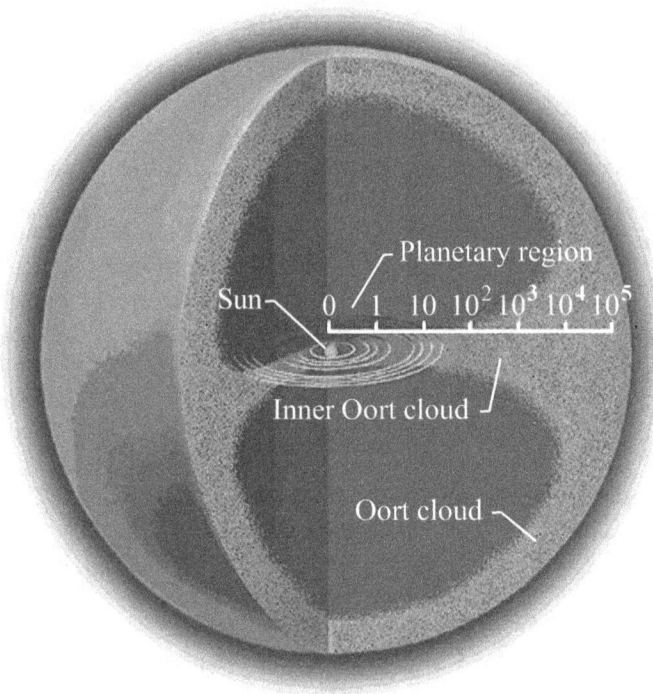

Figure 2. Oort cloud (image courtesy of The Electronic Universe Project, University of Oregon).

the number that should be observed based on the expected steady-state distribution of Oort cloud comets. Dynamic studies generally predict that there should be many more comets than we currently observe, unless some physical mechanism acts to reduce the intrinsic brightness of these objects during subsequent perihelial returns. This phenomenon is often referred to as comet "fading" and has been a vital assumption in comparing the observed and predicted orbits of LPCs (ref. 20)

There are several possible explanations for this discrepancy, and ultimately the answer to this paradox may be that a combination of several mechanisms is responsible. One explanation is that some comets disrupt as they travel through the inner solar system. From an impact hazard standpoint, the important question is whether they disintegrate into very small particles, such as dust, or do the nuclei fragment into smaller, but significant, "cometesimals" that cannot be easily observed with Earth-based telescopes. Comet C/1999 S4 (LINEAR) provided a valuable example of how important space-based observations can be for NEO detection. As it passed through perihelion, intense solar heating triggered a massive disruption of the comet's nucleus around July 26, 2000. Initial Earth-based observations, using 2-m class telescopes, showed a diffuse cloud of dust and gas, indicating that the comet had completely vaporized (ref. 21). On August 5, 2000, the Hubble Space Telescope conclusively showed that the comet has broken into a swarm of more than a dozen minicomets, some tens of meters in diameter (ref. 22). Subsequent observations with the 8.2-m European Southern Observatory Very Large Telescope (VLT) also showed the remnants of the comet nucleus. As the fragments faded from view, it was not possible to ascertain whether or not they continued to disintegrate completely. Comet Shoemaker-Levy 9, which was much larger than Comet C/1999 S4, broke into many large well-defined fragments from tidal stresses as it passed within 100 000 km of Jupiter. Although we have direct evidence that some comets disrupt or fragment, a second possibility is that some LPC nuclei are structurally capable of remaining intact and simply stop outgassing, resulting in either extinction or dormancy. Comets may deplete their volatile gases and evolve into asteroid-like bodies, with some being captured into short-period orbits while others return on long-period trajectories. If this is another possible fate of LPCs, the very real concern exists that there may be impacting objects that have never been previously observed, have extremely low albedos (≈0.02), and never brighten significantly. It is also very possible that both of these mechanisms are responsible for the lack of observed SPCs. If a large cometary body was to disrupt into a collection of significant size fragments that later become extinct or dormant, the ability for any detection system to discover them would be taxed considerably. Because comets and asteroids might form a continuous spectrum of planetary bodies rather than two distinct types, it is very possible that the above mechanisms, as well as others, could explain the lack of observed comets. Finally, another possibility is that the flux of LPCs is not constant (see discussion in next section) and that we are currently in a period of relatively low LPC flux. Drawing definitive conclusions for extremely long-term processes based on a constrained set of observations is always a difficult task.

NEO Impact Flux and Showers

Estimates for the impact flux from comets and asteroids are based on a variety of data, including cratering data (particularly the Moon and Mars), current NEO discovery efforts, and declassified U.S. Department of Defense data from space-based, earthward looking infrared sensors. Between 1975 and 1992, these sensors detected 136 meteoroid fireballs with observed energies between 1 and 10 kilotons (kt) (ref. 23). These objects are small and do not normally survive entry through the atmosphere to reach Earth, but they create upper atmospheric explosions that can be detected. These observations provide important information for bounding the number of asteroids larger than a certain size. The population estimates are based on observational data, so a potential for biases exists in the impact flux for any given size and type of object.

NEOs are removed from the solar system on timescales of 10 to 100 million years (ref. 3). This is accomplished via collisions and gravitational interactions with the planets. As mentioned earlier, the NEA population is constantly being replenished by main belt asteroids perturbed into Earth-approaching orbits. The collision of main belt asteroids and the gravitational influence of Jupiter can send new asteroids toward Earth. Additionally, some NEAs may be extinct comet nuclei so their populations may be altered by several mechanisms.

The flux of LPCs may also vary over time due to tidal forces associated with periodic movement of the solar system through the galactic disk and impulsive perturbations to the Oort cloud. The formation of large impact craters during the last 220 million years appears to have 4 apparent pulses at approximately 2, 35, 65, and 99 million years ago, with a best-fit period of 32 million years. Although the possibility of a periodic astronomical disturbance mechanism exists, the current hypotheses regarding possible sources are either improbable or only cause a weak periodic modulation of the comet flux (ref. 24). These comet showers are most likely random events resulting from the close passage of stars through the Oort cloud. Two significant Earth cratering events, the 100-km crater in Popigai (Siberia) and the 90-km Chesapeake Bay crater (Virginia), occurred nearly synchronously 35.6 million years ago. Analysis of the flux of extraterrestrial helium-3 in pelagic limestone deposits indicates that both impacts likely occurred during a short-lived burst of LPCs in the late Eocene epoch. Helium-3 is an isotope that rarely occurs on Earth but is common in interplanetary dust particles. Although several mechanisms could be responsible for the observed increases in the helium-3 flux, the most likely explanation is that they were the result of a comet shower from an isolated impulsive Oort cloud perturbation (ref. 25). It is likely that comet showers in the last 100 million years have been no more intense than 30 times the combined background flux of comet nuclei and asteroids. Although the flux of LPCs is not expected to change significantly in the foreseeable future, a shower of LPCs could result in the most challenging protection effort imaginable. Large numbers of LPCs entering the inner solar system for the first time each year would require an extremely coordinated detection effort to determine if any objects were a threat. Not only would the probability of an impact during the first perihelial passage be greatly increased, but the threat of impact from smaller cometary fragments would be a problem. The ability to accurately determine the orbits of LPCs would not only assist with the threat of direct collision, but would also permit the more accurate tracking of comet fragments.

The fact that the population and flux of NEOs are not constants emphasizes the need for a continuous detection capability. It is not sufficient to undertake a campaign to search down to a limiting size and then stop observing. This need for continuous monitoring is absolutely critical for LPCs, where we will likely only have one opportunity to identify an impactor.

Unknown Objects and Orbit Classes

Due to limitations in current Earth-based NEO search campaigns, certain classes of minor planets could remain undiscovered and result in an impact without prior warning. Obviously, NEOs too faint (small and/or low albedo) to be observed with ground-based telescopes would be missed by an Earth-based detection system. Additionally, asteroids in fairly long periods or resonant orbits could be missed during a finite search period. One class of hypothesized asteroids could be similar to Atens or the unproven IEAs. IEAs would have aphelia that just reach to the Earth's orbit. Presently, their aphelia could coincide with Earth's aphelion. Therefore, this class of NEOs would not be currently visible in the night sky, and search strategies based on searching near solar opposition would be incapable of detecting them. Hence, we currently have no data regarding the population of this theoretical orbital class. As an IEA's line of apsides precess with respect to Earth's orbit, its orbit could become Earth-crossing. By the time an impacting asteroid in this orbital class became observable, it would allow for little, if any, warning time. An inadequate space-based system could suffer from similar problems as Earth-based systems, but a properly deployed space-based system could more easily identify this class of objects. Two possible approaches for detecting this orbital class would be to make observations near the Sun's observed position, or from an orbit significantly interior to that of the Earth. Placing the system at the Earth-Sun L1 Lagrange point would permit constant monitoring, but placing the system in a heliocentric orbit different than the Earth's orbit would only allow periodic coverage due to the differences in the orbital periods.

Although it is not necessarily fair to use hypothetical objects to assist in the justification of an NEO detection system, astronomy is a field with numerous examples of how the discovery of previously unknown objects has altered long-standing perceptions. It is very likely that as NEO detection systems significantly improve, the increased capability of these systems will identify new classes of asteroids and comets. One recent discovery highlights this point. In 1997, Asteroid 3753 Cruithne (1986TO) was determined to be a dynamical companion of Earth following a complex "horseshoe" orbit when viewed in a heliocentric reference frame corotating with the Earth (ref. 26). When viewed from this corotating vantage point, the asteroid traces out a kidney bean-shaped orbit once each year with the full cycle of the overlapping horseshoe completed in approximately 375 years. Its high inclination (almost 20°) and eccentricity (0.51) make it an unlikely candidate for being a dynamical companion of the Earth. Although 3753 Cruithne neither orbits the Earth nor follows Earth's orbit around the Sun, it is Earth's only known natural companion other than the Moon. Repeated close approaches with the Earth result in gravitational perturbations that increase the object's period slightly. The object's distance from the Earth varies from approximately 0.1 to 2.5 au. Although horseshoe orbits are a well-known feature of the gravitational three-body problem, 3753 Cruithne's orbit has characteristics never before observed or anticipated using theory or computer simulations. The intricate horseshoe orbit includes a spiraling motion, a significant inclination relative to the ecliptic, and an overlap at the end of the horseshoe. Prior to 1997, no other near-Earth asteroid had been discovered in a horseshoe orbit. Although this object had been originally discovered in 1986, it was never sufficiently tracked to determine its unprecedented orbit. The object has an absolute magnitude of 15.1 and an estimated diameter of 5 km. Recently, two additional asteroids of unknown size, 1998 UP1 and 2000 PH5, have been determined to be in similar orbital relationships with Earth. In that 3753 Cruithne's orbit has a fairly high inclination, there is no danger of collision with the Earth when it reaches its closest approach of approximately 0.1 au. However, the discovery of such an asteroid raises the question of how many additional undetected asteroid companions exist that could be impact threats. Understanding the characteristics of 3753 Cruithne's orbit and other possible orbital classes provides tremendous benefit in developing and establishing NEO detection strategies.

The Impact Hazard

It is generally believed by scientists that an impact with an asteroid or comet between 1 and 2 km in diameter would have the ability to disturb Earth's climate on a global scale. Although very infrequent, these impacts could result in explosions with the energy equivalent of a million megatons of TNT. The uncertainty regarding the precise threshold diameter is significant, but it is certain that destruction and loss of life unprecedented in human history would result. The predicted and observed effects of Earth impacts have been documented extensively in various reports, such as "The Spaceguard Survey" (ref. 3), and books, such as *Hazards Due to Comets and Asteroids* (ref. 4) and *Cosmic Catastrophes* (ref. 5). Subsequent sections provide a brief summary describing the effects of Earth impacts and emphasize the hazards over the entire range of threat.

Various statistical analyses have been performed to estimate the probability of death or destruction from the impact of a comet or asteroid of a given size, including the recent computer modeling by John S. Lewis in *Comet and Asteroid Impacts on a Populated Earth: Computer Modeling* (ref. 11). The initial goal of CAPS is to identify what is required to protect against LPCs capable of global devastation and smaller impactors capable of regional hazards. Defending against the smallest impactors ($\approx$10-m diameter) capable of severe local destruction is likely to prove an insurmountable task in the foreseeable future, from both technological and economic standpoints. However, if a viable approach for cataloging smaller NEOs (50- to 100-m diameter) can be developed, it is possible that the approach could be applicable to even smaller objects. A general outline of the comet and asteroid impact hazard is provided subsequently in ascending order of destructive power, followed by a few examples of the indirect hazards that could accompany even a relatively small impact event. It is important to realize that the kinetic energy released during an impact is proportional to the mass of the impacting NEO (which is proportional to the cube of the object's diameter) and the square of its relative velocity. The explosive yield of an impact is usually expressed in megatons (Mt) or kilotons (kt), where 1 Mt = 4.184×10^{15} joules of energy. As mentioned previously, NEO densities can vary by a factor of 40 when comparing metallic asteroids with the lowest density assumed for comet nuclei. The mean velocity of NEAs relative to the Earth is approximately 21 km/s while LPCs have higher velocities with a mean of approximately 55 km/s (ref. 13).

Many estimates have been made in an attempt to quantify the probable fatalities resulting from the impact of an NEO of a particular size, and most have focused on the asteroid part of the impact problem. Figure 3 shows the average fatalities estimated to occur from an asteroid impact event of a given diameter (adapted in ref. 13 and originally from ref. 3). Our knowledge of the population of potential impactors is not complete and is subject to change. Although the efforts to discover NEAs greater than 1 km in

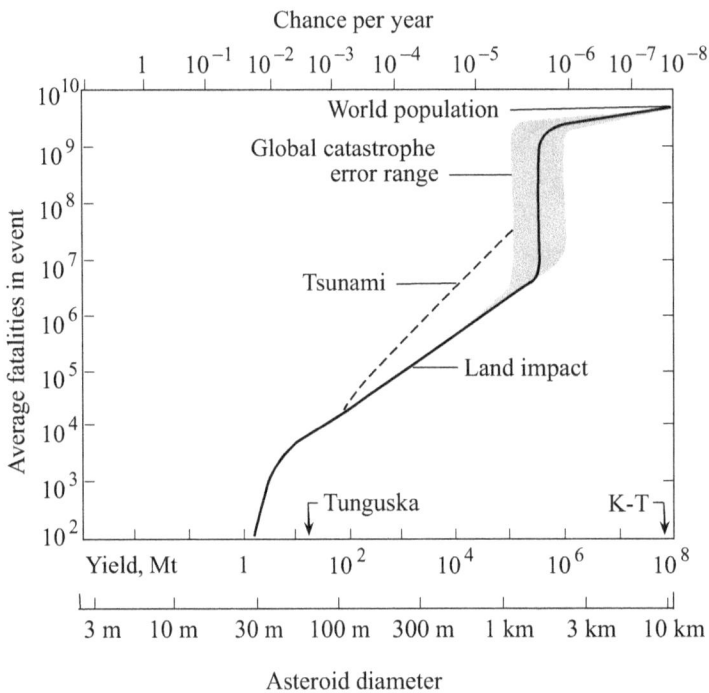

Figure 3. Estimate of averaged fatalities for an asteroid impact versus diameter, explosive yield, and probability per year.

diameter are resulting in many objects smaller than 1 km being detected, only a small percentage of all NEAs have been discovered and cataloged to date. Additionally, the accuracy of the estimated orbits for these known objects varies significantly with the number and quality of the observations. It is the undiscovered asteroids and comets that should warrant much more concern, considering that the likely warning time for one of these objects on an impact trajectory might be as short as a few seconds.

Local Threats

Very small meteorites (stone size) survive entry through Earth's atmosphere and reach the ground on a fairly regular basis. Contrary to popular belief that no person has ever been killed by a meteorite, historical events indicate death may have resulted from an impact. A large number of historical records appear to describe terrestrial impacts or specifically document their occurrences. A table summarizing the deaths, injuries, and damage from these events is provided in reference 11. On average, over the past 20 years one documented case per year is provided where a small meteorite either hit something or nearly hit a person. Considering that much of the Earth's surface is unpopulated, these meteorite falls are quite frequent but only capable of damage on an extremely local scale.

Impacts from meteorites approximately 10 m in diameter can result in significant explosions ($\approx$10- to 100-kt yield) that are capable of destroying a town or small city. The Earth is impacted annually by bodies this size, but most are so fragile that they detonate in the upper atmosphere. The population of NEOs of this size is believed to be on the order of 200 million, and the effort required to identify, catalog, and defend against these objects is difficult to comprehend (ref. 11).

Regional Hazards

Objects with diameters larger than a few tens of meters are capable of releasing enormous amounts of destructive energy (multimegaton-class explosions). Fortunately, impacts with objects of sufficient size and composition to survive the heating and stresses imposed during atmospheric entry are much less common. However, they have occurred in the past and will occur in the future. Impacts with objects larger than 50 m, which is generally taken as the limit for atmospheric breakup, occur on the order of a century or hundreds of years. The 50000-year-old Barringer Meteorite Crater (also known as Meteor Crater) in Arizona provides an enduring reminder of this fact. This crater is about 1.2 km in diameter and 175 m deep and was created by a nickel-iron meteorite with a diameter of approximately 45 m. The impact resulted in an explosion with a yield of around 20 Mt (ref. 27). NEAs between 50 and 100 m in diameter are estimated to number in the millions (see fig. 1). Small LPCs (approximately 100 m or less) are much more common than the 1-km class bodies and possibly 10000 times more abundant. These objects may represent a very significant hazard with the mean interval between lethal impacts estimated to be approximately 600 years (ref. 11). Although damage from an object of this size would still be limited on a global scale, an impact near an urban area or coastline could result in considerable loss of life, extensive damage, and economic disruption.

It is important to realize that an object does not need to reach the ground to be destructive. The Earth's atmosphere can effectively protect us from even megaton-class explosions, which disperse and are vaporized before reaching the lower atmosphere. However, an airburst that results in a 276-hPa (4-psi) overpressure at the surface is capable of producing winds of approximately 70 m/s, or 157 mph (ref. 28). These wind velocities are comparable with those of a category 5 catastrophic hurricane (Saffir-Simpson hurricane scale) and can result in massive destruction on the surface. Stoney objects ≈30 m and LPCs ≈60 m in diameter can penetrate the Earth's atmosphere to an altitude that can create a 4-psi shock at the surface (ref. 28). The Tunguska event of 1908, believed to be an aerial explosion of a stony asteroid or comet fragment approximately 50 m in diameter, released up to 20 Mt of energy, the equivalent of 1600 Hiroshima-size bombs, and devastated 2000 km^2 of Siberian forest. Figure 4 shows the effect that the impact of one of these smaller NEOs could have (adapted from ref. 3). The figure shows the size of the blast field in the Siberian forest compared with Washington, DC. Had the arrival of this object been delayed several hours, it could have impacted in densely populated Europe and had much greater consequences than simply the diminished blast wave that propagated across England. It is not a strict requirement for a comet or asteroid to actually reach the surface to cause extensive destruction, and this event testifies to that fact.

Global Hazard

The size of impactor that would produce deadly and devastating effects of a global scale is not known with great certainty. Many factors, such as where the NEO hits, its relative velocity, and its composition, all contribute to the lethality of the impact. An explosion that releases energy approaching a million megatons of TNT is believed to be capable of producing a global catastrophe. This energy release is achievable with an impact by a comet or asteroid that is between 1 and 2 km in diameter (ref. 3). Even extrapolations based on global nuclear war, our best understood impact analogy, are difficult to quantify when explosions of this magnitude are contemplated. What is known is that the fossil record shows periods of global mass extinctions that have been definitely linked to impact events. The most famous is the Cretaceous-Tertiary (K-T) extinction event 65 million years ago, which resulted from the impact of a 10-km asteroid or comet and created the 170-km Chicxulub crater near the current Yucatan Peninsula. This impact is best known for wiping out the dinosaurs, but it also triggered the extinction of approximately 75 percent of all species on Earth. The massive loss of life was likely caused by a triple punch of

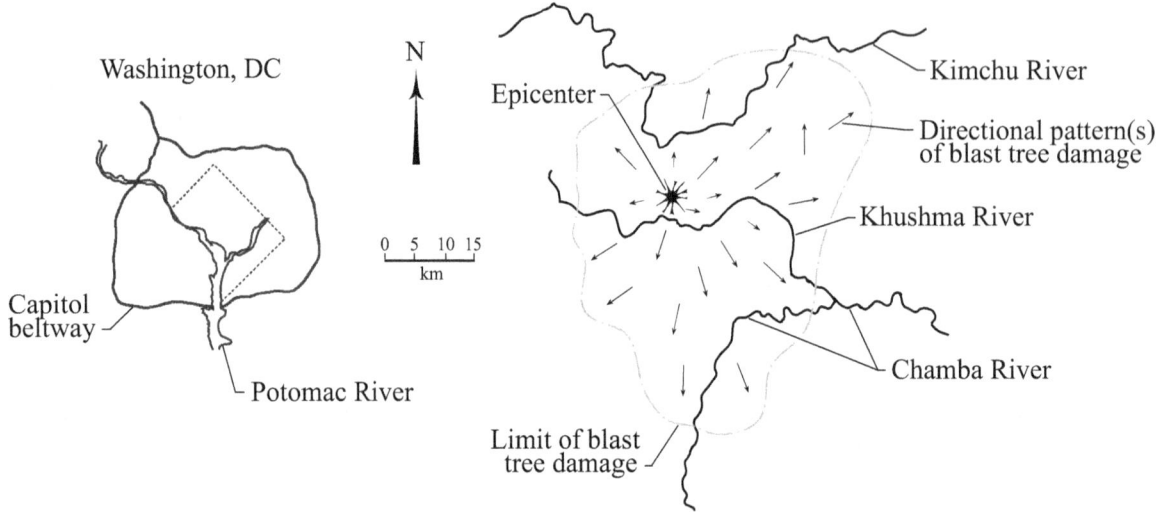

Figure 4. Tunguska impact blast field comparison.

global firestorms, mega-tsunamis, and extensive long-term environmental changes. The shock waves created by such an impact would have triggered massive earthquake and volcanic activity, which would have contributed to the dust and toxic gases released into the atmosphere. In many ways, it is difficult to understand how any species could have survived such conditions. Somehow life on Earth did continue, and this extinction event may very well have allowed for the existence of the human race. Human death and destruction of enormous proportions could be expected from an impact capable of initiating a global catastrophe. Figure 3 shows that a significant fraction of the world's population could die from the direct and indirect effects of a global impact catastrophe.

The Earth is covered with large impact craters. A 1-km-diameter impactor will produce a crater approximately 10 to 15 km in diameter (ref. 3). Currently, there are about 70 known craters larger than 10 km in size, with the largest being the 300-km scar in Vredefort, South Africa (ref. 29). Many more smaller craters have also been identified, but erosion can more easily erase the traces of smaller impact events. Additionally, because approximately 71 percent of the Earth's surface is covered with water, no visible record is available for most impacts.

Indirect Hazards

The enormous heat and acoustical energy directly released from an impact produces destruction that is fairly localized. Depending on the size and location of the impact, the direct effects of an impact may be inconsequential compared with the secondary destruction that could result. Large impacts are expected to affect Earth and its inhabitants on a global scale, but a majority of the damage from smaller impacts may be particularly disproportionate. Outlined in the following sections are some of the possible indirect physical effects of an NEO impact. Additionally, a brief discussion is presented regarding the possibility of lunar surface impacts damaging Earth-orbiting satellites and infrastructure. This is by no means a complete accounting and does not begin to address the very real and important social, economic, and political consequences. The main goal of highlighting some of these secondary destruction mechanisms is to emphasize the breadth of possible effects that need to be considered as part of the impact hazard problem. Because we have not experienced this kind of calamity in modern history, it is possible that the secondary effects could be much more far reaching than we currently realize.

Tsunamis

Tsunamis resulting from the ocean impact of an asteroid or comet may represent the most significant aspect of the impact hazard because most of the Earth's surface is covered with water and a disproportionate number of people inhabit coastal regions. The word tsunami comes from a combination of the Japanese words for harbor (tsu) and wave (nami), because harbors amplify the wave height that reaches the shore. Tsunamis have been recorded for thousands of years and are typically caused by underwater earthquakes, volcanic activity, and landslides. Significant ocean impacts represent a regional hazard that could extend into the realm of global consequences because they can potentially propagate over great distances and wreak havoc on distance shorelines. The United States is particularly vulnerable with both coastlines being highly developed and heavily populated. Additionally, a tsunami can propagate outward from the impact sight at speeds of 750 k/hr (450 mph) and can be barely perceptible in the open ocean, thus making early warning extremely difficult (ref. 30).

There is a great deal of uncertainty about the effects of impact induced tsunamis in the literature. The two most significant issues concern calculating the initial size of the wave and how the wave dissipates as it travels from the impact site. The energy from an explosion decreases as the square of the distance from the impact increases, but the energy from a radiating coherent ocean wave decreases in proportion to the distance. Some analyses indicate that the impact ocean cavity diameter must be several times wider than the ocean depth at the impact site in order to produce a coherent tsunami capable of traveling great distances. The ratio of cavity diameter to impactor diameter ranges from approximately 40 for a 50-m-diameter impactor to approximately 20 for a 1-km object, assuming an impact velocity of 20 km/s and a density of 3000 kg/m^3 (ref. 31). The generated maximum wave heights are highly dependent on the theory and assumptions incorporated in models, and the run-up height (vertical height of the tsunami above sea level at its furthest point) is highly dependent on local topography and the wave's history. Although most simulations confirm that an ocean impact of a 1-km asteroid would generate enormous tsunamis, there is not good agreement on the effect of smaller impacts. Even if the tsunami resulting from a smaller ocean impact dissipates rapidly, an impact near the coast could create devastating local tsunamis, particularly from an LPC with a much higher impact velocity. The run-up height as the wave reaches the coast can be more than an order of magnitude greater than the deepwater wave height, so a 5-m wave could result in a very destructive tsunami over 50 m high (over 15 stories).

Earthquakes and Volcanic Eruptions

Objects of sufficient size and strength to reach the ground are also capable of producing significant seismic disturbances. Even the Tunguska event, which likely did not reach the surface, produced a magnitude 5 earthquake (ref. 28). Although the damage done by an earthquake is highly dependent on the terrain through which the disturbance propagates, it could exceed the blast wave damage. Large impactors, like the K-T event, could conceivably trigger massive earthquakes, which in turn could initiate volcanic eruptions. The combination of these natural disasters could result in much greater global devastation than that which could be attributed directly to the impact explosion.

Misinterpreted Nuclear Strike

Although small impactors (10- to 100-kt yield) would likely cause damage only on a local scale, the possibility exists that the impact could be misinterpreted as an attack using nuclear weapons. Other than the lack of radiation, an impact event would closely resemble a nuclear explosion. If this natural disaster occurred over an area where tense or hostile geopolitical conditions were present, along with presence of nuclear weapons, a retaliatory nuclear exchange could result. Although the probability of such an event is fairly low, it emphasizes that it is not the impact event itself that might be of greatest concern.

Lunar Impacts

The Earth has an atmosphere to effectively protect its surface from smaller impacts, but the Moon lacks such a buffer. Although the Moon is much smaller than the Earth, impacts still occur regularly on the lunar surface. Due to the Moon's low escape velocity (2.376 km/s) and lack of atmosphere, some lunar impact ejecta may leave the Moon's gravitational influence and be expelled into a solar orbit, or captured by the Earth's gravitational field.

The Antarctic Search for Meteorites (ANSMET) Program, which is funded by the Office of Polar Programs of the National Science Foundation, has recovered thousands of meteorites since 1976. An important discovery of ANSMET was that some meteorites were not derived from asteroids, but rather from the lunar and Martian surfaces. Twenty years ago, it was widely believed that any material ejected from a planet-sized body would be vaporized or altered beyond recognition. This paradigm was completely overturned by the discovery of ANSMET meteorite ALH81005, an anorthositic breccia so similar to Apollo lunar highlands samples that all investigators agreed it had to have come from the Earth's Moon. Since that time, ANSMET has recovered six more lunar specimens (ref. 32).

The discovery of lunar meteorites in Antarctica certainly proves that a delivery mechanism exists, but many important questions remain: (1) what size and composition of an impactor is required, (2) how much debris would be captured by the Earth and in what orbits, and (3) how long would an increased debris flux persist. If a significant amount of debris were to be ejected from a lunar impact, it is possible that the resulting debris field could present a significant hazard for satellites orbiting the Earth. Research and analysis in this area of the impact hazard are lacking and require some level of analysis to determine the credibility of this hypothesized hazard. The primary focus for CAPS is Earth defense. However, if lunar impacts are a credible hazard to Earth-orbiting satellites and future in-space infrastructure and lunar outposts, it may be desirable to have at least some CAPS assets located in cislunar space.

References

1. NASA Solar System Exploration Website—Solar System Bodies—Meteors, Meteoroids and Meteorites. <http://solarsystem.nasa.gov/features/planets/meteoroids/meteoroids.html>

2. The Johns Hopkins University Applied Physics Laboratory (APL) Press Release Website—Lunar Impacts Confirmed During Leonid Meteor Storm. <http://iota.jhuapl.edu/lunar_leonid/lunprd09.htm>

3. Morrison, D., Chairperson: *The Spaceguard Survey: Report of the NASA International Near-Earth-Object Detection Workshop.* Jet Propulsion Laboratory, Pasadena, CA, Jan. 25, 1992.

4. Gehrels, T., ed.: *Hazards Due to Comets and Asteroids.* The University of Arizona Press, Tucson, AZ, 1994.

5. Chapman, C. R.; and Morrison, D.: *Cosmic Catastrophes.* Plenum Press, New York, 1989.

6. Binzel, R. P.; Gehrels, T.; and Matthews, M. S., eds.: *Asteroids II.* The University of Arizona Press, Tucson, AZ, 1989.

7. Sloan Digital Sky Survey News Release Website—Sky Survey Lowers Estimate of Asteroid Impact Risk. <http://www.sdss.org/news/releases/20011108.asteroid.html>

8. Science @NASA Headlines Website—The Mysterious Geminid Meteor Shower. <http://science.nasa.gov/newhome/headlines/ast04dec98_1.htm>

9. Remo, J. L.: Classifying and Modeling NEO Material Properties and Interactions, *Hazards Due to Comets and Asteroids*, T. Gehrels, ed. The University of Arizona Press, Tucson, AZ, 1994, pp. 551–596.

10. Lewis, J. S.: *Physics and Chemistry of the Solar System.* Academic Press, San Diego, CA, 1997.

11. Lewis, J. S.: *Comet and Asteroid Impact Hazards on a Populated Earth: Computer Modeling.* Academic Press, San Diego, CA, 2000.

12. Rabinowitz, D.; Helin, E.; Lawrence, K.; and Pravdo, S.: A Reduced Estimate of the Number of Kilometre-Sized Near-Earth Asteroids. *Nature*, vol. 403, no. 6766, Jan. 2000, pp. 165–166.

13. Gold, R. E., Principal Investigator: *SHIELD—A Comprehensive Earth Protection System. A Phase I Report to the NASA Institute for Advanced Concepts.* The Johns Hopkins University Applied Physics Laboratory, 1999.

14. Bowell, E.; Muinonen, K.; and Shoemaker, E. M.: Discovery of Earth-Crossing Asteroids. I. Observing Strategy. Paper presented at the Near-Earth Asteroids Conference, San Juan Capistrano, June 1991.

15. Cornell News Website—Radar Reveals Five Double Asteroid Systems Orbiting Each Other Near Earth, Likely Formed in Close Encounters With Planet. <http://www.news.cornell.edu/releases/April02/Asteroids.Margot.deb.html>

16. Students for the Exploration and Development of Space—Nine Planets Solar System Tour Website—Comet Halley. <http://seds.lpl.arizona.edu/nineplanets/nineplanets/halley.html>

17. Lodders, K; and Fegley, B., Jr.: *The Planetary Scientist's Companion.* Oxford Univ. Press, New York, 1998.

18. Students for the Exploration and Development of Space—Nine Planets Solar System Tour Website—The Kuiper Belt and the Oort Cloud. <http://seds.lpl.arizona.edu/nineplanets/nineplanets/kboc.html>

19. Whipple, F. L.: The Forest and the Trees. *Comets in the Post-Halley Era, Vol. 2* (R. L. Newborn, Jr., et al., eds.), Kluwer Academic Publishers, Dordrecht, The Netherlands, 1991, pp. 1259–1278.

20. Wiegert, P.; and Tremaine, S.: The Evolution of Long-Period Comets. *Icarus*, vol. 137, 1999, pp. 84–121.

21. Science @NASA Headlines Website—"Meltdown! Comet LINEAR continues to disintegrate and could disappear completely within a few days. <http://spacescience.com/headlines/y2000/ast31jul_1m.htm>

22. Hubblesite.org News Center Website—Hubble Discovers Missing Pieces of Comet Linear. <http://oposite.stsci.edu/pubinfo/PR/2000/27/pr-photos.html>

23. Tagliaferri, E., et al.: Detection of Meteoroid Impacts by Optical Sensors in Earth Orbit. *Hazards Due to Comets and Asteroids*, T. Gehrels, ed., The University of Arizona Press, Tucson, AZ, 1994, pp. 199–220.

24. Showmaker, E. M.; and Wolfe, R. F.: Mass Extinctions, Crater Ages and Comet Showers. *The Galaxy and the Solar System*, R. Smoluchowski, et al., eds., The University of Arizona Press, Tucson, AZ, 1986, pp. 338–386.

25. Farley, K. A.; Montanari, A.; and Shoemaker, E. M.: Geochemical Evidence for a Comet Shower in the Late Eocene. *Science*, vol. 280, 1998, pp. 1250–1252.

26. Wiegert, P. A.; Innanen, K. A.; and Mikkola, S.: An Asteroidal Companion to the Earth. *Nature*, vol. 387, 1997, pp. 685–686.

27. The Barringer Meteorite Crater Website—The Science. <http://www.barringercrater.com/science/>

28. Toon, O. B., et al.: Environmental Perturbations Caused by the Impacts of Asteroids and Comets. *Reviews of Geophysics*, vol. 35, 1997, pp. 41–78.

29. Planetary and Space Science Centre at the University of New Brunswick Website—Earth Impact Database. <http://www.unb.ca/passc/ImpactDatabase/CIDiameterSort.html>

30. Lewis, J. S.: *Rain of Fire and Ice.* Addison-Wesley Publ. Co., Inc., Reading, MA, 1996.

31. Ward, S. N.; and Asphaug, E.: Asteroid Impact Tsunami: A Probabilistic Hazard Assessment. *Icarus*, vol. 145, 2000, pp. 64–78.

32. Antarctic Search for Meteorites Program (ANSMET) Website—Frequently Asked Questions. <http://www.cwru.edu/affil/ansmet/faqs.html>

Accurate Determination of Comet and Asteroid Orbits Leading to Collision With Earth[3]

CARLOS M. ROITHMAYR
NASA Langley Research Center

LINDA KAY-BUNNELL
Analytical Mechanics Associates Inc.

DANIEL D. MAZANEK
NASA Langley Research Center

RENJITH R. KUMAR AND HANS SEYWALD
Analytical Mechanics Associates Inc.

MATTHEW A. HAUSMAN
Colorado Center for Astrodynamics Research
University of Colorado, Boulder

Introduction

Movements of the celestial bodies in our solar system inspired Isaac Newton to work out his profound laws of gravitation and motion; with one or two notable exceptions, all of those objects move as Newton said they would. But normally harmonious orbital motion is accompanied by the risk of collision, which can be cataclysmic. The Earth's moon is thought to have been produced by such an event, and we recently witnessed magnificent bombardments of Jupiter by several pieces of what was once Comet Shoemaker-Levy 9. Other comets or asteroids may have met the Earth with such violence that dinosaurs and other forms of life became extinct; it is this possibility that causes us to ask how the human species might avoid a similar catastrophe, and the answer requires a thorough understanding of orbital motion.

The two red square flags with black square centers displayed in figure 1 are internationally recognized as a warning of an impending hurricane. Mariners and coastal residents who know the meaning of this symbol and the signs evident in the sky and ocean can act in advance to try to protect lives and property; someone who is unfamiliar with the warning signs or chooses to ignore them is in much greater jeopardy. Although collisions between Earth and large comets or asteroids occur much less frequently than landfall of a hurricane, it is imperative that we learn to identify the harbingers of such collisions by careful examination of an object's path.

An accurate determination of the orbit of a comet or asteroid is necessary in order to know if, when, and where on the Earth's surface a collision will occur. Generally speaking, the longer the warning time, the better the chance of being able to plan and execute action to prevent a collision. The more accurate the determination of an orbit, the less likely such action will be wasted effort or, what is worse, an effort that increases rather than decreases the probability of a collision. Conditions necessary for a collision to occur are discussed, and warning times for long-period comets and near-Earth asteroids are presented.

Orbit determination is the process of using a collection of measurements obtained by observation to calculate a set of orbital elements, six quantities that give (either implicitly or explicitly) the position and

---

[3]Chapter nomenclature available in chapter notes, p. 217.

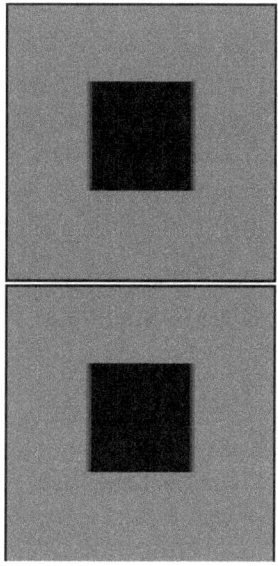

Figure 1. Hurricane Warning Flag.

velocity of an object at a particular instant of time. The classical methods of Laplace, Gauss, and Olbers utilize six angular measurements, obtained two at a time from each of three optical observations, to make a preliminary determination of orbital elements. Additional measurements can be employed together with the iterative method of least squares to improve accuracy of the elements. With all these methods, the quality of the result is affected by errors in the measurements and by the spatial and temporal spacing of observations which are in turn related to the geometry resulting from the orbits of the object and the observatory. The method of least squares makes it possible to take advantage of multiple measurements made from a single observatory or from two or more observatories spread throughout the solar system; the accuracy of orbital elements so obtained is thus a function of the number and spacing of observations, and the number and placement of observatories.

A study of the aforementioned factors and their effects on orbit determination is undertaken in order to identify trends and make a preliminary determination as to the number, placement, and resolution of instruments required to form an effective system for determining orbits of comets and asteroids. In addition, an idea can be obtained of the number and timing of observations that yield the best results, leading to a strategy for using the system to perform observations. In order to keep this stage of the analysis from becoming unnecessarily complicated it is assumed that measurements can be made at will, but of course this cannot always be the case. Orbit determination of long-period comets using a batch filter and sequential filters is examined, and a batch filter is applied to the study of near-Earth asteroids.

Collisions

This analysis is facilitated by the assumption that a collision occurs when the heliocentric distance of a comet or asteroid is identical with that of the Earth; in the case where the object's orbit is not coplanar with the Earth's orbit, the distances must be identical at a point where the object passes through the ecliptic, the plane in which the Earth orbits the Sun. Warning times are studied with the aid of coplanar orbits, and a condition required for a collision in the more general situation of noncoplanar orbits is examined.

Warning Time

The amount of warning before a collision is the interval between the time the object is first discovered (or, to be more precise, the time the orbit becomes known accurately) and the time of collision. Although several orbital periods may elapse during this time, we concern ourselves here with the worst situation in which the comet or asteroid is discovered less than one orbital period before a collision. Unlike large asteroids with relatively short orbital periods, long-period comets (LPCs) do not present themselves for observation over multiple orbits, making it much more difficult to predict a collision decades in advance. Smaller asteroids may escape detection until less than one orbit remains before a collision. Warning time can be obtained through a straightforward application of time-of-flight equations.

Figure 2 shows the Earth E in a circular orbit of radius $r_k = 1$ astronomical unit (au) about the Sun S, and a comet (or asteroid) C in a coplanar elliptical orbit. The axes s_1 and s_2 lie in the ecliptic, with s_1 in the direction of vernal equinox. It is assumed that discovery of C occurs after aphelion and before perihelion, at the point where the red line intersects the orbit of C, and E is assumed to be at the wrong place at the wrong time so that E collides with C before C reaches perihelion, at the point where the blue line intersects the orbits of both objects; the time of flight of C between these two points is to be determined. The motion of S and C is regarded as being governed by two-body orbit mechanics; that is, S and C are each treated as a particle, or a sphere whose mass is distributed uniformly, and the only forces exerted on S and C are those of mutual gravitation. The circular orbit of E about S also proceeds as two-body motion; however, the gravitational force exerted by E and C on each other is left out of account, as are all other perturbing forces acting on E, C, and S.

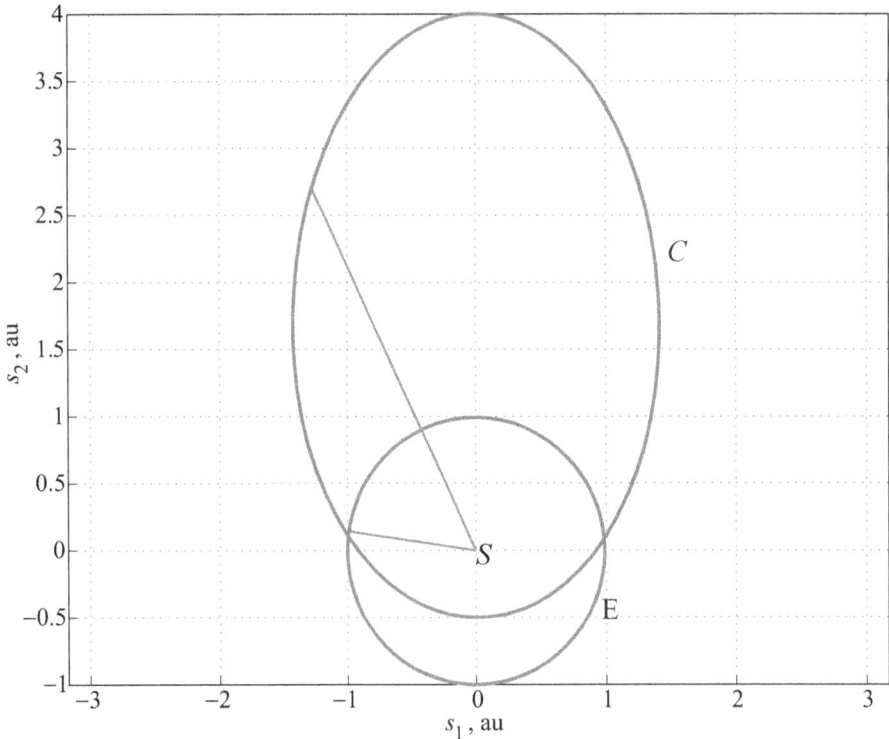

Figure 2. Orbits of C and E.

The eccentric anomaly E at any point of an elliptical orbit obeys the relationship in equation (4.2-14) of reference 1:

$$\cos E = \frac{1 - r/a}{e} \qquad (1)$$

where a and e are, respectively, the semimajor axis and the eccentricity of the orbit, and r is the distance between the centers of S and C at the point of interest. If E designates the eccentric anomaly at the point of discovery where the heliocentric distance is r, and E_k denotes the eccentric anomaly at the point of collision where the heliocentric distance is $r_k = 1$ au, then the time of flight for an interval less than one orbital period is obtained via Kepler's equation, as given by equation (4.2-9) of reference 1:

$$t - t_k = \sqrt{\frac{a^3}{\mu}}\left[(E - e \sin E) - (E_k - e \sin E_k)\right] \qquad (2)$$

where μ is the gravitational parameter of the primary, in this case S.

Now, the principal values of the inverse cosine function are $0 \leq \cos^{-1} x \leq \pi$; E is defined to be 0 at perihelion and π at aphelion. If E and E_k are to have the correct values for the quadrants as illustrated in figure 2, the sign of the right hand member of equation (1) would have to be changed, as would the signs of E, E_k, $\sin E$, and $\sin E_k$ in equation (2); however, the absolute value of the time of flight $t - t_k$ would remain unaltered.

Results

Times of flight until collision for LPCs are calculated with the aid of equations (1) and (2). A family of orbits is constructed with perihelia r_p equal to 0.1 and 1 au, and aphelia r_a of 15, 20, 25, ..., 100, 200, 300, ..., 1000, 2000, ..., 50×10^3 au. An aphelion of 15 au corresponds to an orbital period of about 20 years, whereas 50×10^3 au reaches the middle of the Oort cloud and corresponds to a period of 4×10^6 years. (LPCs are often regarded as those with periods greater than 200 years.) The distance r at which the orbit becomes known takes on the values 5, 6, and 7 au; corresponding time of flight, regarded as warning time, is presented in figure 3 as a function of r_a. Warning time does not change appreciably for aphelia in the range $1000 \leq r_a \leq 50 \times 10^3$ au, and a reduction in r_p by a factor of 10 reduces warning time by approximately the same amount as a reduction of 1 au in r. For the orbits studied, warning times range from 2 years in the case of a 20-year comet with $r_p = 1$ au, detected at a distance of 7 au from the Sun, to 9.5 months in the case of a comet coming from the Oort cloud with $r_p = 0.1$ au, detected at 5 au.

Near-Earth asteroids (NEAs) are studied in a similar manner, with r_p equal to 0.2 or 0.9 au, and r_a = 1.4, 1.5, ..., 3.0 au; the associated orbital periods range between approximately 9 months, and 2 years and 9 months. The detection distance takes on the values 1.1, 1.2, and 1.3 au, and warning time as a function of r_a is presented in figure 4. Warning times range from approximately 90 days for an asteroid in a 0.9×1.4 au orbit, detected at a heliocentric distance of 1.3 au, to 7 days in the case of an asteroid with a 0.2×3.0 au orbit spotted when it is 1.1 au from the Sun.

Although it has been assumed thus far that the orbits of C and E are coplanar, a collision is possible also when the orbit plane of C is inclined to the ecliptic, so long as an ascending or descending node of C has a heliocentric distance of 1 au. Under these conditions, together with the two-body assumptions set forth in the discussion of warning time, the time of flight is unchanged.

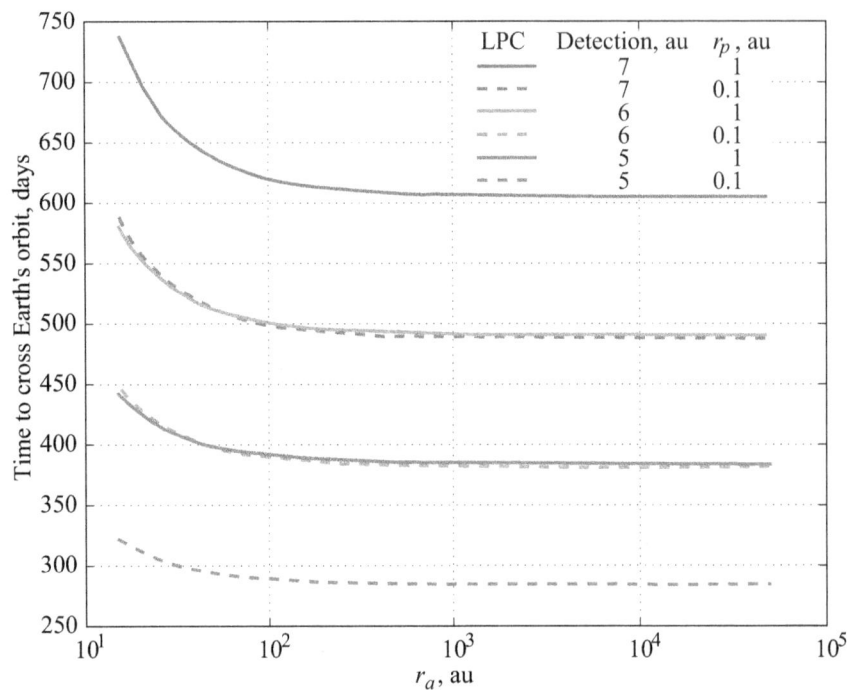

Figure 3. Times of flight to collision, long-period comets.

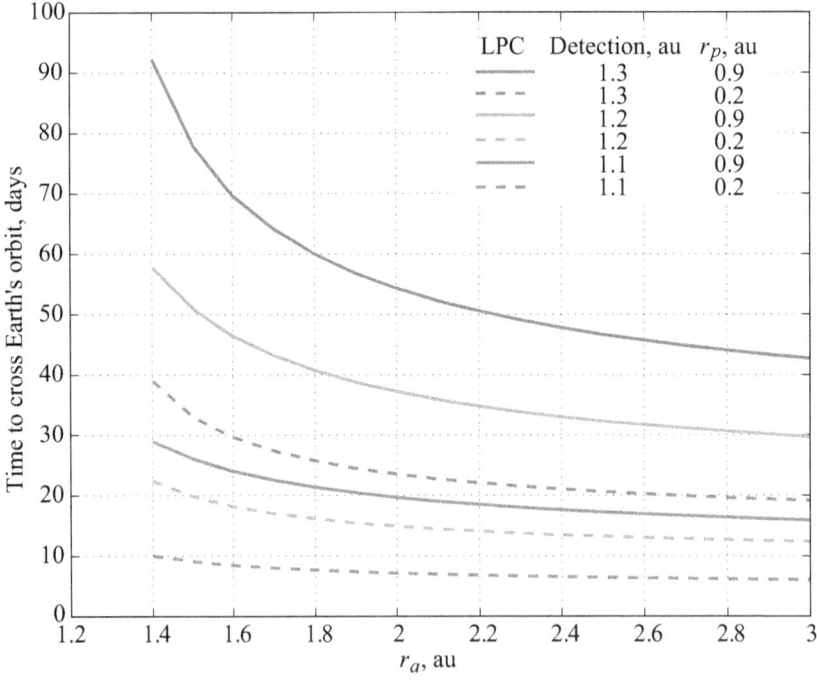

Figure 4. Times of flight to collision, near-Earth asteroids.

Designing a Collision

Forthcoming batch filter analyses rest on the foregoing supposition, namely that a collision with Earth occurs when the object passes through the ecliptic plane and the object's heliocentric distance is equal to r_k, the radius at collision, or 1 au. By design, all of the orbits to be studied meet these conditions; this requires a certain relationship, to be developed presently, between the argument of periapsis on the one hand and, on the other, r_p, r_k, and e.

With the ecliptic chosen as the reference or fundamental plane, an object whose orbit plane is inclined to the ecliptic is, by definition, either at the ascending node or the descending node of the orbit when it passes through the ecliptic. Equations (1.5-4) and (1.5-7) of reference 1 allow us to write

$$r = \frac{r_p(1+e)}{1+e \cos \nu} \tag{3}$$

where r_p is the radius of periapsis, e is the eccentricity of the orbit, and ν is the true anomaly measured in the plane of orbit from the periapsis. At the ascending node $\nu = -\omega$, where ω is known as the argument of periapsis, and at the descending node $\nu = \pi - \omega$. The requirement that $r = r_k$ at one of the nodes therefore can be expressed as

$$r_k = \frac{r_p(1+e)}{1 \pm e \cos \omega} \tag{4}$$

where the positive sign preceding $\cos \omega$ means the condition is imposed at the ascending node, and the negative sign is associated with the descending node. This relationship can be rearranged,

$$\pm \cos \omega = \frac{r_p}{r_k}\left(1+\frac{1}{e}\right) - \frac{1}{e} \tag{5}$$

Now, any member of equation (5) must of course remain between -1 and 1 (inclusive) if it is to be solved for ω,

$$-1 \leq \pm \cos \omega \leq 1 \tag{6}$$

however, one can simply work with the positive sign because it can be seen that the requirement is the same no matter which sign is used. Substitution from equation (5) into (6) gives

$$-1 \leq \frac{r_p}{r_k}\left(1+\frac{1}{e}\right) - \frac{1}{e} \leq 1 \tag{7}$$

or

$$\frac{1-e}{1+e} \leq \frac{r_p}{r_k} \leq 1 \tag{8}$$

and the right hand inequality yields the expected restriction on r_p, namely

$$r_p \leq r_k \tag{9}$$

One can deal with the left hand inequality in (8) by appealing to equation (1.7-4) of reference 1 and substituting $(r_a - r_p)/(r_a + r_p)$ for e, which yields a rather sensible result involving the radius at apoapsis,

$$r_k \leq r_a \tag{10}$$

Thus, as long as the inequalities (9) and (10) are satisfied, one is able to solve equation (5) for a value of argument of perihelion that meets the condition necessary for a collision to take place at the ascending or descending node, according to the choice of sign. (Changing the sign of the argument of the function $\cos^{-1}$ yields the supplementary angle: $\cos^{-1}(x) + \cos^{-1}(-x) = \pi$.) The analyst is presented with a second choice of sign because $\cos(\omega) = \cos(-\omega)$; the result of $\cos^{-1}[\cos(\omega)]$ provided by a calculating machine typically lies in the range of principal values of the inverse function, namely $0 \leq \omega \leq \pi$, but a solution on the interval $-\pi \leq \omega \leq 0$ is also correct. Because ω is always measured from the ascending node, $\omega > 0$ always corresponds to a periapsis that is to the north side of the reference plane, and $\omega < 0$ always implies the periapsis is on the south side. Each of the four possible combinations of the two choices is associated with a pre- or postperihelion collision as indicated in table 1.

Table 1. Position of Collision

Node of collision	Perihelial hemisphere	
	North ($\omega > 0$)	South ($\omega < 0$)
Ascending	Preperihelion	Postperihelion
Descending	Postperihelion	Preperihelion

Batch Filter for Long-Period Comets

As mentioned in the introduction, the design of a system for performing observations can be guided by a quantitative study of the way in which orbit determination is influenced by the quality, number, and timing of the observations, as well as the number and location of the observatories. Because the orbital parameters of LPCs are distinct from those of asteroids, the investigation is divided accordingly with comets examined in the present section and asteroids in the following section. It would be highly desirable if a single system could serve effectively in observing the two classes of objects; however, it remains to be seen if this is in fact possible.

In order to gauge the effects of the aforementioned factors on orbit determination, it is necessary to have in hand a way to judge the quality of orbital parameters obtained from observations. To this end the concept referred to as "erroneous predicted miss distance" is introduced.

The study of LPCs begins with an introduction to the method of weighted least squares. The requirement for collision expressed in equation (5) is used to construct a large number of orbital parameter sets, which then yield "true" values of measurements to which errors are introduced intentionally. The measurements thus corrupted account for limits in resolution that can be furnished by an actual telescope and are used to determine or estimate the orbit and obtain an associated erroneous predicted miss distance. The quality of preliminary orbit determination obtained from three observations is examined, followed by a study of the effects of multiple (more than three) observations and observations taken from two or more observatories.

Erroneous Predicted Miss Distance

Orbital elements determined on the basis of observations generally differ from the true orbital elements, that is to say at a particular value of time t, the true position **r** and velocity **v** differ from the position **r′** and velocity **v′** resulting from the process of orbit determination. In practice the true orbital elements are unknown; however, they are to be specified in analysis that follows. The two sets of orbital elements can be compared in order to judge the quality of the result of orbit determination; however, it is more convenient to compare a single parameter if at all possible instead of six scalar values associated with position and velocity (or, for that matter, six classical orbital elements). Such a parameter ε, hereafter referred to as an "erroneous predicted miss distance," is now introduced. When ε vanishes, the orbit determined from observations results in a collision at the specified position and at the designated time. (Other metrics can be found in the literature, but seem to require a comparison of two scalars.)

An object C travels along a true orbit, shown in figure 5 with a solid red arc, designed to collide with the Earth E at time t_k, at a point K where the object passes through the ecliptic plane at a heliocentric distance of $r_k = 1$ au. The orbit of C is generally not identical to an orbit determined from measurements, shown with a dashed red arc, and associated with an object C'. An orbit solution is obtained for some epoch t, at which time the true position vector from S to C is $\mathbf{r}(t)$, and the position determined from observation is shown as $\mathbf{r}'(t)$. When the true orbit is propagated to the time of collision t_k, $\mathbf{r}(t_k)$ is of course the position vector from S to K. The orbit of C' can also be propagated to t_k, yielding a position vector $\mathbf{r}'(t_k)$; the magnitude ε of the difference $\mathbf{r}'(t_k) - \mathbf{r}(t_k)$ is defined to be the "erroneous predicted miss distance."

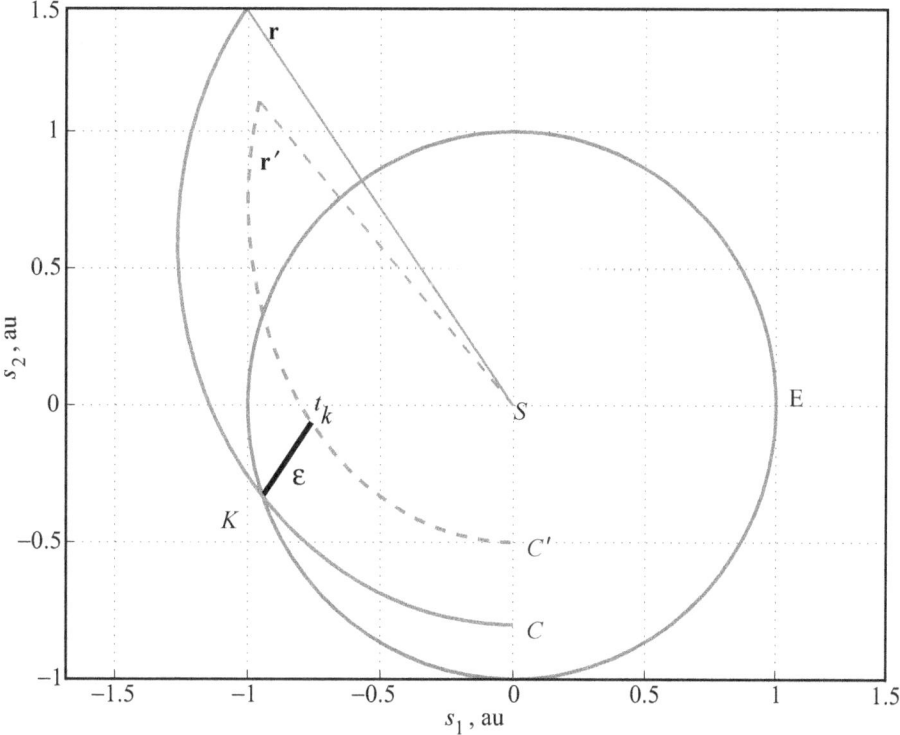

Figure 5. True orbit and orbit determined from observations.

There is one drawback to using ε to judge the quality of an orbit solution: ε may vanish even though $\mathbf{r}'(t) \ne \mathbf{r}(t)$ and $\mathbf{v}'(t) \ne \mathbf{v}(t)$. A solution to the Gauss problem, described in section 5.2 of reference 1, yields the velocities $\mathbf{v}(t_1)$ and $\mathbf{v}(t_2)$ at two specified positions $\mathbf{r}(t_1)$ and $\mathbf{r}(t_2)$, where the time of flight $t_2 - t_1$ between the two positions is also specified, as is the "direction of motion" (i.e., whether the orbit proceeds from $\mathbf{r}(t_1)$ to $\mathbf{r}(t_2)$ via an angular displacement less than or greater than π). Consequently, for any position solution $\mathbf{r}'(t)$ and time of flight $t_k - t$, there exists some velocity $\mathbf{v}'(t)$ that results in C' passing through K at precisely t_k, thus $\varepsilon = 0$ even though the orbit of C' differs from that of C. If by unlucky coincidence an orbit determination algorithm happened to settle on just such a combination of $\mathbf{r}'(t)$ and $\mathbf{v}'(t)$, ε would vanish, giving a false impression that the orbit solution was identical to the true orbit. The effects of such coincidences will be diminished by introducing errors in the measurements in a random fashion, and averaging ε over a large number of measurement sets.

For each orbit examined in the sequel, ε is obtained by the following steps:

1. A set of classical orbital elements is constructed in accordance with equation (5) and used together with orbit propagation software to produce time histories of $\mathbf{r}$ and $\mathbf{v}$ for an interval of time over which optical measurements are to be taken.

2. A time of flight to collision $t_k - t$ is determined according to equation (2), whereupon Lagrange coefficients F and G (p. 179, ref. 2) are computed and used to obtain the position of collision $\mathbf{r}(t_k)$.

3. Several values of time t_i ($i = 1, 2, 3, \ldots$) are selected for obtaining observations. For each such value a position of an observatory is constructed and subtracted from $\mathbf{r}(t_i)$ to yield the position vector from the observatory to the object of interest; this vector is in turn used to calculate a longitude and latitude of the object with respect to a set of inertially fixed heliocentric-ecliptic axes.

4. For each observation, errors are intentionally introduced in the longitude and latitude to reflect the limits in angular resolution of a telescope; the resulting angles serve as measurements.

5. A guess is made as to position and velocity at the time t of one of the observations. The object of the process of orbit determination is to adjust the guess and obtain $\mathbf{r}'(t)$ and $\mathbf{v}'(t)$ that are as consistent as possible with the measurements.

6. The time of flight obtained in step 2 is used together with $\mathbf{r}'(t)$ and $\mathbf{v}'(t)$ to obtain $\mathbf{r}'(t_k)$.

7. The erroneous predicted miss distance ε is the magnitude of $\boldsymbol{\varepsilon} = \mathbf{r}'(t_k) - \mathbf{r}(t_k)$.

Method of Weighted Least Squares

Precision orbit determination can be performed with the linear method of weighted least squares formulated by Carl Friedrich Gauss in 1809, sketched out in references 3 and 4, and explained in detail in reference 5. We have used this method to obtain preliminary determination of orbits using three or four observations as described subsequently, as well as improved orbit determination from many observations and multiple observatories.

The method of weighted least squares can be described briefly as follows: Let $X_1, \ldots, X_6$ be six orbital elements, for example, six scalars associated at a particular instant of time with position $\mathbf{r}(t)$ and velocity

v(*t*), to be determined from measurements of longitude $\tilde{\phi}(t_1), \ldots, \tilde{\phi}(t_n)$ and latitude $\tilde{\lambda}(t_1), \ldots, \tilde{\lambda}(t_n)$ obtained at the times $t_1, \ldots, t_n$ of n observations. Using estimates $\hat{X}_1, \ldots, \hat{X}_6$ of the orbital parameters, one may compute corresponding values of longitude $\overline{\phi}(\hat{X}_1, \ldots, \hat{X}_6, t_1), \ldots, \overline{\phi}(\hat{X}_1, \ldots, \hat{X}_6, t_n)$ and latitude $\overline{\lambda}(\hat{X}_1, \ldots, \hat{X}_6, t_1), \ldots, \overline{\lambda}(\hat{X}_1, \ldots, \hat{X}_6, t_n)$, and form $2n$ residuals by subtracting the computed values from the measurements,

$$y_j(\hat{X}_1, \ldots, \hat{X}_6) = \begin{cases} \tilde{\phi}(t_j) - \overline{\phi}(\hat{X}_1, \ldots, \hat{X}_6, t_j) & (j = 1, \ldots, n) \\ \tilde{\lambda}(t_{j-n}) - \overline{\lambda}(\hat{X}_1, \ldots, \hat{X}_6, t_{j-n}) & (j = n+1, \ldots, 2n) \end{cases} \quad (11)$$

The purpose of the method of weighted least squares is to find the values $\hat{X}_1^*, \ldots, \hat{X}_6^*$ that minimize the sum of the squares of the weighted residuals,

$$Q = \sum_{j=1}^{2n} w_j y_j^2 \quad (12)$$

where weights $w_1, \ldots, w_{2n}$ are chosen to give more weight or importance to the measurements obtained with a better resolution; $w_j^{-1/2}$ is the assumed accuracy of the jth measurement. Now, for values of $X_1, \ldots, X_6$ in the neighborhood of $\hat{X}_1, \ldots, \hat{X}_6$, the computed values can be expanded in a Taylor series, for example,

$$\overline{\phi}(X_1, \ldots, X_6, t_j) = \overline{\phi}(\hat{X}_1, \ldots, \hat{X}_6, t_j) + \sum_{i=1}^{6} (X_i - \hat{X}_i) \frac{\partial \overline{\phi}}{\partial X_i}(\hat{X}_1, \ldots, \hat{X}_6, t_j) + \cdots \quad (j = 1, \ldots, n) \quad (13)$$

which can be used to write approximate expressions for residuals

$$y_j(X_1, \ldots, X_6) = \tilde{\phi}(t_j) - \overline{\phi}(X_1, \ldots, X_6, t_j)$$

$$\approx y_j(\hat{X}_1, \ldots, \hat{X}_6) - \sum_{i=1}^{6}(X_i - \hat{X}_i) \frac{\partial \overline{\phi}}{\partial X_i}(\hat{X}_1, \ldots, \hat{X}_6, t_j) \quad (j = 1, \ldots, n) \quad (14)$$

If one defines arrays

$$\{y\} \triangleq \begin{bmatrix} y_1(X_1, \ldots, X_6) & y_2(X_1, \ldots, X_6) & \cdots & y_{2n}(X_1, \ldots, X_6) \end{bmatrix}^T \quad (15)$$

$$\{\hat{y}\} \triangleq \begin{bmatrix} y_1(\hat{X}_1, \ldots, \hat{X}_6) & y_2(\hat{X}_1, \ldots, \hat{X}_6) & \cdots & y_{2n}(\hat{X}_1, \ldots, \hat{X}_6) \end{bmatrix}^T \quad (16)$$

$$\{x\} \triangleq \begin{bmatrix} (X_1 - \hat{X}_1) & (X_2 - \hat{X}_2) & \cdots & (X_6 - \hat{X}_6) \end{bmatrix}^T \quad (17)$$

and matrices $[H]$ and $[W]$, whose elements are defined as

$$H_{ji} = \begin{cases} \dfrac{\partial \overline{\phi}}{\partial X_i}\left(\hat{X}_1, ..., \hat{X}_6, t_j\right) & (j=1, ..., n;\ i=1, ..., 6) \\[1em] \dfrac{\partial \overline{\lambda}}{\partial X_i}\left(\hat{X}_1, ..., \hat{X}_6, t_{j-n}\right) & (j=n+1, ..., 2n;\ i=1, ..., 6) \end{cases} \qquad (18)$$

$$W_{ji} = \begin{cases} w_j & (i=j;\ j=1, ..., 2n) \\ 0 & (i \ne j;\ i=1, ..., 2n) \end{cases} \qquad (19)$$

then Q can be expressed as

$$Q = \sum_{j=1}^{2n} w_j y_j^2 = \{y\}^T [W]\{y\} \approx (\{\hat{y}\}-[H]\{x\})^T [W](\{\hat{y}\}-[H]\{x\})$$

$$= \{\hat{y}\}^T [W]\{\hat{y}\} - \{\hat{y}\}^T [W][H]\{x\} - \{x\}^T [H]^T [W]\{\hat{y}\} + \{x\}^T [H]^T [W][H]\{x\} \qquad (20)$$

The value of $\{x\}$ that minimizes Q is obtained by setting $\partial Q/\partial\{x\} = \{0\}$, a 6×1 array of zeros, which yields

$$\{x\} = \left([H]^T[W][H]\right)^{-1}[H]^T[W]\{\hat{y}\} \qquad (21)$$

as long as $([H]^T[W][H])$ is nonsingular; this quantity is known as the normal or information matrix and is equal to the inverse of the covariance matrix describing the accuracy of $\{x\}$. When the normal matrix is singular, the orbit cannot be determined uniquely from the measurements and the orbit is said to be unobservable (ref. 5, sec. 4.12). A new estimate of the orbital parameters is formed by adding the adjustment $\{x\}$ to the old estimate $\{\hat{x}\} = [\hat{x}_1\ \hat{x}_2\ \hat{x}_3\ \hat{x}_4\ \hat{x}_5\ \hat{x}_6]^T$, and the process is repeated, leading to the values of $\hat{x}_1^*, ..., \hat{x}_6^*$ that minimize Q.

As discussed in reference 5, each row of the $2n \times 6$ mapping matrix $[H]$ can be regarded as the product $[\tilde{H}][\Phi(t_j,t)]$, where $[\tilde{H}]$ is a 1×6 matrix of partial derivatives of a computed value ($\overline{\phi}$ or $\overline{\lambda}$) with respect to $X_1, ..., X_6$, and $[\Phi(t_j,t)]$ is the 6×6 state transition matrix for the time of the measurement t_j and the time t at which the orbit is to be determined. Both matrices are evaluated with the values $\hat{X}_1, ..., \hat{X}_6$. Although the rows of $[H]$ are often computed with these products, it is also possible to form $[H]$ numerically and this is one option available with the MATLAB® function LSQNONLIN. With this software, a search direction can be determined with the Levenberg-Marquardt method, a hybrid of the Gauss-Newton approach and the method of steepest descent (ref. 6, pp. 2-16 to 2-21). Parameter options that we set specifically for orbit determination include the minimum change in variables for finite differencing, DiffMinChange $= 5 \times 10^{-4}$, the termination tolerance on the function value, TolFun $= 1 \times 10^{-10}$, and LargeScale is set to off to select the Levenberg-Marquardt method rather than a large-scale optimization algorithm. The elements of $\{\hat{x}\}$ are scaled so that the first three are in units of au and the last three are in units of km/s. The residuals are rectified so that $-\pi \le y_j \le \pi$ and are then multiplied by 1×10^6 to avoid difficulties when the residuals are on the scale of the numerical precision of the computer, and where round off errors can cause a serious loss of accuracy.

Preliminary Orbit Determination

The method of least squares has been used together with the seven-step procedure described in connection with the erroneous predicted miss distance, and MATLAB® software, to determine how the accuracy of preliminary orbit determination based on three observations is affected by the time interval between successive observations, by the distance r at which the object is detected, and by the resolution of the telescope.

Proceeding with the first step, a set of classical orbital elements is specified as $r_a = 10$ au, $r_p = 0.5$ au, $i = 16°$, and $\Omega = 36.5°$, where i is the inclination of the object's orbital plane to the ecliptic, and Ω is the longitude of ascending node measured from the vernal equinox. The argument of perihelion ω is computed according to equation (5) to be 93.017°. Observations are to be performed when the object's heliocentric distance r is in the neighborhood of some specified value, so an initial value of true anomaly $\nu(t_0)$ is obtained by solving equation (3) for ν, chosen ≤ 0 so that the object is traveling toward perihelion. With this set of orbital elements, time histories of $\mathbf{r}$ and $\mathbf{v}$ are recorded over an interval of 98 days.

Continuing with step 3, the observatory is assumed to travel in the ecliptic plane in a circular orbit of radius 1 au; however, *the observatory is not necessarily coincident with the Earth*. At the beginning of the 98-day observation period, the observatory has an initial true longitude $L_0 = 180°$, measured in the ecliptic from the direction of vernal equinox. The times of the three observations are assumed to have equal spacing, $(t_3 - t_2) = (t_2 - t_1)$, ranging from 5 to 49 days. (For example, measurements obtained at the beginning of the 1st, 50th, and 99th days have an interval of 49 days between successive observations, and the total span of the data arc is 98 days.)

Figure 6 illustrates the comet's orbit with a blue curve, and the red portion of the orbit indicates the interval over which observations are made, beginning at $\nu(t_0) = -159.7°$. The green plane represents the ecliptic and the asterisk denotes the Sun. The three black points indicate the positions of the observatory at the beginning, middle, and end of the observation period, where the beginning is marked by the leftmost point.

As described in step 4, each of the six measurements is regarded as the sum of a true angle and an error, which is now assumed to be no greater than the telescope resolution ρ. The MATLAB® function RAND is employed to produce pseudorandom errors uniformly distributed between the limits of the telescope's resolution, $\pm\rho$. In a Monte Carlo approach, an orbit is determined 100 times using a different random number seed in each trial, and an average erroneous predicted miss distance is reported. The assumed distribution of the error is not actually an important consideration when using the method of least squares since this method does not account for such statistics. A uniform distribution is simple to generate and just as valid as any other in this case.

Figure 7 shows, on logarithmic scale of base 10, the average value of ε as a function of the interval between observations, $(t_3 - t_2)$ or $(t_2 - t_1)$, where the time of the second observation t_2 is taken to be the 50th day of the 98-day data arc. The heliocentric distance r at t_2 is varied between 5, 6, and 7 au by using initial values $\nu(t_0) = -155.3°$, $-159.7°$, and $-163.7°$, respectively. Each point on a line represents the average value $\bar{\varepsilon}$ of ε in units of lunar distance from Earth (384 400 km), from a set of 100 measurement errors formed with $\rho = 0.2$ seconds of arc (arcsec). Accuracy of orbit determination becomes better as $r(t_2)$ decreases; however, the most dramatic improvements are obtained by increasing the observation interval to 49 days, at which point all three values of $\bar{\varepsilon}$ are less than 2 lunar distances. Using the laws of logarithms, it is easily shown that a relationship of the form $\bar{\varepsilon} = \bar{\varepsilon}_0 (t_2 - t_1)^m$, where $\bar{\varepsilon}_0$ is a constant,

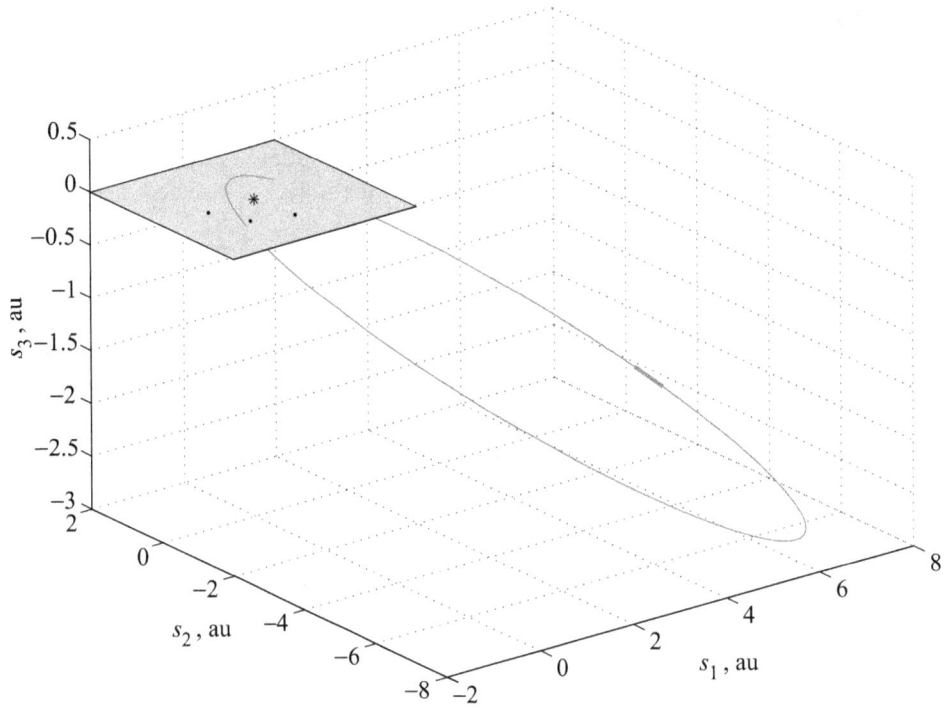

Figure 6. Observation arc.

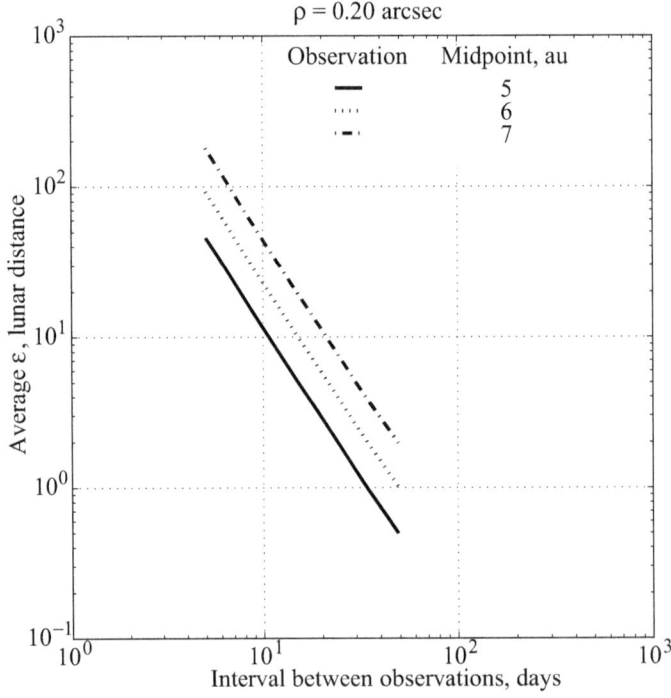

Figure 7. Average ε for various detection distances.

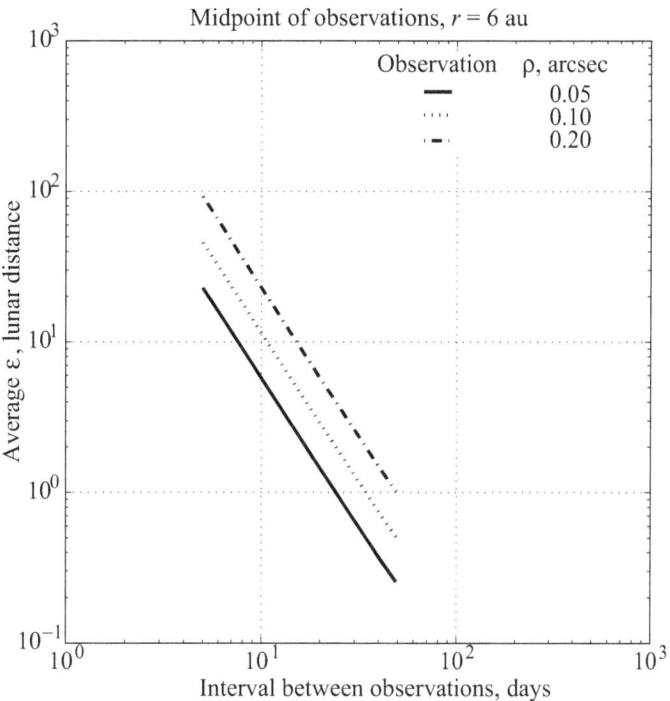

Figure 8. Average ε for various angular resolutions.

becomes $\log_{10} \bar{\varepsilon} = m \log_{10}(t_2 - t_1) + \log_{10} \bar{\varepsilon}_0$, which is the equation of a line with slope m. In each line of figure 7, $\bar{\varepsilon}$ decreases by two powers of 10 for every power of 10 increase in $(t_2 - t_1)$; therefore $m = -2$, meaning the accuracy of the preliminary orbit improves with the square of the observation interval.

Figure 8 displays similar information, with $r(t_2)$ held fixed at 6 au and $\rho = 0.05$, 0.10, and 0.20 arcsec. Telescope resolution is seen to have an effect on preliminary orbit determination accuracy, but the length of the observation interval is again the most significant factor. With an observation interval of 49 days, all values of $\bar{\varepsilon}$ are below 1 lunar distance. Unfortunately, longer observation intervals yield shorter warning times.

In the same vein, an extensive analysis of preliminary determination of LPC orbits is carried out with the following additional details.

In connection with the first step, 1008 sets of classical orbital elements are constructed with $r_a = 10$, 40, 70, and 100 au, $r_p = 0.1$, 0.4, 0.7, and 1.0 au, $i = 10°$, 30°, 50°, ..., 170°, and $\Omega = 0°$, 30°, 60°, ..., 180°. Time of flight is nearly constant with respect to r_a for $r_a > 100$ au (see fig. 3); therefore, such orbits are not considered. For each set, the argument of perihelion ω is computed according to equation (5). A 98-day data arc begins with $r(t_0) = 6.5$ au, and $v(t_0)$ is chosen accordingly.

As before, the observatory is assumed to travel in the ecliptic plane in a circular orbit of radius 1 au, and it is worth remembering that *the observatory is not necessarily coincident with Earth*. The effect of observatory position on orbit determination is studied by allowing the initial true longitude L_0 of the observatory to take on four values, each differing by 90°. Thus, a total of $4 \times 1008 = 4032$ cases are examined. The times of the three observations are equally spaced, with 33 days between subsequent observations.

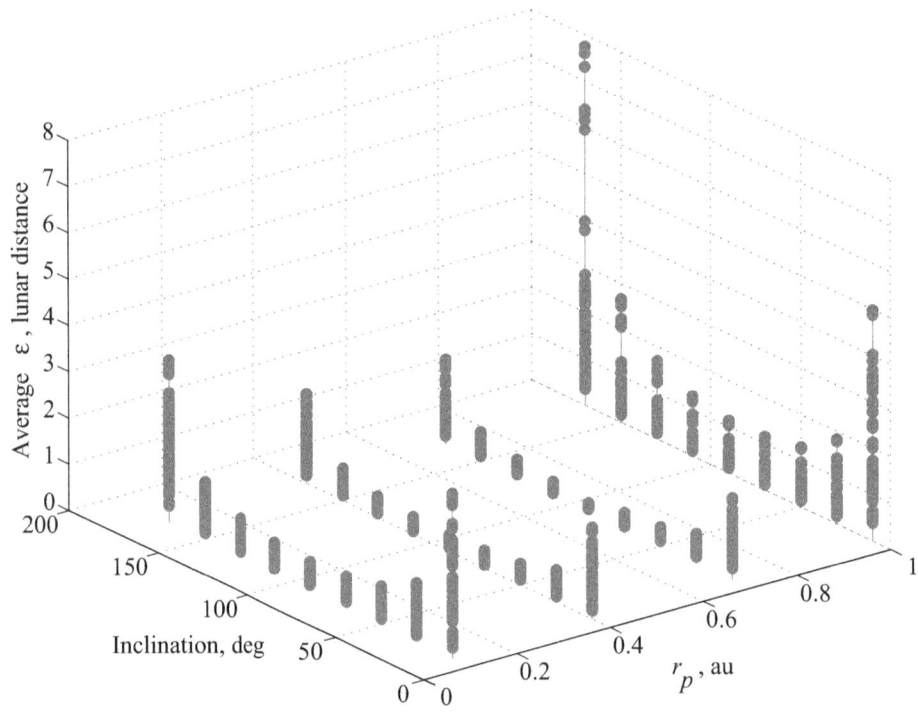

Figure 9. Accuracy of comet orbits, as a function of i and r_p.

For step 4, a telescope resolution of $\rho = 0.1$ arcsec is used to form 100 sets of measurement errors for each orbit to be examined, and the average value of erroneous predicted miss distance is recorded.

The results are displayed in figure 9 as functions of i and r_p; each data point represents $\bar{\varepsilon}$ for one case, and a family of 112 cases (in which r_a, Ω, and L_0 vary) form one stack of points. The best orbit determination is obtained when the orbital planes of the observatory and comet are perpendicular, whereas nearly coplanar orbits yield the poorest determinations. This relationship to inclination stems from the fact that, with three observations, an orbit is unobservable when it is coplanar with the orbit of the observatory. One may form analytic expressions for the elements of the matrix $[\tilde{H}]$ discussed in connection with the method of weighted least squares which are partial derivatives of longitude and latitude. In addition, one may determine which elements of the state transition matrix $[\Phi]$ are zero when two-body mechanics are assumed; a reference orbit coplanar with that of the observatory is used and $[\Phi]$ is written according to equations (9.84)–(9.87) of reference 2. It can then be shown that for three observations, the 6 × 6 matrix $[H]$ is singular, but for four observations the 6 × 6 matrix $[H]^T[H]$ is not singular. A study of figure 9 shows that with $r_p = 1$ au, retrograde orbits are harder to determine accurately than prograde orbits. Orbit determination accuracy with $r_p = 0.1$ au is noticeably poorer than with $r_p = 0.4$ or 0.7 au, and even poorer with $r_p = 1$ au. In figure 10 the results are shown as functions of r_a and r_p; each stack contains 252 values of $\bar{\varepsilon}$ associated with varied i, Ω, and L_0. Orbit determination is the most difficult when r_a has the smallest value, 10 au.

Results are not presented as functions of Ω, ω, and $v(t_0)$ because there is limited value in doing so. No remarkable relationship appears to exist between $\bar{\varepsilon}$ and Ω. The angle $v(t_0)$ simply determines the point of the object's orbit at which observations begin. Finally, argument of perihelion ω is dependent upon r_a and r_p through equation (5).

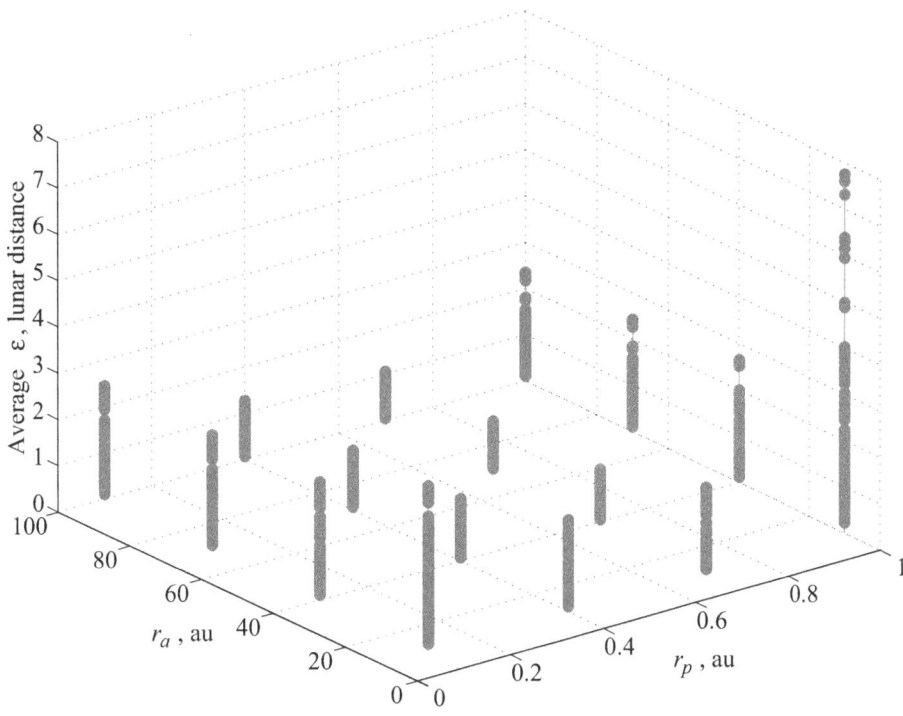

Figure 10. Accuracy of comet orbits, as a function of r_a and r_p.

The 4032 average values of ε are sorted from largest to smallest, and the worst and best cases are identified. The effect of the length of observation interval is then studied for these two cases, with times of the three observations again given equal spacing, ranging from 5 to 49 days between subsequent observations. The worst case, numbered 246, is associated with the orbital parameters $r_a = 10$ au, $r_p = 1$ au, $i = 170°$, $\Omega = 0°$, $\omega = 0°$, and $\nu(t_0) = -151.68°$, and $L_0 = 180°$ for the observatory. Curves for the cases with the same orbital parameters and other values of L_0 are also shown in figure 11. Observatory location affects the accuracy of orbit determination, but not to the same extent as length of the data arc. In all four cases, $\bar{\varepsilon}$ is less than 4 lunar distances with an observation interval of 49 days.

The best case is numbered 3946, with orbital parameters of $r_a = 100$ au, $r_p = 0.7$ au, $i = 110°$, $\Omega = 120°$, $\omega = 66.69°$, and $\nu(t_0) = -142.86°$, and $L_0 = 90°$ for the observatory. Curves for the cases with the same orbital parameters and other values of L_0 are shown in figure 12; long observation intervals once again result in the best preliminary orbit determination. With an observation interval of 49 days, the value of $\bar{\varepsilon}$ associated with each observatory is less than 0.1 lunar distance, or about 6 times the Earth's radius. A slope of $m = -2$ is evident in figures 11 and 12, confirming the inverse square relationship between $\bar{\varepsilon}$ and observation interval established in figures 7 and 8.

The foregoing results involving preliminary orbit determination corroborate statements made in references 3 and 4, pointing out that the length of the data arc is the single most important factor in determining the accuracy of the orbit solution. The number and precision of the observations, the object's proximity to the observatory when measurements are obtained, and even the use of radar measurements are all secondary to the length of the data arc.

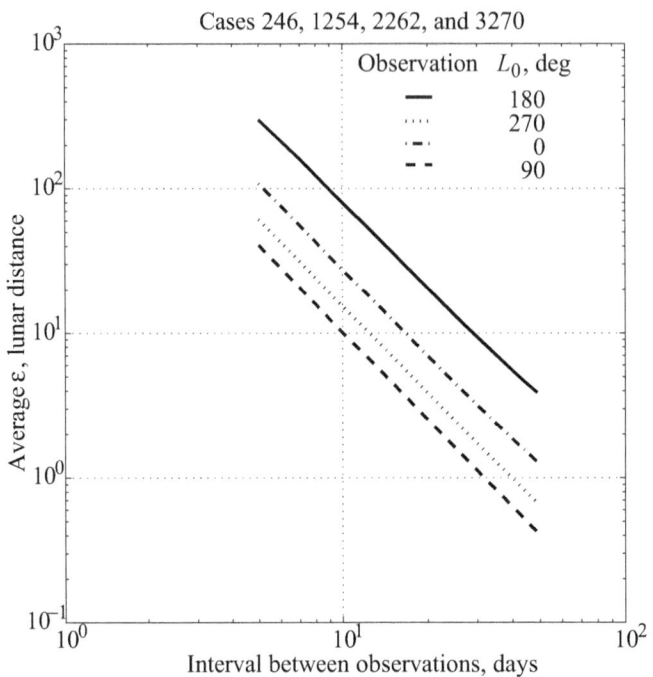

Figure 11. Worst orbit, various observatory longitudes.

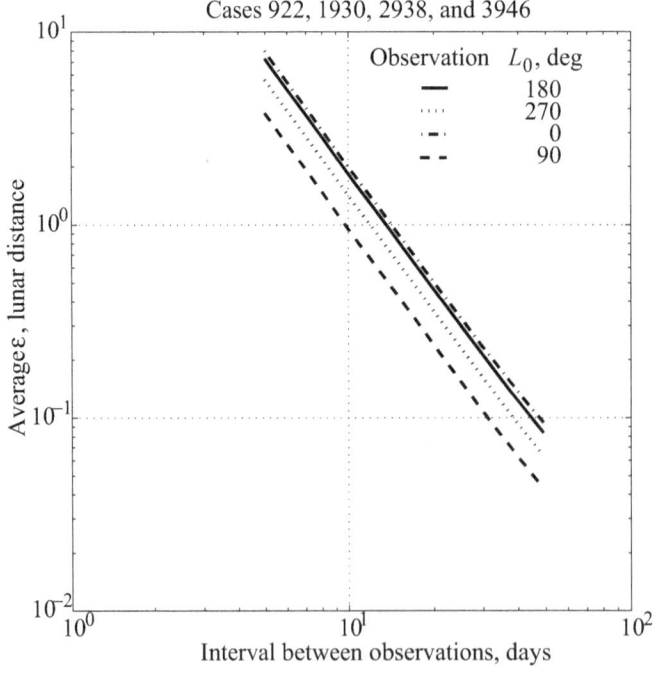

Figure 12. Best orbit, various observatory longitudes.

Multiple Observations

As opposed to the classical methods of Laplace, Gauss, and Olbers, the method of least squares permits the use of a great number of measurements, which can be obtained from one or more observatories. We now proceed to study the effects of multiple observations and observatories on orbit determination of LPCs. A number of observations, ranging from 3 to 99, are taken in equally timed increments over a period of 98 days. We have chosen to obtain an orbit solution for the time of the first observation. Differential correction begins with an estimate $\{\hat{x}\}$ corresponding to an arbitrary day, typically a day after the end of the 98-day observation period.

The forthcoming results, shown in figures 13 through 18, are reported first for a single observatory, and then for multiple observatories.

Single observatory. The orbits identified in the discussion of preliminary orbit determination as resulting in the largest and smallest average value $\bar{\varepsilon}$ of ε, cases 246 and 3946, respectively, are revisited. Figures 13 and 15 show results calculated for various numbers of observations obtained from a single observatory. In each case, a solid curve shows the average erroneous predicted miss distance $\bar{\varepsilon}$ from 100 sets of measurements, each of which is produced with a telescope resolution ρ of 0.1 arcsec.

Results for the worst orbit are contained in figure 13. With 3 observations spaced 49 days apart, $\bar{\varepsilon}$ is nearly 3.8 lunar distances, whereas 99 observations spaced 1 day apart reduce $\bar{\varepsilon}$ to less than 0.22 of a lunar distance, or about 13 Earth radii. In the case of the best orbit, the quality of the result is evidenced by the use of Earth radius (6378 km) as the unit for measuring average erroneous predicted miss distance. The data in figure 15 show that with 3 observations $\bar{\varepsilon}$ is approximately 2.5 Earth radii; 99 observations improve the measure to less than 0.5 Earth radius. In the worst case there is a marked improvement from increasing the number of observations from 3 to 5, followed by a more gradual slope of approximately $m = -1/2$ in going from 10 to 99 observations. In the best case, the slope is also approximately $-1/2$. As discussed previously, this is an indication that the accuracy of the orbit as measured by $\bar{\varepsilon}$ improves as the square root of the number of observations.

As one might expect, better telescope resolution (indicated by a smaller value of ρ) improves orbit determination accuracy. In order to quantify the improvement available with very high resolution measurements (obtained, for example, with interferometry) an analysis is made with 11 observations taken over a 98-day data arc involving a mixture of resolutions; $\rho = 0.01$ arcsec for initial observations, and $\rho = 0.0001$ arcsec for some number of final observations. As described earlier, weights are assigned using the value of ρ with which the corresponding measurements are made, thus giving more weight or importance to the measurements obtained with a better resolution ($w_j = 1/\rho_j^2$). In figures 17 and 18, improvements in $\bar{\varepsilon}$ obtained for the worst orbit (Case 246) and the best orbit (Case 3946), respectively, are shown. The number of final observations made with the better resolution is indicated on the abscissa. A comparison of figures 13 and 17 reveals $\bar{\varepsilon}$ is made better by a factor of 10 when the resolution of 11 observations is improved from 0.1 to 0.01 arcsec. The same conclusion is reached by comparing figures 15 and 18. In each case, a further improvement by a factor of 10 is obtained when the last 4 observations are made with $\rho = 0.0001$ arcsec.

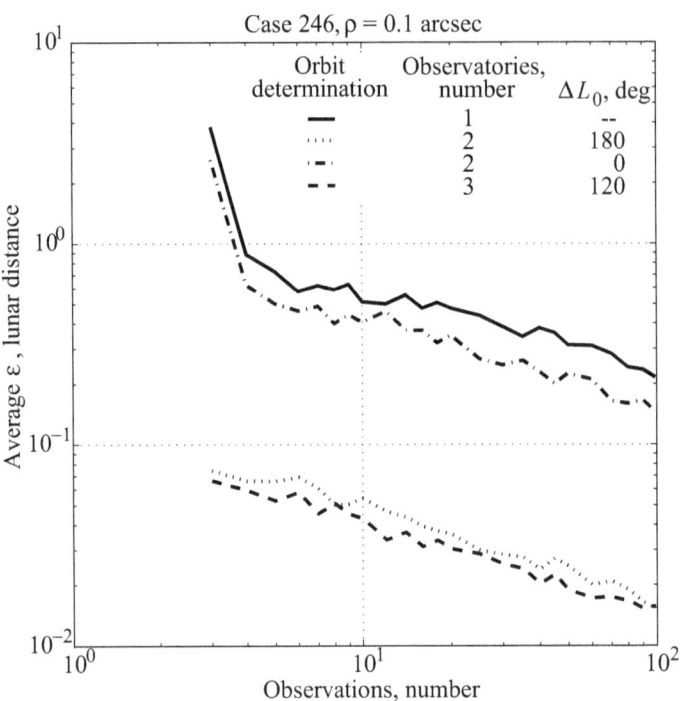

Figure 13. Worst orbit, various observatory configurations.

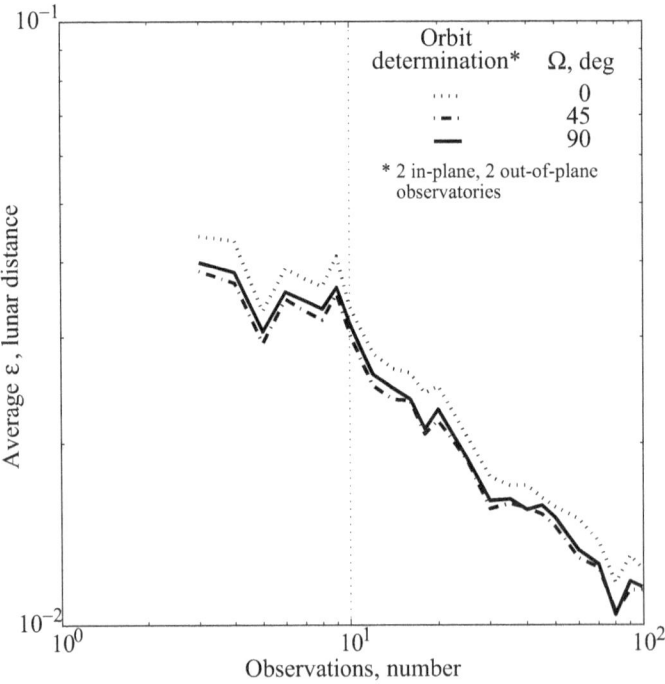

Figure 14. Worst orbit, various observatory configurations (detailed view).

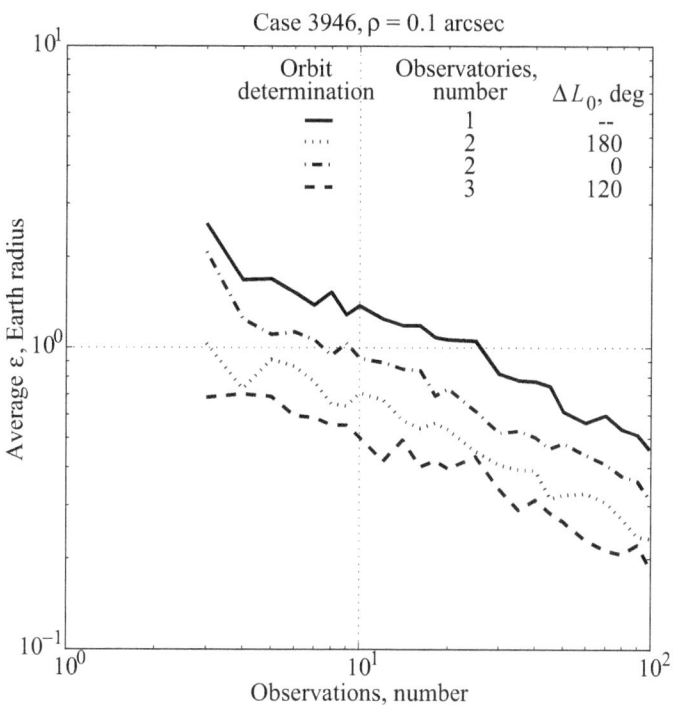

Figure 15. Best orbit, various observatory configurations.

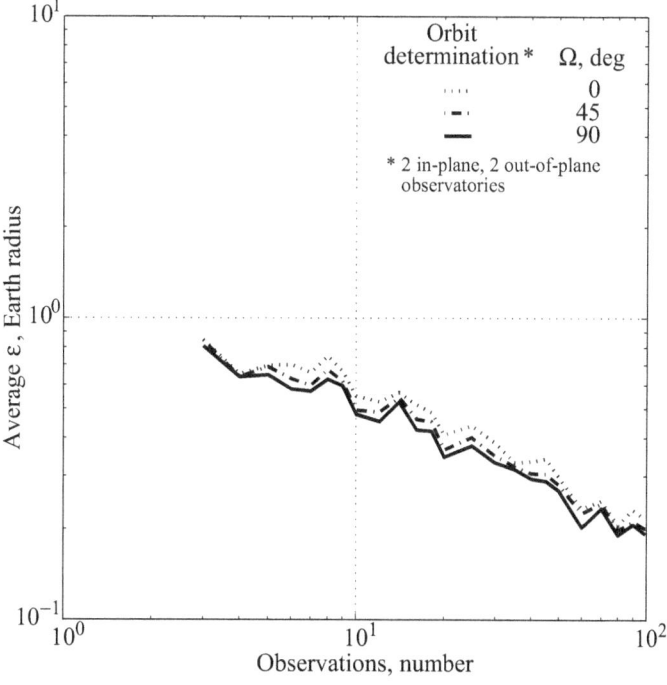

Figure 16. Best orbit, various observatory configurations (detailed view).

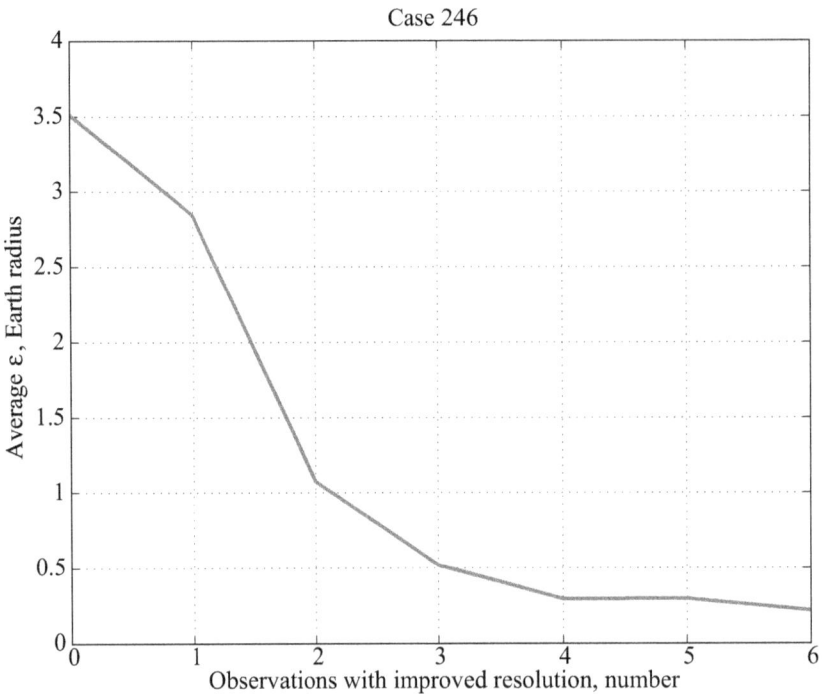

Figure 17. Worst orbit, observations with mixed resolutions.

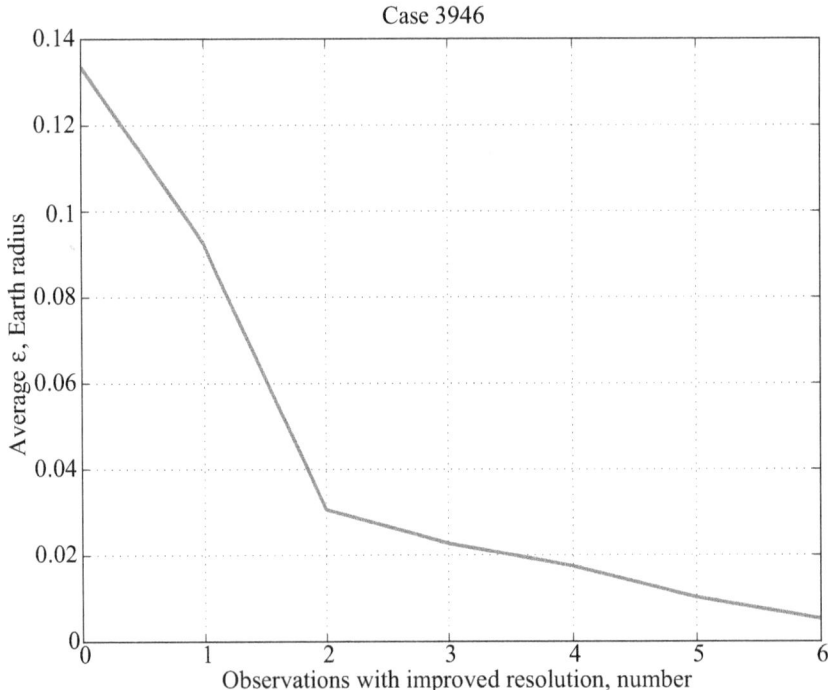

Figure 18. Best orbit, observations with mixed resolutions.

Multiple observatories. As previously mentioned, the method of least squares permits orbit determination to take advantage of measurements not only from additional observations, but also from additional observatories. The benefits of configurations of 2, 3, and 4 observatories are studied with the aid of the worst and best LPC orbits.

The positions of two observatories are constructed such that they have heliocentric circular orbits of radius 1 au in the ecliptic plane, and their true longitudes L_0 are always 180° apart. Three observatories are placed in similar orbits, with their true longitudes phased by 120°. In the case of four observatories, the first two have orbits identical to the configuration of two observatories just described, and the remaining two are in similar coplanar orbits perpendicular to the ecliptic with the ascending node Ω taking on values of 0°, 45°, and 90° in order to determine what effect, if any, Ω has on orbit determination.

In connection with the worst comet orbit, figures 13 and 14 display $\bar{\varepsilon}$ as a function of the number of observations for all observatory configurations described heretofore. It is evident that two or more observatories offer an improvement in $\bar{\varepsilon}$ of nearly a factor of 10 over that from a single observatory. With 99 observations each, two observatories yield $\bar{\varepsilon}$ less than 0.016 lunar distance, slightly less than 1 Earth radius. Three observatories are only marginally better than two observatories. Four observatories are not substantially better than three, and $\bar{\varepsilon}$ is relatively insensitive to the value of Ω for the members of the four-observatory configuration that have orbits normal to the ecliptic. Similarly, figures 15 and 16 show that in the best case substantial improvement in $\bar{\varepsilon}$ is obtained by employing two observatories phased by 180° instead of a single observatory, but the addition of a third or fourth observatory does not appear to be cost effective for this orbit and observatory heliocentric distances of 1 au.

The advantage of two observatories over one may stem from a doubling in the number of observations, or from parallax; in order to gauge the contribution of each, one may form a configuration of two observatories that are coincident with one another, thus eliminating parallax. Results obtained from this configuration are represented with a dash-dot curve in figures 13 and 15. The first of these shows that in the worst case, the improvement obtained from a pair of observatories stems more from parallax than from doubling the number of observations. Figure 15 shows that in the best case, the benefit of two observatories has as much to do with parallax as it does with the availability of twice as many observations.

Of the multiobservatory configurations examined here, two observatories in circular heliocentric orbits of radius 1 au in the ecliptic plane, phased 180° apart, provide the best balance of cost and accuracy in determining the orbits of LPCs. Thirty to forty observations equally spaced over a 98-day interval appear to give results nearly as good as 99 observations taken 1 day apart.

Sequential Filter for Long-Period Comets

The method of weighted least squares, described previously and used to produce all of the results presented earlier, is referred to as a batch processor because it employs observations collected together in a single set; the larger the batch, the more arithmetical operations are required to carry out the matrix multiplications indicated in equation (21). Moreover, when new observations are obtained, they must be added to the previous batch to form an even larger set, and the entire orbit determination procedure must be repeated. Another shortcoming of the method is that there are no provisions for treating the state and observation deviations as random processes and using any associated statistical information to improve the orbit solution.

Remedies for these problems exist in the Kalman filter, a recursive or sequential algorithm in which measurements are used continuously, as they become available, to refine the solution for the orbit. The recursive property of the procedure allows the previous solution to be used as the starting point for the solution to follow. The computational expense of the sequential filter can be less than that of the batch filter because the matrix requiring inversion in the former case can be smaller—when only one measurement is processed at a time, the matrix is reduced to a scalar.

We describe briefly two variations of the Kalman filter, the conventional and the extended algorithms. A method for obtaining the a priori information needed to start the filter is also discussed. Warning time offered by a sequential filter with high resolution optical measurements from a single observatory is compared with that derived from measurements of lower resolution from two observatories. The advantages of measurements of range and range-rate are illustrated. Effects of comet outgassing are considered, followed by a discussion of the construction and use of the body-targeting plane or B-plane in orbit determination analysis. Finally, the calculation of probability and probability of collision are taken up.

The Kalman Filter

The conventional Kalman filter is developed from differential equations for the state, and expressions that relate the measurements to the state, both of which have been linearized about a reference state time history. To reduce the effects of the errors introduced in the process of linearization, the reference state can be replaced after each observation with the improved estimate of the state; this procedure is referred to as the extended Kalman filter. The two varieties of the filter are discussed in turn in what follows.

Conventional Kalman filter. The conventional Kalman filter is described in detail in section 4.7.1 of reference 5 in terms similar to those used earlier in connection with the batch algorithm.

The $n \times 1$ column matrix $\{X\}$ contains n ($n = 6$) orbital elements, dot products of the position $\mathbf{r}(t)$ and velocity $\mathbf{v}(t)$ with three right-handed, mutually perpendicular unit vectors. The state deviation matrix $\{x\}$ is defined as the difference between the true elements $\{X\}$ and the elements $\{X^*\}$ associated with a reference orbit,

$$\{x\} \stackrel{\Delta}{=} \{X\} - \{X^*\} \tag{22}$$

The $m \times 1$ observation deviation, or residual matrix $\{y\}$, is defined as the difference between m measurements $\{Y\}$ and the values $\{Y^*\}$ calculated from the reference orbit,

$$\{y\} \stackrel{\Delta}{=} \{Y\} - \{Y^*\} \tag{23}$$

where the measurements $\{Y\}$ are regarded as the sum of nonlinear functions $\{G(\{X\}, t)\}$ of the state and time, and errors $\{e\}$,

$$\{Y\} = \{G(\{X\},t)\} + \{e\} \tag{24}$$

Equations for the filter time update are given as equations (4.7.18) and (4.9.50) of reference 5:

$$\{\bar{x}_j\} = [\Phi(t_j, t_{j-1})]\{\hat{x}_{j-1}\} \tag{25}$$

$$[\overline{P}_j] = [\Phi(t_j, t_{j-1})][P_{j-1}][\Phi(t_j, t_{j-1})]^{\mathrm{T}} + [Q_{j-1}] \tag{26}$$

where $\{\hat{x}_{j-1}\}$ and $\{\overline{x}_j\}$ are, respectively, the estimated values of $\{x\}$ following a measurement update at an instant of time t_{j-1}, and prior to a measurement update at t_j. The $n \times n$ state transition matrix $[\Phi(t_j, t_{j-1})]$ that relates the state deviations at times t_{j-1} and t_j is discussed presently. The $n \times n$ estimate error variance-covariance matrices (or simply covariance matrices) of the state deviations $\{\hat{x}_{j-1}\}$ and $\{\overline{x}_j\}$ are denoted by $[P_{j-1}]$ and $[\overline{P}_j]$, respectively. The $n \times n$ matrix Q_{j-1} is the covariance at t_{j-1} of a zero-mean, white sequence known as process noise or state noise.

The $m \times 1$ observation deviation $\{y_j\}$, $m \times n$ observation-state mapping matrix $[\tilde{H}_j]$, and the $n \times m$ Kalman gain $[K_j]$ are assembled in preparation for the measurement update:

$$\{y_j\} = \{Y_j\} - \left\{G\left(\{X_j^*\}, t_j\right)\right\} \tag{27}$$

$$[\tilde{H}_j] = \frac{\partial\left\{G\left(\{X_j^*\}, t_j\right)\right\}}{\partial\{X\}} \tag{28}$$

$$[K_j] = [\overline{P}_j][\tilde{H}_j]^{\mathrm{T}} \left([\tilde{H}_j][\overline{P}_j][\tilde{H}_j]^{\mathrm{T}} + [R_j]\right)^{-1} \tag{29}$$

where $\{Y_j\}$ contains the actual measurements at t_j, and where the $m \times m$ matrix $[R_j]$ is the covariance at t_j of $\{e\}$, assumed to have a normal distribution and a mean of zero. Expressions for the elements of $[\tilde{H}_j]$ associated with measurements of longitude and latitude are developed presently, and measurements of range and range-rate are taken up subsequently.

Equations for the filter measurement update are set forth in equations (4.7.16) and (4.7.12) of reference 5,

$$\{\hat{x}_j\} = \{\overline{x}_j\} + [K_j]\left(\{y_j\} - [\tilde{H}_j]\{\overline{x}_j\}\right) \tag{30}$$

$$[P_j] = \left([I] - [K_j][\tilde{H}_j]\right)[\overline{P}_j] \tag{31}$$

where $\{\hat{x}_j\}$ contains the estimated values of the state deviation following a measurement update at time t_j, and $[P_j]$ is the corresponding covariance matrix. After any update performed with measurements obtained at t_j, the best estimate $\{X'_j\}$ of the orbital elements can be formed by adding $\{\hat{x}_j\}$ to the elements $\{X_j^*\}$ associated with the reference orbit at that time, $\{X'_j\} = \{\hat{x}_j\} + \{X_j^*\}$.

In the forthcoming discussion of Gauss's method we take up the matter of obtaining the a priori information needed to start the filtering process, namely the initial values $\{X_0^*\}$ of the elements of the reference orbit, the initial state deviation estimate $\{\hat{x}_0\}$, and a consistent initial covariance matrix $[P_0]$.

The state transition matrix $[\Phi(t_j, t_{j-1})]$ required in equations (25) and (26) for the time update is formed on the assumption that motion of the Sun and the object is governed by two-body gravitational mechanics. The position vector $\mathbf{r}^*_{j-1}$ at time t_{j-1} from the mass center of the Sun to the mass center of the object traveling on the reference orbit, and the inertial time derivative $\mathbf{v}^*_{j-1}$ of that position vector, are related to the relative position $\mathbf{r}^*_j$ and velocity $\mathbf{v}^*_j$ at another time t_j by equations (9.68) of reference 2,

$$\mathbf{r}^*_j = F \mathbf{r}^*_{j-1} + G \mathbf{v}^*_{j-1} \tag{32}$$

$$\mathbf{v}^*_j = F_t \mathbf{r}^*_{j-1} + G_t \mathbf{v}^*_{j-1} \tag{33}$$

where F, G, F_t, and G_t are known as Lagrangian coefficients, which can be calculated with Battin's universal variables as indicated in equations (9.69) of reference 2. The state transition matrix is partitioned into four 3×3 parts,

$$[\Phi(t_j, t_{j-1})] = \begin{bmatrix} [\tilde{R}(t_j, t_{j-1})] & [R(t_j, t_{j-1})] \\ [\tilde{V}(t_j, t_{j-1})] & [V(t_j, t_{j-1})] \end{bmatrix} \tag{34}$$

where the superscript * is omitted because it is used by Battin to indicate the adjoint of a matrix, but it is to be understood that the partitions are evaluated with $\mathbf{r}^*_{j-1}$, $\mathbf{v}^*_{j-1}$, $\mathbf{r}^*_j$, and $\mathbf{v}^*_j$ of the reference orbit according to equations (9.84)–(9.87) of reference 2. The universal variables U_0, U_1, U_2, and U_3 required to evaluate the Lagrangian coefficients are obtained straightforwardly from the relationships given in problem 4-21 of reference 2. The coefficient C, required to form $[\tilde{V}(t_j, t_{j-1})]$ according to equation (9.87), is expressed in equation (9.74) in terms of U_2, U_4, and U_5. The second of these, U_4, is easily obtained from U_1, U_2, and U_3 with the aid of equation (4.108). The universal variable U_5 can be expressed as in problem 4-30 in terms of U_1 and U_3, except when $\chi = U_1 = 0$, in which case $U_0 = 1$, and $U_2 = U_3 = U_4 = 0$; the variable u also vanishes in view of equation (4.100), which leads to $q = 0$ by way of equation (4.104), and to $U_5(2\chi) = 0$ from equation (4.112). Finally, the first of equations (4.113) yields $U_5(\chi) = U_5(0) = 0$.

In connection with measurements of longitude ϕ and latitude λ, the observation-state mapping matrix $[\tilde{H}_j]$ defined in equation (28) is derived as follows. The position vector from the Sun S to the object C can be written as

$$\mathbf{r} = \mathbf{r}^{SC} = X_1 \hat{\mathbf{s}}_1 + X_2 \hat{\mathbf{s}}_2 + X_3 \hat{\mathbf{s}}_3 \tag{35}$$

where $\hat{\mathbf{s}}_1$, $\hat{\mathbf{s}}_2$, and $\hat{\mathbf{s}}_3$ are a set of right-handed, mutually orthogonal unit vectors fixed in an inertial or Newtonian reference frame N. Unit vectors $\hat{\mathbf{s}}_1$ and $\hat{\mathbf{s}}_2$ lie in the ecliptic plane, with $\hat{\mathbf{s}}_1$ in the direction of vernal equinox, and the direction of $\hat{\mathbf{s}}_3$ is north of the ecliptic. Similarly, the position vector from S to an observatory O can be expressed as

$$\mathbf{r}^{SO} = O_1 \hat{\mathbf{s}}_1 + O_2 \hat{\mathbf{s}}_2 + O_3 \hat{\mathbf{s}}_3 \tag{36}$$

and the position vector **d** from O to C is thus

$$\mathbf{d} \triangleq \mathbf{r}^{OC} = \mathbf{r}^{SC} - \mathbf{r}^{SO} = (X_1 - O_1)\hat{\mathbf{s}}_1 + (X_2 - O_2)\hat{\mathbf{s}}_2 + (X_3 - O_3)\hat{\mathbf{s}}_3 \tag{37}$$

A unit vector $\hat{\mathbf{d}}$ that has the same direction as **d** can be brought into a general orientation in N by first giving it the same direction as $\hat{\mathbf{s}}_1$ and then subjecting it to a body-two, 3-2-3 rotation sequence with angles of ϕ (longitude), $-\lambda$ (negative of latitude), and 0 (zero). Consequently, $\hat{\mathbf{d}}$ can be expressed as

$$\hat{\mathbf{d}} = \cos\phi\cos\lambda\hat{\mathbf{s}}_1 + \sin\phi\cos\lambda\hat{\mathbf{s}}_2 + \sin\lambda\hat{\mathbf{s}}_3 \tag{38}$$

and relationships for longitude and latitude follow,

$$\tan\phi = \frac{\hat{\mathbf{d}}\cdot\hat{\mathbf{s}}_2}{\hat{\mathbf{d}}\cdot\hat{\mathbf{s}}_1} = \frac{\mathbf{d}\cdot\hat{\mathbf{s}}_2}{\mathbf{d}\cdot\hat{\mathbf{s}}_1} \tag{39}$$

$$\sin\lambda = \hat{\mathbf{d}}\cdot\hat{\mathbf{s}}_3 = \frac{\mathbf{d}\cdot\hat{\mathbf{s}}_3}{d} \tag{40}$$

where d is the magnitude of **d**, $d = (\mathbf{d}\cdot\mathbf{d})^{1/2}$.

Each observation of an object with an optical telescope yields measurements of ϕ and λ; the corresponding matrix $[\tilde{H}]_j$ is then

$$[\tilde{H}]_j = \left\{ \begin{bmatrix} \frac{\partial\phi}{\partial X_1} & \frac{\partial\phi}{\partial X_2} & \frac{\partial\phi}{\partial X_3} & \frac{\partial\phi}{\partial X_4} & \frac{\partial\phi}{\partial X_5} & \frac{\partial\phi}{\partial X_6} \\ \frac{\partial\lambda}{\partial X_1} & \frac{\partial\lambda}{\partial X_2} & \frac{\partial\lambda}{\partial X_3} & \frac{\partial\lambda}{\partial X_4} & \frac{\partial\lambda}{\partial X_5} & \frac{\partial\lambda}{\partial X_6} \end{bmatrix} \right\} \tag{41}$$

The partial derivatives of ϕ are obtained by noting that $\frac{\partial}{\partial x}\tan^{-1}u = \frac{1}{1+u^2}\frac{\partial u}{\partial x}$; therefore, in view of equation (39),

$$\frac{\partial\phi}{\partial X_i} = \frac{(\mathbf{d}\cdot\hat{\mathbf{s}}_1)^2}{(\mathbf{d}\cdot\hat{\mathbf{s}}_1)^2 + (\mathbf{d}\cdot\hat{\mathbf{s}}_2)^2} \frac{\partial}{\partial X_i}\left[\frac{(\mathbf{d}\cdot\hat{\mathbf{s}}_2)}{(\mathbf{d}\cdot\hat{\mathbf{s}}_1)}\right]$$

$$= \frac{(\mathbf{d}\cdot\hat{\mathbf{s}}_1)^2}{(\mathbf{d}\cdot\hat{\mathbf{s}}_1)^2 + (\mathbf{d}\cdot\hat{\mathbf{s}}_2)^2}\left[\frac{1}{(\mathbf{d}\cdot\hat{\mathbf{s}}_1)}\frac{\partial\mathbf{d}}{\partial X_i}\cdot\hat{\mathbf{s}}_2 - \frac{(\mathbf{d}\cdot\hat{\mathbf{s}}_2)}{(\mathbf{d}\cdot\hat{\mathbf{s}}_1)^2}\frac{\partial\mathbf{d}}{\partial X_i}\cdot\hat{\mathbf{s}}_1\right]$$

$$= \frac{1}{(\mathbf{d}\cdot\hat{\mathbf{s}}_1)^2 + (\mathbf{d}\cdot\hat{\mathbf{s}}_2)^2}\left[(\mathbf{d}\cdot\hat{\mathbf{s}}_1)\frac{\partial\mathbf{d}}{\partial X_i}\cdot\hat{\mathbf{s}}_2 - (\mathbf{d}\cdot\hat{\mathbf{s}}_2)\frac{\partial\mathbf{d}}{\partial X_i}\cdot\hat{\mathbf{s}}_1\right] \quad (i = 1, \ldots, 6) \tag{42}$$

With the aid of equation (37), it can be seen that $\partial \mathbf{d}/\partial X_i = \hat{\mathbf{s}}_i$ for $i = 1, 2, 3$, and $\partial \mathbf{d}/\partial X_i = \mathbf{0}$ for $i = 4, 5, 6$; thus,

$$\frac{\partial \phi}{\partial X_1} = \frac{-(\mathbf{d} \cdot \hat{\mathbf{s}}_2)}{(\mathbf{d} \cdot \hat{\mathbf{s}}_1)^2 + (\mathbf{d} \cdot \hat{\mathbf{s}}_2)^2} \qquad (43)$$

$$\frac{\partial \phi}{\partial X_2} = \frac{(\mathbf{d} \cdot \hat{\mathbf{s}}_1)}{(\mathbf{d} \cdot \hat{\mathbf{s}}_1)^2 + (\mathbf{d} \cdot \hat{\mathbf{s}}_2)^2} \qquad (44)$$

$$\frac{\partial \phi}{\partial X_3} = 0 \qquad (45)$$

$$\frac{\partial \phi}{\partial X_4} = \frac{\partial \phi}{\partial X_5} = \frac{\partial \phi}{\partial X_6} = 0 \qquad (46)$$

The partial derivatives of λ are obtained by noting that $\frac{\partial}{\partial x} \sin^{-1} u = \frac{1}{\sqrt{1-u^2}} \frac{\partial u}{\partial x}$; therefore, in view of equation (40),

$$\frac{\partial \lambda}{\partial X_i} = \frac{d}{\sqrt{d^2 - (\mathbf{d} \cdot \hat{\mathbf{s}}_3)^2}} \frac{\partial}{\partial X_i}\left[\frac{(\mathbf{d} \cdot \hat{\mathbf{s}}_3)}{d}\right]$$

$$= \frac{d}{\sqrt{d^2 - (\mathbf{d} \cdot \hat{\mathbf{s}}_3)^2}} \left[\frac{1}{d} \frac{\partial \mathbf{d}}{\partial X_i} \cdot \hat{\mathbf{s}}_3 - \frac{(\mathbf{d} \cdot \hat{\mathbf{s}}_3)}{2d^3} 2\mathbf{d} \cdot \frac{\partial \mathbf{d}}{\partial X_i}\right]$$

$$= \frac{1}{\sqrt{(\mathbf{d} \cdot \hat{\mathbf{s}}_1)^2 + (\mathbf{d} \cdot \hat{\mathbf{s}}_2)^2}} \left[\frac{\partial \mathbf{d}}{\partial X_i} \cdot \hat{\mathbf{s}}_3 - \frac{(\mathbf{d} \cdot \hat{\mathbf{s}}_3)}{d^2} \mathbf{d} \cdot \frac{\partial \mathbf{d}}{\partial X_i}\right] \quad (i = 1, \ldots, 6) \qquad (47)$$

from which we obtain

$$\frac{\partial \lambda}{\partial X_1} = \frac{-(\mathbf{d} \cdot \hat{\mathbf{s}}_1)(\mathbf{d} \cdot \hat{\mathbf{s}}_3)}{d^2 \sqrt{(\mathbf{d} \cdot \hat{\mathbf{s}}_1)^2 + (\mathbf{d} \cdot \hat{\mathbf{s}}_2)^2}} \qquad (48)$$

$$\frac{\partial \lambda}{\partial X_2} = \frac{-(\mathbf{d} \cdot \hat{\mathbf{s}}_2)(\mathbf{d} \cdot \hat{\mathbf{s}}_3)}{d^2 \sqrt{(\mathbf{d} \cdot \hat{\mathbf{s}}_1)^2 + (\mathbf{d} \cdot \hat{\mathbf{s}}_2)^2}} \qquad (49)$$

$$\frac{\partial \lambda}{\partial X_3} = \frac{\sqrt{(\mathbf{d} \cdot \hat{\mathbf{s}}_1)^2 + (\mathbf{d} \cdot \hat{\mathbf{s}}_2)^2}}{d^2} \qquad (50)$$

$$\frac{\partial \lambda}{\partial X_4} = \frac{\partial \lambda}{\partial X_5} = \frac{\partial \lambda}{\partial X_6} = 0 \qquad (51)$$

Extended Kalman filter. The extended form of the Kalman filter is presented in section 4.7.2 of reference 5; as mentioned previously, it serves to reduce the effects of errors resulting from linearization upon which the conventional filter rests. The error reduction leads to faster convergence of the extended filter in comparison with the conventional form and is accomplished by replacing the reference orbit with the current, best estimate of the true orbit after each observation. As a consequence, the estimate $\{\hat{x}_{j-1}\}$ of the state deviation following a measurement update vanishes, and it becomes unnecessary to perform the time update of the state deviation indicated in equation (25). The time update of the covariance, equation (26), and preparation for the measurement update, equations (27)–(29), proceed as before. It can be seen that, with $\{\bar{x}_j\} = \{0\}$, the measurement update for the state deviation, equation (30), gives way to

$$\{\hat{x}_j\} = [K_j]\{y_j\} \tag{52}$$

The measurement update of the covariance, equation (31), remains the same, and the aforementioned replacement can be expressed as

$$\{X_j^*\} \leftarrow \{X_j^*\} + [K_j]\{y_j\} \tag{53}$$

$$\{\hat{x}_j\} \leftarrow \{0\} \tag{54}$$

A Priori Information Obtained by the Method of Gauss

The venerable method of Gauss for preliminary determination of orbits from optical measurements has been in use for two centuries; the occasion for its creation was the discovery (and subsequent disappearance behind the Sun) in 1801 of the first minor world, Ceres. The asteroid was recovered through the application of the Gaussian method with observations that spanned only 1 month, and Carl Friedrich Gauss became recognized immediately as the premier mathematician in all of Europe.

The method requires only three observations consisting of two angular measurements each, which is in most cases the minimum number of observations needed. (As discussed previously, four observations are required when the orbits of the object and observatory are coplanar.) Gauss's method is used to obtain the a priori information needed to start the filtering sequence, as described in what follows.

A variant of Gauss's method presented (but not labeled as such) in section 5.8 of reference 1 is brought to bear, with the Sun substituted for Earth in the role of primary body, and ecliptic longitude and latitude used in place of right ascension and declination. Battin's expressions for the Lagrange coefficients F and G in terms of universal functions (eqs. (4.84), ref. 2) are used in place of the infinite series suggested in reference 1. The heart of the procedure is an iterative solution of a system of six linear algebraic equations, yielding estimates of the position $\mathbf{r}'(t_2)$ and velocity $\mathbf{v}'(t_2)$ at the time t_2 of the second of three observations. Such a solution is obtained with these steps:

1. Three pairs of measurements of longitude $\tilde{\phi}_i$ and latitude $\tilde{\lambda}_i$ for times t_i ($i = 1, 2, 3$) are produced according to steps 1, 3, and 4 of the procedure given in the discussion of erroneous predicted miss distance. The intentional errors to be added ($\{e\}_i$ in eq. (24)) are produced with the MATLAB® function RANDN, yielding a normal distribution with zero mean (in accordance with the Kalman filter assumption) and a standard deviation of 1. Errors are then multiplied by the telescope resolution ρ, but subsequently limited in magnitude; errors less than -2ρ are replaced with -2ρ, and errors greater than 2ρ are replaced with 2ρ.

2. The measurements constructed in step 1 are used to form three unit vectors $\hat{\mathbf{d}}_i$ with the aid of equation (38).

3. An initial guess for the vector $\mathbf{r}'(t_2) = \mathbf{r}^{SO}(t_2) + d\hat{\mathbf{d}}_2$ is formed with the known observatory position $\mathbf{r}^{SO}(t_2)$ and a guess for the value of d, such as 5 au.

4. An initial guess of the velocity $\mathbf{v}'(t_2)$ is produced by assuming that the object is in a circular orbit and therefore has a magnitude of $\sqrt{\mu/r_2}$, where r_2 is the magnitude of $\mathbf{r}'(t_2)$. The direction of $\mathbf{v}'(t_2)$ is given by a unit vector normal to $\mathbf{r}'(t_2)$, having the same direction as $(\hat{\mathbf{d}}_1 \times \hat{\mathbf{d}}_3) \times \mathbf{r}'(t_2)$, where the cross product $\hat{\mathbf{d}}_1 \times \hat{\mathbf{d}}_3$ is approximately normal to the orbital plane.

5. Lagrange coefficients F_1 and G_1 that relate $\mathbf{r}'(t_1)$ to $\mathbf{r}'(t_2)$ and $\mathbf{v}'(t_2)$ are constructed from these two vectors and the time difference $t_1 - t_2$. Coefficients F_3 and G_3 corresponding to t_3 are obtained similarly.

6. An iterative solution of equations (5.8-10) in reference 1 is terminated when the magnitudes of all six components of $\mathbf{r}'(t_2)$ and $\mathbf{v}'(t_2)$ change by less than 1×10^{-14} percent from one pass to the next, or after 50 passes, whichever occurs first. It is asserted in reference 1 that this process converges quickly if the time intervals $t_3 - t_2$ and $t_2 - t_1$ are "not too large."

7. Initial elements $\{X_0^*\}$ for a reference orbit are simply dot products of $\mathbf{r}'(t_2)$ and $\mathbf{v}'(t_2)$ with $\hat{\mathbf{s}}_1$, $\hat{\mathbf{s}}_2$, and $\hat{\mathbf{s}}_3$. Initial values of a state deviation $\{\hat{x}_0\}$ are given by dot products of the same unit vectors with the vector differences $\mathbf{r}'(t_2) - \mathbf{r}(t_2)$ and $\mathbf{v}'(t_2) - \mathbf{v}(t_2)$, where $\mathbf{r}(t_2)$ and $\mathbf{v}(t_2)$ are the true position and velocity at t_2.

This procedure is used 100 times with 100 sets of observations to obtain 100 different matrices $\{X_0^*\}$ and $\{\hat{x}_0\}$. Convergence was achieved in all cases. An initial covariance matrix $[P_0]$ describing the uncertainties in $\{\hat{x}_0\}$ is calculated from the 100 state deviation matrices using the MATLAB® function COV. This a priori information is used to start a Kalman filter sequence. When Monte Carlo simulations are performed, each of the 100 pairs of associated matrices $\{X_0^*\}$ and $\{\hat{x}_0\}$ are used in a separate simulation, but the same matrix $[P_0]$ is used in each simulation.

Single Observatory, and Two Observatories

The results presented in connection with multiple observations indicate that it may be possible to make reliable forecasts with two observatories whose angular resolutions are on the order of 0.1 arcsec, or with a single observatory whose resolution is better by 1 to 3 orders of magnitude. An extended Kalman filter is used to compare the amount of warning time offered by the two alternatives, where warning time is defined (somewhat more rigorously than before) to be the interval between the time a filter yields an erroneous predicted miss distance less than 1 Earth radius, and the designed time of collision. The task is facilitated by introducing a new term, "watch time," defined to be the interval between the time ε is calculated to be less than or equal to 1 lunar distance, and the designed time of collision. A watch is issued prior to a warning for hurricanes and other reasonably predictable natural calamities, and so it would be in the case of a celestial collision.

Single observatory. We assume the single observatory achieves a resolution of 0.0001 arcsec by means of interferometry; two telescopes, each having a resolution of 0.01 arcsec, are placed some distance apart in a circular heliocentric orbit in the ecliptic plane with a radius of 1 au. Observations are obtained at 7-day intervals from only one of the telescopes until $\varepsilon \leq 1$ lunar distance, and the watch time is calculated. Subsequently, both telescopes are used together to produce measurements at 1-day intervals with $\rho = 0.0001$ arcsec, and the filter continues to process those measurements until $\varepsilon \leq 1$ Earth radius, at which point a warning time is computed.

Two observatories. The configuration of two observatories is the same as that previously described: two telescopes are placed in a circular heliocentric orbit in the ecliptic plane, with a radius of 1 au, and phased 180° apart. For the purpose of making the comparison, one of the telescopes is assumed to be coincident with one of the instruments in the single-observatory configuration. Each telescope has a resolution of 0.01 arcsec. For any one observation, this configuration produces twice as many measurements as the single observatory. Furthermore, two observatories provide parallax that cannot be obtained with the single observatory. Measurements are taken at 7-day intervals until $\varepsilon \leq 1$ lunar distance, at which time the frequency increases to once per day until $\varepsilon \leq 1$ Earth radius.

All measurements contain Gaussian noise and are constructed as described in step 1 of the procedure for obtaining a priori information.

Results. Watch and warning times have been computed for both configurations, using the 1008 hypothetical LPC orbits given previously, and four values of the observatory initial true longitude L_0. The results obtained for $L_0 = 180°$ are presented in figure 19. Computational expense prevents us from performing Monte Carlo analysis and presenting averages; each data point represents a single set of random measurement errors for that orbit.

In almost all cases two observatories, enjoying the advantages of parallax and twice as many measurements, provide a greater watch time than a single observatory taking measurements at the same resolution. The advantage passes to the single observatory after it begins to obtain interferometric measurements; the improvement in resolution by a factor of 100 quickly reduces ε to 1 Earth radius or less, giving a greater warning time in general than two observatories.

Sixteen groups can be identified in the watch and warning times given in figure 19. The first, second, third, and fourth sets of four groups correspond to $r_a = 10, 40, 70$, and 100 au, respectively. The four groups within each set are associated with $r_p = 0.1, 0.4, 0.7$, and 1.0 au. The results for other values of L_0 are similar to those shown in figure 19.

When a conventional filter is used in place of the extended algorithm, it ceases to converge on the correct orbit and yields a warning time of zero in approximately 100 of the 1008 orbits, rendering it unsuitable for obtaining watch and warning times. An example of this behavior is exhibited in figure 20. The lack of an update to the reference trajectory in a conventional filter is the source of this defect.

Range and Range-Rate Measurements

The accuracy of orbit determination can be improved greatly with measurements of range and range-rate obtained from radar or lidar instruments. Range is simply the distance d between the observatory and the object, or the magnitude of the vector **d** given by equation (37). Range-rate, $\dot{d}$, is the time derivative of d. Measurements are performed by analyzing the returning echo from an electromagnetic pulse that

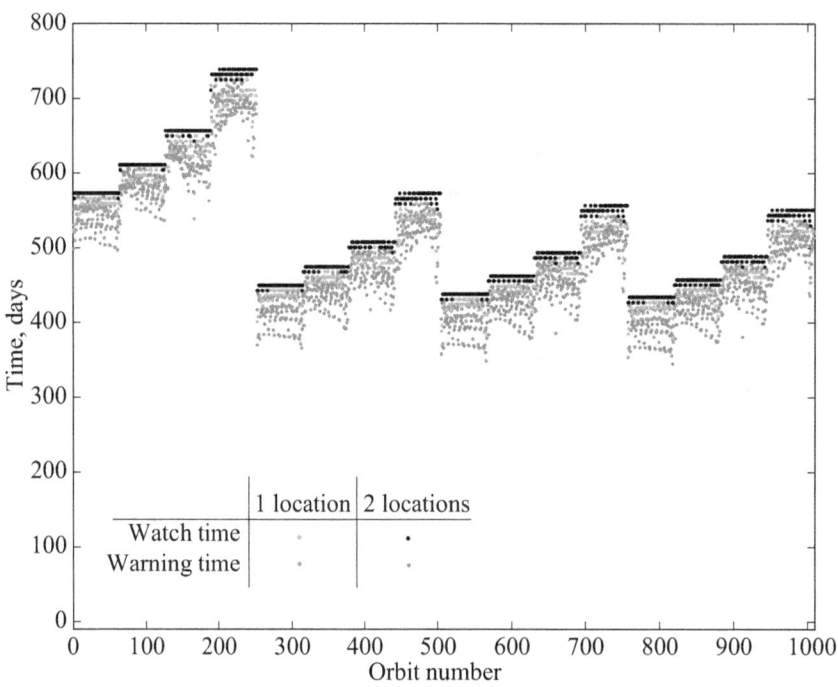

Figure 19. Watch and warning times.

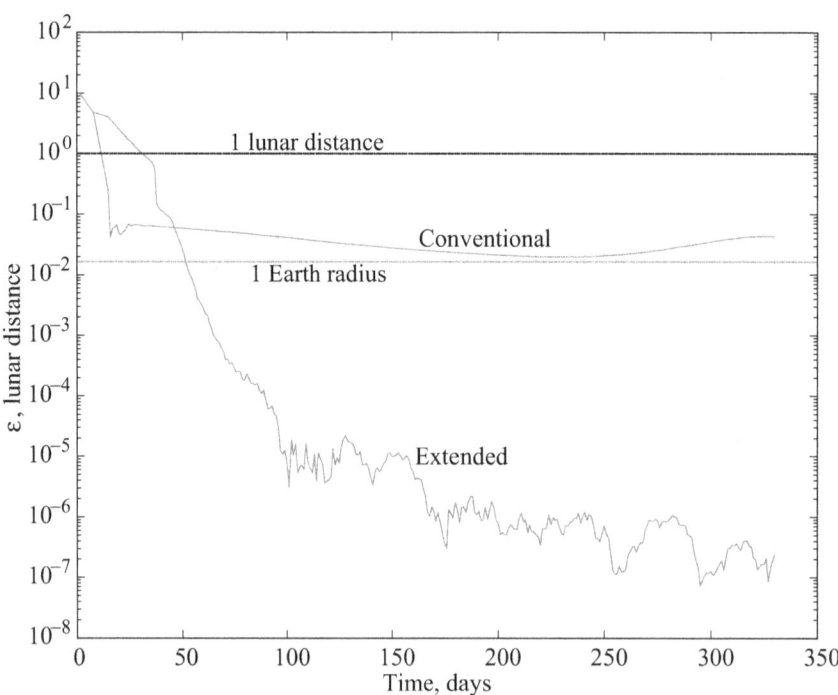

Figure 20. Lack of convergence with a conventional filter.

has been directed toward the object. The distance d is related to the wave's round trip time, and $\dot{d}$ is related to the Doppler shift in the frequency.

As pointed out in the discussion of the Kalman filter, the observation-state mapping matrix $[\tilde{H}_j]$ is dimensioned $m \times n$, the measurement covariance matrix $[R_j]$ is $m \times m$, and the observation deviation matrix $\{y_j\}$ is $m \times 1$, where m is the number of measurements and n is the size of the state, in our case $n = 6$. So far, only angular measurements of longitude and latitude have been considered, corresponding to $m = 2$. The addition of measurements of range and range-rate increases the value of m to 3 and 4, respectively. The covariance matrix $[R_j]$ is diagonal; with $m = 4$, the first two nonzero elements of $[R_j]$ contain the square of the resolution of the optical telescope, the third is the square of the resolution of range measurements, and the fourth is the square of the resolution of range-rate. When observing an asteroid at 20 times the lunar distance, current terrestrial radar systems can achieve resolutions of 10 m and 0.1 mm/s for range and range-rate, respectively.

Expressions for the elements of $[\tilde{H}_j]$ associated with measurements of longitude and latitude are developed in equations (43)–(46) and (48)–(51). The third and fourth rows of $[\tilde{H}_j]$, associated respectively with measurements of range and range-rate, are derived as follows:

$$\frac{\partial d}{\partial X_i} = \frac{\partial}{\partial X_i}(\mathbf{d}\cdot\mathbf{d})^{1/2} = \frac{1}{2}(\mathbf{d}\cdot\mathbf{d})^{-1/2} 2\mathbf{d}\cdot\frac{\partial \mathbf{d}}{\partial X_i}$$

$$= \frac{\mathbf{d}}{d}\cdot\frac{\partial \mathbf{d}}{\partial X_i} \qquad (i = 1,\ldots,6) \tag{55}$$

As mentioned earlier, it is evident from equation (37) that $\partial \mathbf{d}/\partial X_i = \hat{\mathbf{s}}_i$ for $i = 1, 2, 3$, and $\partial \mathbf{d}/\partial X_i = \mathbf{0}$ for $i = 4, 5, 6$; therefore,

$$\frac{\partial d}{\partial X_i} = \frac{\mathbf{d}\cdot\hat{\mathbf{s}}_i}{d} \qquad (i = 1, 2, 3) \tag{56}$$

$$\frac{\partial d}{\partial X_i} = 0 \qquad (i = 4, 5, 6) \tag{57}$$

Range-rate can be expressed as

$$\dot{d} = \frac{d}{dt}[(\mathbf{d}\cdot\mathbf{d})^{1/2}] = \frac{1}{2}(\mathbf{d}\cdot\mathbf{d})^{-1/2} 2\mathbf{d}\cdot\frac{{}^N d}{dt}\mathbf{d} = \frac{\mathbf{d}}{d}\cdot\frac{{}^N d}{dt}\mathbf{d} \tag{58}$$

where ${}^N d\mathbf{d}/dt$ indicates differentiation of $\mathbf{d}$ with respect to time in N, a Newtonian reference frame in which unit vectors $\hat{\mathbf{s}}_i$ ($i = 1, 2, 3$) are fixed. Strictly speaking, specification of a reference frame is not required because the left hand member of the equation is the derivative of a scalar, but the choice of N leads right away from equation (37) to

$$\frac{{}^N d}{dt}\mathbf{d} = (\dot{X}_1 - \dot{O}_1)\hat{\mathbf{s}}_1 + (\dot{X}_2 - \dot{O}_2)\hat{\mathbf{s}}_2 + (\dot{X}_3 - \dot{O}_3)\hat{\mathbf{s}}_3 \tag{59}$$

Now,

$$\frac{\partial \dot{d}}{\partial X_i} = \frac{1}{d}\frac{\partial \mathbf{d}}{\partial X_i}\cdot\frac{{}^N\mathbf{d}}{dt}\mathbf{d} + \frac{\mathbf{d}}{d}\cdot\frac{\partial}{\partial X_i}\left(\frac{{}^N\mathbf{d}}{dt}\mathbf{d}\right) - \frac{1}{d^2}\frac{\partial d}{\partial X_i}\mathbf{d}\cdot\frac{{}^N\mathbf{d}}{dt}\mathbf{d}$$

$$= \frac{1}{d}\frac{\partial \mathbf{d}}{\partial X_i}\cdot\frac{{}^N\mathbf{d}}{dt}\mathbf{d} + \frac{\mathbf{d}}{d}\cdot\frac{\partial}{\partial X_i}\left(\frac{{}^N\mathbf{d}}{dt}\mathbf{d}\right) - \frac{1}{d^2}\left(\frac{\mathbf{d}}{d}\cdot\frac{\partial \mathbf{d}}{\partial X_i}\right)\mathbf{d}\cdot\frac{{}^N\mathbf{d}}{dt}\mathbf{d} \qquad (i = 1, \ldots, 6) \qquad (60)$$

The last three elements of $\{X\}$ are simply

$$X_i = \dot{X}_{i-3} \qquad (i = 4, 5, 6) \qquad (61)$$

therefore, in view of equation (59), $\frac{\partial}{\partial X_i}({}^N\mathbf{d}\,\mathbf{d}/dt) = \mathbf{0}$ for $i = 1, 2, 3$, and $\frac{\partial}{\partial X_i}({}^N\mathbf{d}\,\mathbf{d}/dt) = \hat{\mathbf{s}}_{i-3}$ for $i = 4, 5, 6$. Together with the values for $\partial \mathbf{d}/\partial X_i$ used previously, this leads to

$$\frac{\partial \dot{d}}{\partial X_i} = \frac{\hat{\mathbf{s}}_i}{d}\cdot\frac{{}^N\mathbf{d}}{dt}\mathbf{d} - \frac{\mathbf{d}\cdot\hat{\mathbf{s}}_i}{d^3}\left(\mathbf{d}\cdot\frac{{}^N\mathbf{d}}{dt}\mathbf{d}\right) \qquad (i = 1, 2, 3) \qquad (62)$$

$$\frac{\partial \dot{d}}{\partial X_i} = \frac{\mathbf{d}\cdot\hat{\mathbf{s}}_{i-3}}{d} \qquad (i = 4, 5, 6) \qquad (63)$$

The benefits of measurements of range and range-rate are illustrated in figure 21, with curves showing a reduction in average erroneous predicted miss distance $\bar{\varepsilon}$ in units of Earth radius (R_E) as a function of the number of observations, taken once a day. The average is taken from 100 extended Kalman filter simulations, each with a different set of normally distributed measurement errors, and a priori information as discussed earlier. The analysis involves a hypothetical LPC designed to collide with Earth, with orbital elements obtained as described for designing a collision: $r_a = 100$ au, $r_p = 0.7$ au, $i = 50°$, $\Omega = 60°$, $\omega = 66.68597°$, and $v_0 = -137.06483°$ (corresponding to $r = 5$ au). The orbital period of this comet is 1839.4 years.

All measurements are obtained at a single observatory. The least accurate determination of the orbit, shown with the solid line, is obtained from measurements that are strictly angular with a resolution of 0.01 arcsec. The slope of the line is approximately –2.3; therefore $\bar{\varepsilon}$ varies inversely with the number of observations to the power 2.3. Greater accuracy is obtained when the angular measurements are supplemented with range measurements having a resolution of 10^3 km, as shown with the dashed curve, and the inclusion of range-rate measurements with a resolution of 1 m/s leads to the most accurate orbit determination, as displayed with the dash-dot curve. The usefulness of range and range-rate measurements is most pronounced over a short data arc of 15 to 30 days; they do not appear to be necessary over an arc of 80 or 90 days.

Modeling Comet Outgassing

A visible comet in the night sky typically appears as a fuzzy object; light is reflected diffusely from a cloud, or coma, consisting of dust evaporated by solar heating from the comet's core, or nucleus. Solar

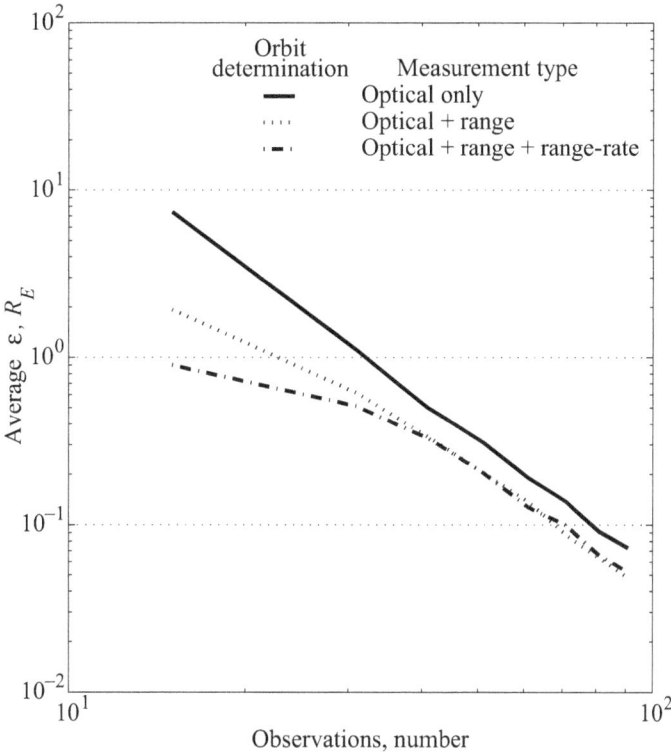

Figure 21. Benefit of range and range-rate measurements.

radiation pressure and the solar wind move the dust and ionized particles away from the nucleus, and a tail is formed. The composition of a comet is not well-known, but the nucleus is generally thought to be a "dirty snowball" made mostly of frozen water and organic and silicate compounds. When the comet is close enough to the Sun that the heat causes these volatile materials to boil off and carry solid particles with them, the comet is said to be "outgassing." Consequently, force is exerted on the comet, and the resulting perturbations to the orbit should be taken into account when attempting to calculate a trajectory precisely enough to determine if a collision with Earth will occur.

The most widely accepted method for modeling force due to outgassing was developed by Marsden, Sekanina, and Yeomans, as described in reference 7, and employed in reference 8, based on the assumption that the comet is an icy conglomerate—an object consisting mostly of frozen water that holds together bits and pieces of rock-like material. The contribution of outgassing to force per unit mass is expressed as

$$\mathbf{f}_g = g(r)[A_1\hat{\mathbf{a}}_1 + A_2\hat{\mathbf{a}}_2 + A_3\hat{\mathbf{a}}_3] \tag{64}$$

where the function $g(r)$ of heliocentric distance r is given by

$$g(r) = \alpha\left(\frac{r}{r_0}\right)^{-m}\left[1+\left(\frac{r}{r_0}\right)^n\right]^{-l} \tag{65}$$

The parameters other than r are constants obtained from studies of vaporization rates of comet nucleus material. The normalizing constant α is defined such that $g(1) = 1$, and r_0 is the scale heliocentric distance of high outgassing activity with a value equal to 2.808 au for frozen water. The remaining constants are reported to be $m = 2.15$, $n = 5.093$, and $l = 4.6142$; substituting these values into equation (65) and solving for the normalizing constant yields $\alpha = 0.1113$.

A right-handed, mutually perpendicular set of unit vectors $\hat{\mathbf{a}}_1$, $\hat{\mathbf{a}}_2$, and $\hat{\mathbf{a}}_3$ is defined such that $\hat{\mathbf{a}}_1$ is in the direction of the position vector from S to C, $\hat{\mathbf{a}}_2$ lies in the orbital plane, and $\hat{\mathbf{a}}_3$ is normal to the orbital plane in the direction of the specific inertial angular momentum of C relative to S. Numerical values of the constant coefficients A_1 and A_2 are calculated by studying changes in comet orbital periods, and are listed for more than 20 comets in table I of reference 7. The coefficient A_3 is usually neglected because the force normal to the orbit plane has no detectable effect on the period, therefore values for A_3 are not given. However, a normal force will influence a comet's final Earth-encounter distance, so this component must be included in simulations involving prediction of a collision.

Before undertaking such simulations, it is important to understand the magnitude of the perturbation to the comet's trajectory caused by outgassing. This is accomplished with the aid of dynamical equations formed by adding the perturbing force per unit mass to the right hand member of the equations governing two-body motion,

$$\frac{{}^N d^2}{dt^2}\mathbf{r} = -\frac{\mu}{r^3}\mathbf{r} + g(r)[A_1\hat{\mathbf{a}}_1 + A_2\hat{\mathbf{a}}_2 + A_3\hat{\mathbf{a}}_3] \tag{66}$$

where $\mathbf{r}$ is the position vector from S to C, and ${}^N d^2\mathbf{r}/dt^2$ denotes the second derivative in frame N of $\mathbf{r}$ with respect to time. By comparing numerical solutions obtained through integration of equations (66) with $[g(r) \neq 0]$ and without $[g(r) = 0]$ the contribution of the force of expelled gas, one can quantify its effect on the time history of $\mathbf{r}$. Integration is performed in connection with the hypothetical LPC whose orbital elements are given at the conclusion of the material dealing with range and range-rate measurements. Equations (66) are integrated from t_0 to t_k, at which times $r = 5$ au and 1 au, respectively. The maximum values of the coefficients reported in reference 7 are used: $A_1 = 3.61$ au/10^8 day^2 = 5.4 km/day^2, and $A_2 = 0.3269$ au/10^8 day^2 = 0.49 km/day^2. Because no values for A_3 are given, A_3 is set equal to A_2.

Figure 22 includes time histories of the differences in $\mathbf{r} \cdot \hat{\mathbf{s}}_i$ ($i = 1, 2, 3$) caused by outgassing, as well as the difference in the magnitude of $\mathbf{r}$, which is approximately 4730 km by the time $r = 1$ au. However, the more important result is that differences in position are negligible until about 250 days, when r is approximately 2.2 au; therefore, a two-body trajectory furnishes a good reference orbit until this point, warranting the linearization about such a reference orbit performed in constructing the state transition matrix of equation (34). The state transition matrix should account for outgassing if the filter is to be used with observations of a comet made with r less than, say, 2.5 au, but need not do so for r greater than this number.

The uncertainty that results from outgassing is accounted for with a white state noise of zero mean and time-varying diagonal covariance $[Q(t_j)] \triangleq [Q_j]$, the elements of which are determined by assuming the perturbing force per unit mass is constant during the interval between observations. The uncertainty in

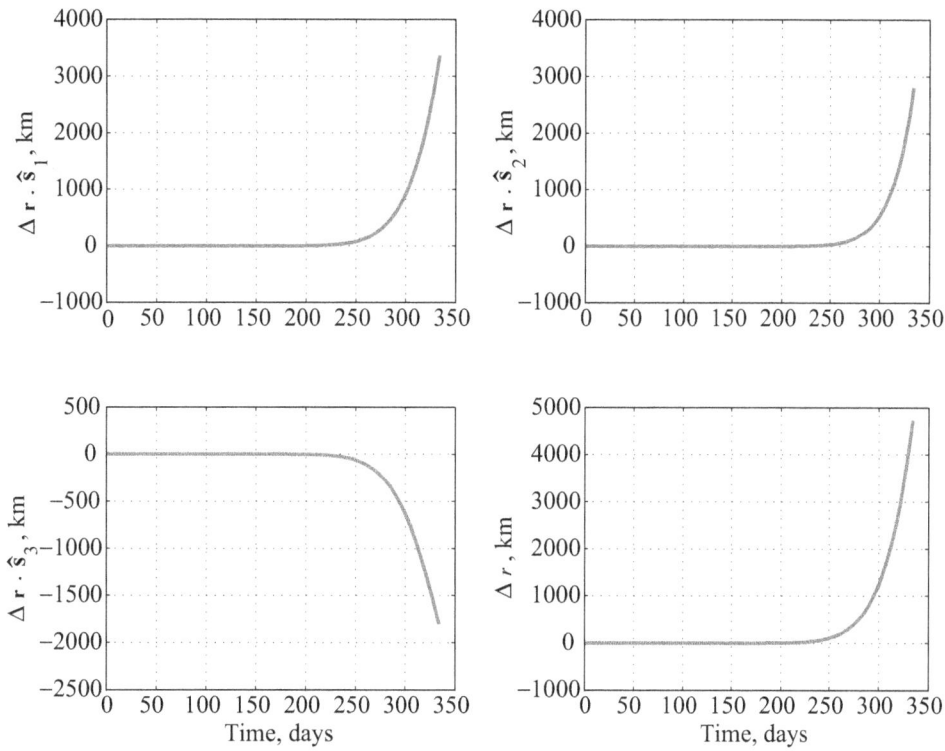

Figure 22. Change in position due to outgassing for a hypothetical comet.

the force per unit mass is reflected in the standard deviations of the coefficients listed in reference 7: σ_{A_1} = 2.096 km/day^2, and σ_{A_2} = σ_{A_3} = 0.085 km/day^2. The elements of $[Q_j]$ are given by

$$Q_{ii}(t_j) = \left[\frac{\mathbf{f}_g(t_j)\cdot\hat{\mathbf{s}}_i}{2}(t_j - t_{j-1})^2\right]^2 \qquad (i = 1, 2, 3) \qquad (67)$$

$$Q_{ii}(t_j) = \left\{[\mathbf{f}_g(t_j)\cdot\hat{\mathbf{s}}_{i-3}](t_j - t_{j-1})\right\}^2 \qquad (i = 4, 5, 6) \qquad (68)$$

where $\mathbf{f}_g(t_j)$ is evaluated at t_j with σ_{A_i} in place of A_i (i = 1, 2, 3) in equation (64). As heliocentric distance r decreases, the elements of $[Q_j]$ grow along with $g(r)$, as do the elements of the Kalman gain matrix $[K_j]$, and the filter places increasing emphasis on the measurements in determining the best estimate of the state deviation.

Once a best estimate $\{X'_j\}$ of the orbital elements is obtained from the sequential filter, an erroneous predicted miss distance is calculated by using the elements of $\{X'_j\}$ as initial values for the differential equations (66) and by performing numerical integration with the MATLAB® function ODE45 from the time t_j of the final observation to the designed time of collision t_k.

As it is not possible to know ahead of time the values of the outgassing coefficients for a newly discovered comet, a Monte Carlo analysis is performed with a set of random values uniformly distributed between the minimum and maximum values of A_1 and A_2 reported in reference 7. Random values for A_3 are obtained from the same distribution used for A_2. The results of this analysis are presented next.

The Body-Targeting Plane

Body-plane (B-plane) targeting is a method commonly used in the preliminary design of interplanetary missions to determine a distance of closest approach of a spacecraft to a celestial body of interest. A miss distance B calculated with this method can furnish a metric for quantifying orbit determination accuracy in connection with hypothetical comets that collide with Earth. The uncertainty in B can be expressed in terms of error ellipses constructed on the B-plane and is actually the more important result when analyzing the accuracy with which an orbit has been determined.

A vector **B**, and a plane containing **B**, are constructed as follows. First, one neglects the gravitational force exerted by the Earth E on the hypothetical object C; therefore, E does not perturb the heliocentric orbit of C, and the velocity $\mathbf{v}^{C/E^*}$ in N of C relative to the mass center E^* of E is regarded as constant in N during the time C is within the Earth's sphere of influence. The magnitude of $\mathbf{v}^{C/E^*}$ is denoted by v_∞. A unit vector $\hat{\mathbf{S}}$ having the same direction as $\mathbf{v}^{C/E^*}$ is shown in figure 23, and is often considered to be parallel to the incoming asymptote of the geocentric hyperbolic trajectory (assuming C is not captured) resulting from gravitational attraction of the Earth that must in fact exist. In the absence of gravitational

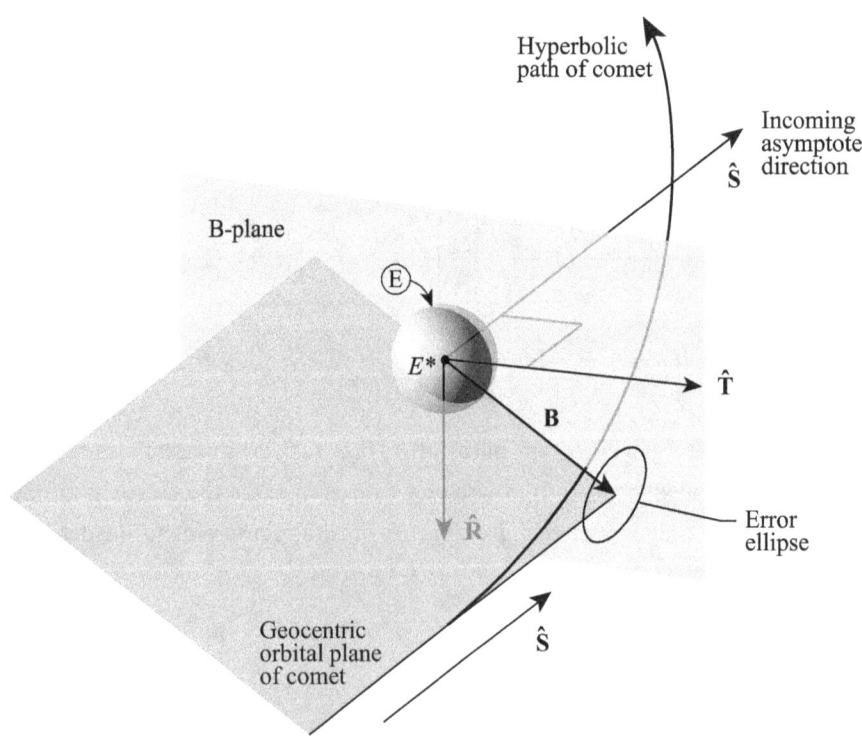

Figure 23. The B-plane.

force exerted by E, C continues along the incoming asymptote. The unit vector $\hat{\mathbf{S}}$ can be obtained with the expression

$$\hat{\mathbf{S}} = \frac{\mathbf{v}^{C/E^*}}{v_\infty} = \frac{\mathbf{v}^{C/S^*}(t_k) - \mathbf{v}^{E^*/S^*}(t_k)}{v_\infty} \qquad (69)$$

where $\mathbf{v}^{C/S^*}(t_k)$ and $\mathbf{v}^{E^*/S^*}(t_k)$ are, respectively, the velocities in N of C and E^*, relative to the Sun's mass center S^*, at the designed time of collision t_k.

The B-plane is defined such that it contains the point E^* and is perpendicular to $\hat{\mathbf{S}}$; the plane contains two unit vectors $\hat{\mathbf{T}}$ and $\hat{\mathbf{R}}$ that are mutually perpendicular to each other and to $\hat{\mathbf{S}}$. The unit vector $\hat{\mathbf{T}}$ is chosen to be in the ecliptic plane; therefore, it can be obtained from the relationship

$$\hat{\mathbf{T}} = \frac{\hat{\mathbf{S}} \times \hat{\mathbf{s}}_3}{\left[(\hat{\mathbf{S}} \times \hat{\mathbf{s}}_3) \cdot (\hat{\mathbf{S}} \times \hat{\mathbf{s}}_3)\right]^{1/2}} \qquad (70)$$

where, as one will recall, $\hat{\mathbf{s}}_3$ is perpendicular to the ecliptic plane. Unit vector $\hat{\mathbf{R}}$ is then simply

$$\hat{\mathbf{R}} = \hat{\mathbf{S}} \times \hat{\mathbf{T}} \qquad (71)$$

The vector $\mathbf{B}$ denotes the position from E^* to the point at which the incoming asymptote intersects the B-plane, which is the point of closest approach when one neglects the gravitational attraction of E (or, equivalently, neglects the mass of E). The magnitude B of $\mathbf{B}$ is the corresponding distance of closest approach.

As previously mentioned, the geocentric velocity $\mathbf{v}^{C/E^*} = v_\infty \hat{\mathbf{S}}$ is assumed to remain constant in N for the purpose of constructing $\mathbf{B}$; therefore, C must necessarily travel on a straight line between the point of closest approach and the position at the designed time of collision, as shown in figure 24. Hence, the specific angular momentum $\mathbf{h}$ in N of C relative to E^* is given by

$$\mathbf{h} = \mathbf{B} \times v_\infty \hat{\mathbf{S}} = \boldsymbol{\varepsilon} \times v_\infty \hat{\mathbf{S}} \qquad (72)$$

where $\boldsymbol{\varepsilon}$ is the position vector from E^* to C at the designed time of collision, obtained from the process of orbit determination. From this relationship, an expression for $\mathbf{B}$ can be derived through premultiplication with $\hat{\mathbf{S}}$ in a cross product,

$$\begin{aligned}\hat{\mathbf{S}} \times \mathbf{h} &= \hat{\mathbf{S}} \times (\boldsymbol{\varepsilon} \times v_\infty \hat{\mathbf{S}}) = \hat{\mathbf{S}} \times (\mathbf{B} \times v_\infty \hat{\mathbf{S}}) \\ &= v_\infty [(\hat{\mathbf{S}} \cdot \hat{\mathbf{S}})\mathbf{B} - (\hat{\mathbf{S}} \cdot \mathbf{B})\hat{\mathbf{S}}] \\ &= v_\infty \mathbf{B}\end{aligned} \qquad (73)$$

because $\hat{\mathbf{S}} \cdot \hat{\mathbf{S}} = 1$, and $\hat{\mathbf{S}} \cdot \mathbf{B} = 0$ by definition. Hence,

$$\mathbf{B} = \frac{\hat{\mathbf{S}} \times \mathbf{h}}{v_\infty} = \hat{\mathbf{S}} \times (\boldsymbol{\varepsilon} \times \hat{\mathbf{S}}) \qquad (74)$$

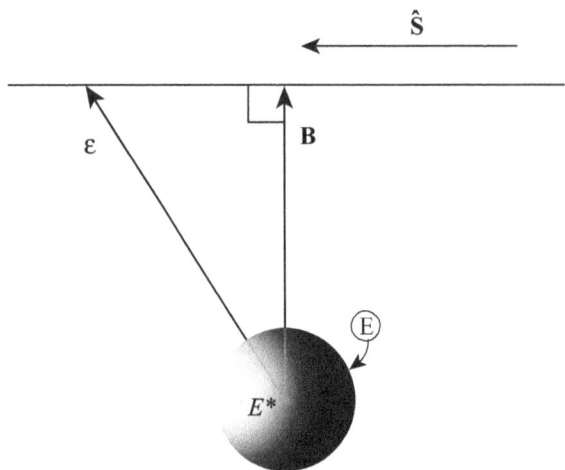

Figure 24. Erroneous predicted miss position, and closest approach.

and, because **B** lies in the plane formed by $\hat{\mathbf{R}}$ and $\hat{\mathbf{T}}$, it can always be expressed as

$$\mathbf{B} = B_R \hat{\mathbf{R}} + B_T \hat{\mathbf{T}} \tag{75}$$

As described previously, the orbit of C is designed to pass through the ecliptic at a heliocentric distance of 1 au, a point that is assumed to be coincident with E^*. A perfect determination of the orbit must necessarily yield $\boldsymbol{\varepsilon} = \mathbf{B} = \mathbf{0}$; however, inevitable random noise present in observations will lead to a prediction that $\boldsymbol{\varepsilon}$ and $\mathbf{B}$ each differ from $\mathbf{0}$. If the magnitude of either of these position vectors is determined to be, say, less than or equal to 1 Earth radius (R_E), one would be justified in predicting a collision unless there is a large uncertainty in the orbit solution as quantified by the covariance of $\boldsymbol{\varepsilon}$ or $\mathbf{B}$, which bears the following relationship to the estimate error covariance matrix $[P_j]$ in equation (31).

A time update according to equation (26) is performed in order to obtain the covariance matrix at the designed time of collision t_k, from the covariance matrix $[P_j]$ after a measurement update at t_j,

$$[\overline{P}_k] = [\Phi(t_k, t_j)][P_j][\Phi(t_k, t_j)]^T + [Q_j] \tag{76}$$

The effects of outgassing can be reflected in $[Q_j]$ as set forth in equations (67) and (68) for short intervals of time. Alternatively, one may divide a large time interval into shorter ones over which $\mathbf{f}_g$ is regarded as constant in N. When 1 year of warning time separates t_k and t_j, dividing the interval into 20 equal parts ensures that $[Q_j]$ is not dependent on the propagation time step and that the constant outgassing force assumption is still valid. The covariance time update for the shorter intervals (each approximately 2 weeks in length) is modified,

$$[\overline{P}_{j+1}] = [\Phi(t_{j+1}, t_j)][P_j][\Phi(t_{j+1}, t_j)]^T + [\overline{Q}_j] \tag{77}$$

to employ the average value of the state noise covariance

$$[\overline{Q}_j] \triangleq \frac{[Q_j] + [Q_{j+1}]}{2} \tag{78}$$

Equation (77) is applied successively until t_{j+1} becomes the time of collision t_k, yielding a 6 × 6 covariance matrix $[\overline{P_k}]$.

The upper left 3 × 3 partition of $[\overline{P_k}]$ deals with position and is therefore of the most interest in what follows; it is denoted by the symbol $[\overline{P_r}]$. As indicated in connection with the modeling of comet outgassing, the state transition matrix should be constructed according to reference 8 when outgassing becomes a significant perturbation; in lieu of this, $[\overline{P_r}]$ can be supplemented by adding a covariance matrix assembled from a difference in position associated with uncertainty resulting from outgassing. Equation (66) is numerically integrated from the time of the first observation t_1 to the time of collision t_k, first with $\mathbf{f}_g = \mathbf{0}$, and then three times, with one of σ_{A_i} used in place of A_i ($i = 1, 2, 3$) in equation (64). The difference in $\mathbf{r}(t_k)$ obtained with σ_{A_i} and with $\mathbf{f}_g = \mathbf{0}$, is expressed in terms of unit vectors $\hat{\mathbf{s}}_1$, $\hat{\mathbf{s}}_2$, and $\hat{\mathbf{s}}_3$ and presented as a column matrix $\{\delta\}_i$, allowing one to form the 3 × 3 covariance matrix

$$[\overline{P_S}] \triangleq [\overline{P_r}] + \sum_{i=1}^{3} \{\delta\}_i \{\delta\}_i^{\mathrm{T}} \tag{79}$$

This matrix (like the rest of $[\overline{P_k}]$) is associated with unit vectors $\hat{\mathbf{s}}_1$, $\hat{\mathbf{s}}_2$, and $\hat{\mathbf{s}}_3$, but it is more convenient to work with a covariance that describes the uncertainties in the directions marked by $\hat{\mathbf{S}}$, $\hat{\mathbf{T}}$, and $\hat{\mathbf{R}}$. The transformed covariance, indicated with $[\overline{P_B}]$, is formed as

$$[\overline{P_B}] = [{}^S C^B]^{\mathrm{T}} [\overline{P_S}] [{}^S C^B] \tag{80}$$

where the elements of the direction cosine matrix $[{}^S C^B]$ are defined as ${}^S C^B_{i1} \triangleq \hat{\mathbf{s}}_i \cdot \hat{\mathbf{S}}$, ${}^S C^B_{i2} \triangleq \hat{\mathbf{s}}_i \cdot \hat{\mathbf{T}}$, and ${}^S C^B_{i3} \triangleq \hat{\mathbf{s}}_i \cdot \hat{\mathbf{R}}$ ($i = 1, 2, 3$).

As discussed in section 4.16 of reference 5, the function

$$(\{x_k\} - \{\overline{x}_k\})^{\mathrm{T}} [P_B]^{-1} (\{x_k\} - \{\overline{x}_k\}) = l^2 \tag{81}$$

describes a three-dimensional ellipsoid, where the portion of the state deviation associated with position at t_k is $\{x_k\}$, having covariance $[\overline{P_B}]$ and mean $\{\overline{x}_k\}$, and where l is a constant. The ellipse corresponding to a given value of l, formed by the intersection of the ellipsoid with the B-plane (shown in fig. 23), is obtained from the 2 × 2 partition of $[\overline{P_B}]$ associated with $\hat{\mathbf{T}}$ and $\hat{\mathbf{R}}$, represented by $[\overline{P_b}]$, which can be diagonalized

$$[\overline{P_p}] = [{}^B C^P]^{\mathrm{T}} [\overline{P_b}] [{}^B C^P] \tag{82}$$

where $[{}^B C^P]$ is a matrix containing the normalized eigenvectors of $[\overline{P_b}]$, with elements defined as ${}^B C^P_{1i} \triangleq \hat{\mathbf{T}} \cdot \hat{\mathbf{p}}_i$ and ${}^B C^P_{2i} \triangleq \hat{\mathbf{R}} \cdot \hat{\mathbf{p}}_i$ ($i = 1, 2, 3$). The diagonal covariance matrix $[\overline{P_p}]$ contains the eigenvalues of $[\overline{P_b}]$ associated with the eigenvectors of unit length $\hat{\mathbf{p}}_1$ and $\hat{\mathbf{p}}_2$, which are parallel to the principal axes of the error ellipse in the B-plane. The orientation of the principal axes of the ellipse with respect to $\hat{\mathbf{T}}$ and $\hat{\mathbf{R}}$ can be obtained as $\theta = \arctan\left({}^B C^P_{21} / {}^B C^P_{11}\right)$. The square roots of the two nonzero elements of $[\overline{P}_p]$ are the semimajor and semiminor axes of the ellipse, denoted as σ_1 and σ_2, respectively. The difference in time that it takes for the comet to travel between the positions marked by **B** and **ε** is assumed to be negligible, and the error ellipse therefore describes the uncertainty in the estimate that the comet will pass through the position indicated by **B**. The bivariate Gaussian probability density function can be expressed in invariant form according to equation (2.2-39) of reference 9, or equation (A.19.1) of reference 5,

$$f(x_1, x_2) = \frac{1}{2\pi |P|^{1/2}} e^{-\frac{1}{2}(\{x\}-\{\overline{x}\})^T [P]^{-1}(\{x\}-\{\overline{x}\})} \tag{83}$$

where $[P]$ can be either of the covariance matrices $[\overline{P}_b]$ or $[\overline{P}_p]$, $|P|$ denotes the determinant of $[P]$, and $\{x\}^T = [x_1 \ x_2]$ contains the random variables of the state deviation at t_k associated with $\hat{\mathbf{T}}$ and $\hat{\mathbf{R}}$ in the case of $[\overline{P}_b]$, or with $\hat{\mathbf{p}}_1$ and $\hat{\mathbf{p}}_2$ in the case of $[\overline{P}_p]$. The random variables have a mean of $\{\overline{x}\}$, a column matrix containing the two dot products $\mathbf{B} \cdot \hat{\mathbf{T}} = B_T$ and $\mathbf{B} \cdot \hat{\mathbf{R}} = B_R$, or $\mathbf{B} \cdot \hat{\mathbf{p}}_1$ and $\mathbf{B} \cdot \hat{\mathbf{p}}_2$. Contours of constant f are ellipses.

Inspection of results from a Monte Carlo analysis shows the validity of the assumption of linearity inherent in the development of the sequential filter. The object of interest is the hypothetical comet described previously at the conclusion of the section dealing with range and range-rate measurements, without the perturbation induced by outgassing. The extended Kalman filter is employed to produce 100 orbit solutions from 91 optical observations spaced 1 day apart, made from a single observatory whose resolution ρ is 0.01 arcsec. Each solution begins with a preliminary orbit determined by the method of Gauss, and processes a different set of measurements of longitude and latitude, whose random errors are formed according to step 1 of the procedure set forth for obtaining a priori information. The procedure yields 100 values of **B** and associated covariance matrices $[\overline{P}_b]$ (or $[\overline{P}_p]$), each of which corresponds to an ellipse whose center is the position given by **B** and indicated with a point in figure 25. The average values of σ_1, σ_2, θ, B_T, and B_R are given in the second column of table 2. Counterparts to the first three of these parameters may be obtained by forming the covariance of the 100 state deviations $\{x\}$ produced by the filter after the final measurement update, and mapping this single covariance matrix to t_k. Error ellipse parameters obtained in this way are given in the third column of table 2, and the 1σ ellipse centered on the average values of B_T and B_R is shown in figure 25 to be a reasonable representation of the distribution of the individual intersection points.

The time of closest approach t_b can be determined from the designed time of collision t_k by referring to figure 24 and recalling the assumption that the geocentric velocity $\mathbf{v}^{C/E*} = v_\infty \hat{\mathbf{S}}$ is assumed to remain constant in N for the purpose of constructing **B**. The time $t_k - t_b$ required for C to travel from the position marked by **B** to the position marked by **ε** is given by

$$t_k - t_b = \frac{1}{v_\infty}(\boldsymbol{\varepsilon} - \mathbf{B}) \cdot \hat{\mathbf{S}} = \frac{\boldsymbol{\varepsilon} \cdot \hat{\mathbf{S}}}{v_\infty} \tag{84}$$

The standard deviation σ_t of this time interval obtained from the 100 orbit solutions is given in the fourth column of table 2.

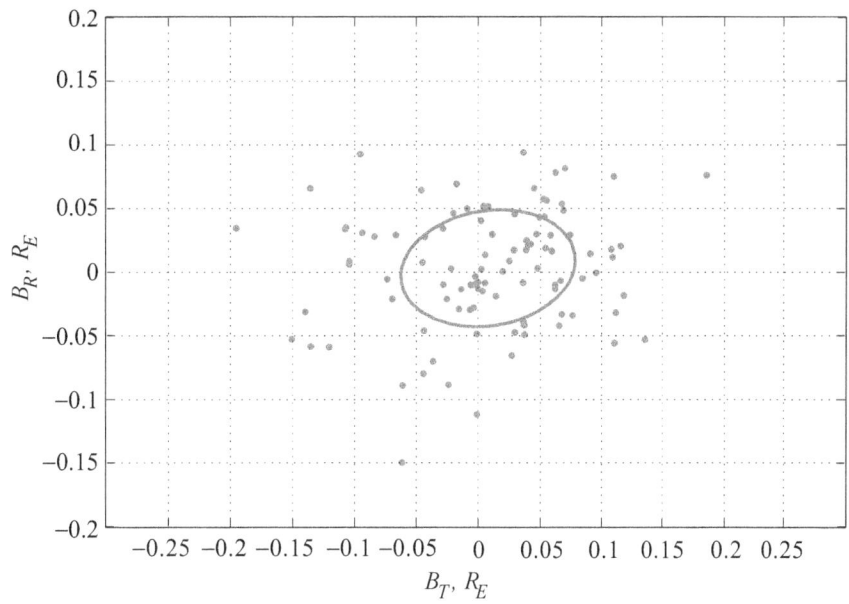

Figure 25. 1σ error ellipse for long-period comet.

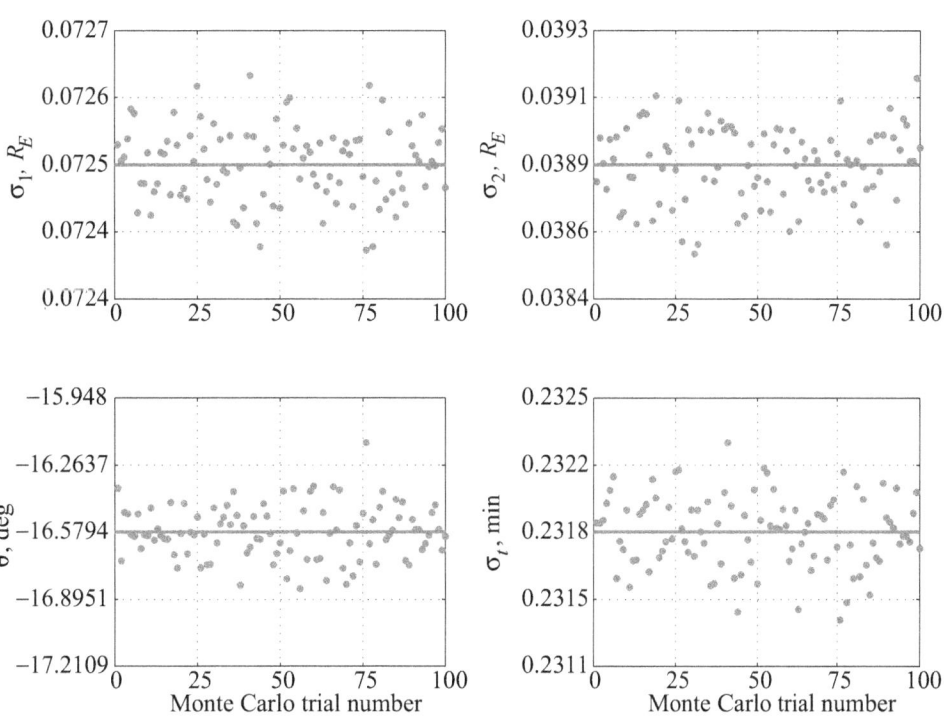

Figure 26. Variation in 1σ error ellipse for long-period comet.

Alternatively, the accuracy of the estimate of the time at which the comet will arrive at the position marked by **B** is indicated by the extent of the three-dimensional error ellipsoid normal to the B-plane. Letting σ_S denote the square root of the element of $[\overline{P_B}]$ associated with $\hat{\mathbf{S}}$, the uncertainty in time of arrival at the B-plane is given simply by

$$\sigma_t = \sigma_S / v_\infty \qquad (85)$$

Table 2 contains the average values of σ_S and time of arrival standard deviation in the second column and the values obtained from the single covariance matrix in the third column. Values of σ_1, σ_2, σ_S, or σ_t corresponding to 2σ and 3σ ellipses are calculated easily by multiplying the 1σ value in table 2 by 2 or 3.

The similarity of σ_t in the fourth column to the values recorded in the second and third columns shows the errors in time of arrival have the same standard deviation as the travel times between the positions marked by **B** and **ε**. The close agreement in the values of all parameters in the second and third columns of table 2 is one indication the assumption of linearity made in deriving the sequential filter is justified. Additional support is furnished in figure 26: the individual error ellipse parameters for each of the 100 Monte Carlo simulations are virtually the same, with only slight variations distributed on either side of the mean (shown with a solid blue line). The similarity in these error ellipse parameters is evidence of linearity; in other words, the covariance of the state deviation at the designed time of collision is virtually independent of the point at which the orbit intersects the B-plane. Consequently, there is no need to perform additional Monte Carlo analysis.

The foregoing conclusions remain unchanged after including the effect of outgassing, as indicated in table 3. A comparison of tables 2 and 3 reveals that comet outgassing increases the size of the 1σ error ellipse, signaling a less accurate determination of the orbit.

Table 2. 1σ Error Ellipse Parameters for Sample Comet

Parameter	Average	Single	Eq. (84)
σ_1, R_E	0.072535	0.071114	
σ_2, R_E	0.038867	0.045221	
θ, deg	−16.579420	−8.787936	
B_T, R_E	0.008113		
B_R, R_E	0.002959		
σ_S, R_E	0.078602	0.077102	
σ_t, min	0.231822	0.227399	0.227400

Table 3. 1σ Error Ellipse Parameters for Sample Outgassing Comet

Parameter	Average	Single	Eq. (84)
σ_1, R_E	0.287882	0.287184	
σ_2, R_E	0.064515	0.069989	
θ, deg	−31.912291	−31.814645	
B_T, R_E	−0.210096		
B_R, R_E	−0.108374		
σ_S, R_E	0.334448	0.334101	
σ_t, min	0.986446	0.985425	0.774256

Probability Associated With B

The probability that an object passes through the B-plane within a certain distance of the position indicated by **B** is found by integrating the probability density function over the area of a circle. The result can be used in two ways, provided the assumption of linearity is warranted. First, in the unique circumstance that occurs because the orbit of our hypothetical object is known precisely, the probability quantifies the accuracy of the orbit determined by the sequential filter. Second, for the general case in which the true orbit is unknown, a probability of collision between the object and Earth can be obtained.

In what follows, numerical examples are created with the hypothetical comet described in the discussion regarding range and range-rate measurements; again, the filter processes 91 optical observations from a single observatory, taken 1 day apart, and measurement errors are normally distributed, based upon a resolution of $\rho = 0.01$ arcsec. In view of the demonstration in the preceding section, the assumption of linearity is justified, and we dispense with Monte Carlo analysis; furthermore, the eigenvalues and eigenvectors of the covariance matrix $[\bar{P}_b]$ are considered independent of **B** (the mean of the state deviation) obtained from the sequential filter in the orbit determination process. The actual orbital parameters of the hypothetical object are adopted to produce the reference trajectory for the sequential algorithm, and a conventional filter is put into service so that the reference orbit remains unaltered.

The bivariate probability density function given by equation (83) is integrated numerically over the area of a circle whose center is given by the coordinate pair $(\bar{x}_1, \bar{x}_2)$, the elements of which are used to form the matrix $\{\bar{x}\}$. The center of the circle varies over several points in a coarsely spaced grid in the neighborhood of (0, 0) in the B-plane. For each center point, the function $f(x_1, x_2)$ is evaluated at every point of a fine, evenly spaced grid that approximates a circle, and then multiplied by the area enclosed by four neighboring grid points (arranged in a square); the products thus obtained are summed to yield the probability that the object passes through the B-plane, somewhere within the circle.

The radius of the circle is not critical for the purpose of gauging orbit determination accuracy. However, a particular radius is required for calculating probability of collision; the area of the circle must be equal to the effective collision cross section of the planet, which is larger than the physical cross section in order to account for Earth's gravitational attraction that has been ignored up to now. The effective collision radius r_c is given by equation (8.3-30) in reference 1,

$$r_c = \frac{R_E}{v_\infty} \sqrt{v_\infty^2 + \frac{2\mu_E}{R_E}} \tag{86}$$

where R_E is the physical radius of Earth (regarded as a sphere), and μ_E is the Earth's gravitational parameter.

The probability that the object passes through a circle of radius r_c centered at the Earth's mass center E^*, (0, 0), is found to be 99.9 percent. The probability of passing through circles with other centers is likewise computed, and each concentric ellipse in figure 27 is a locus of points for which the probability is equal to 1 in 10^z ($z = 1, 3, 6$). Because the actual orbit is designed to result in **B** = **0**, it can be said that the higher the probability, the more accurate the orbit determination. With the luxury of knowing the actual orbit corresponds to $\varepsilon = \mathbf{B} = \mathbf{0}$, the information presented in figure 27 gives reason to be 100 times more confident in a filter solution when **B** marks a point on the contour labeled −1 than when it marks a point on the −3 contour.

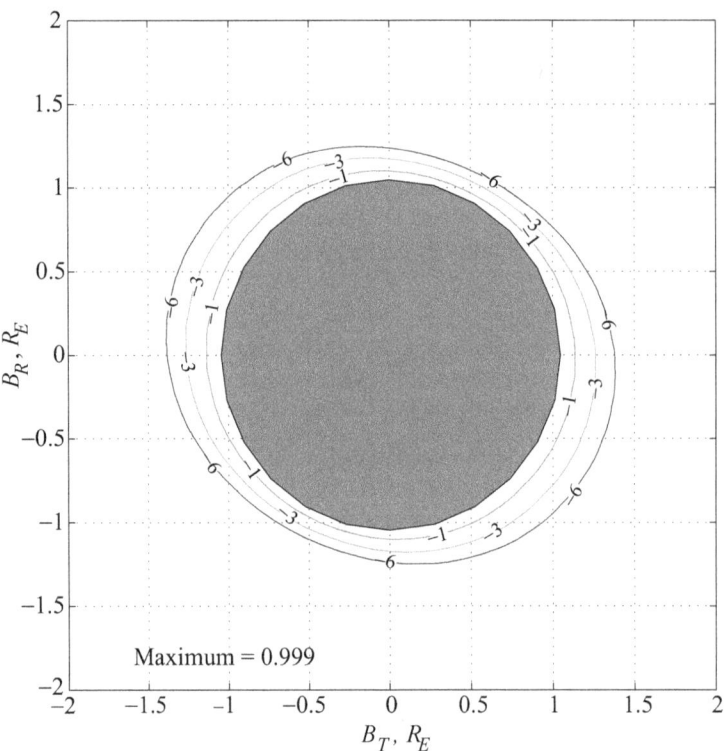

Figure 27. Probability of passing through a circle of radius r_c.

In addition to gauging orbit determination accuracy, figure 27 can be used to indicate probability of collision. In the event the filter solution yields $\mathbf{B} = \mathbf{0}$ (rather unlikely), the probability of collision is 99.9 percent, and the probability is one in a million that the object crosses the B-plane through a circle of radius r_c centered at, say, $(1, -1)$. On the other hand, if the filter solution gives $\mathbf{B} = 1\hat{\mathbf{T}} - 1\hat{\mathbf{R}}$ (a more likely result than $\mathbf{B} = \mathbf{0}$), the probability that the comet passes within a distance r_c of this point is 99.9 percent; furthermore, there is one chance in a million that the comet will instead cross the B-plane at $(0, 0)$ and collide with Earth. This interpretation of figure 27 rests on the assumption of linearity and the concomitant concept that the uncertainty in $\mathbf{B}$ is described by the same error ellipse, no matter what the direction and magnitude of $\mathbf{B}$. In other words, the same error ellipse can be translated or shifted to any point on the B-plane.

The presence of outgassing can be expected to reduce the probability of passing through $\mathbf{B} = \mathbf{0}$; measurements of range and range-rate can be expected to increase it. The orbit solution based on angular measurements with $\rho = 0.01$ arcsec from 91 observations at a single observatory is so accurate that neither of these effects are observed; however, varying the number of observations brings them to light in figure 28. The plot at the top shows the increase in central probability with number of observations. The second and third plots show the respective collision probabilities associated with solutions of $\mathbf{B} = 1\hat{\mathbf{T}}$ and $\mathbf{B} = 1\hat{\mathbf{R}}$. There are six curves in each plot corresponding to three combinations of observation types. The first consists of two angular measurements at each observation, the second includes range measurements at a resolution of 10^3 km, and the third adds measurements of range-rate at a resolution of 1 m/s. For each combination, results with outgassing are compared with those in which outgassing is absent.

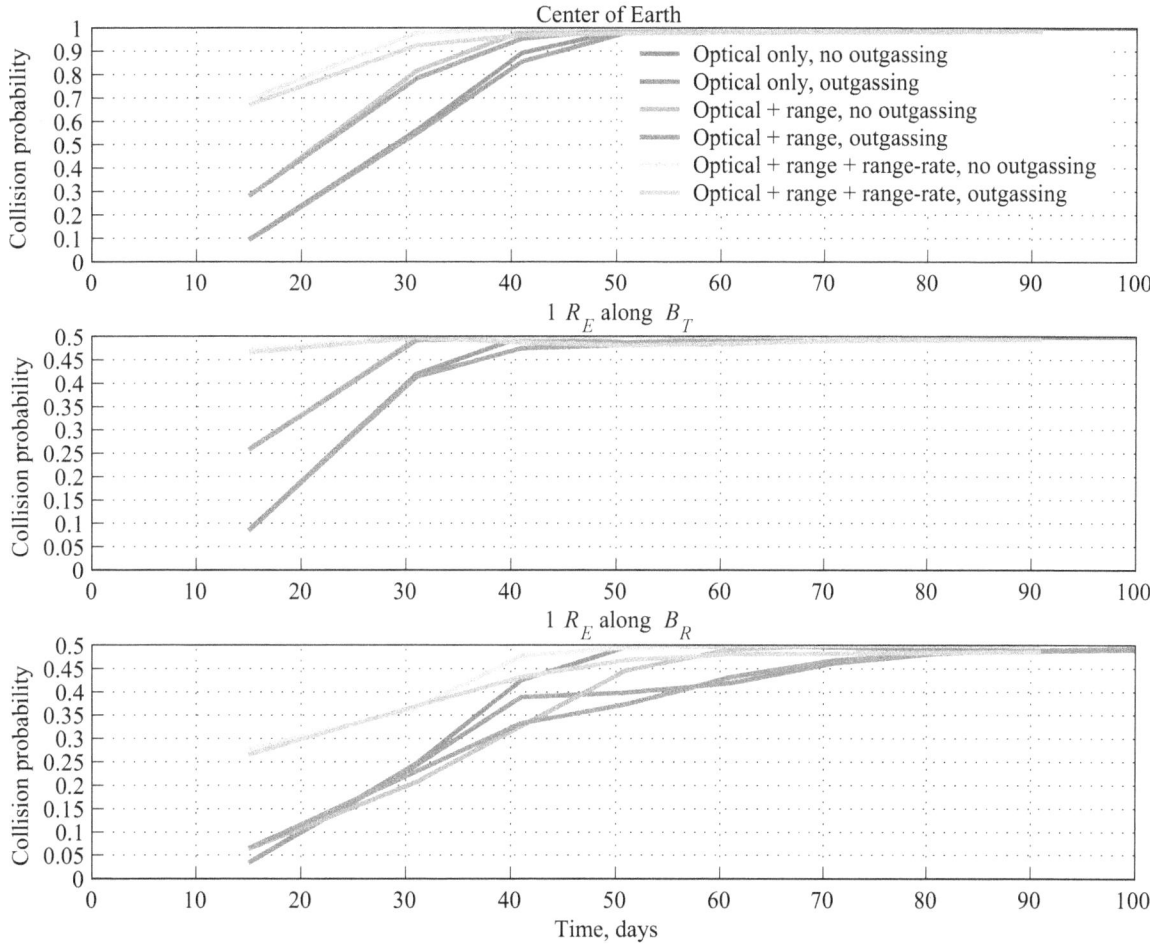

Figure 28. Probability, outgassing, and measurements of range and range-rate.

Batch Filter for Near-Earth Asteroids

Having examined orbit determination of LPCs in some detail, we turn our attention now to potentially dangerous objects whose orbital parameters are rather different, NEAs that are so named when they have perihelial distances less than 1.3 au. These asteroids are further classified by dividing them into groups based upon values of semimajor axis a, aphelial distance r_a, and perihelial distance r_p. Earth-crossing asteroids with $a < 1$ au and $r_a > 0.983$ au are members of the Aten class. Earth-crossing asteroids with $a > 1$ au and $r_p < 1.017$ au belong to the Apollo class. Earth-approaching asteroids with $a > 1$ au and $1.017 < r_p < 1.3$ au are classified as Amors (these asteroid orbits do not actually cross Earth's orbit but come close enough that a perturbation could cause a collision). It is possible that there exists another class of Earth-approaching asteroids for which $r_a < 1$ au, although no asteroids of this type have been detected to date.

The following discussion of analysis is facilitated by the definition of two types of Earth-crossing asteroids that are more general than the foregoing classes. "Interior" asteroids are said to be those with $r_p < 1$ au and $r_a = 1$ au, while "exterior" asteroids have $r_p \leq 1$ au and $r_a > 1$ au. Figures 29 and 30 contain examples of these two types of orbits, where the Earth E is shown in a circular orbit of radius 1 au

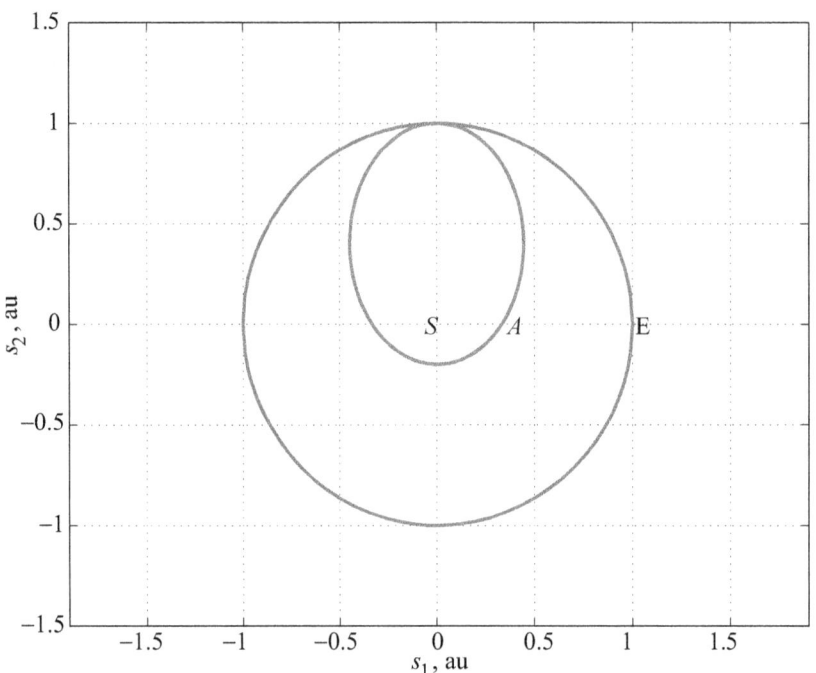

Figure 29. Example of interior asteroid orbit.

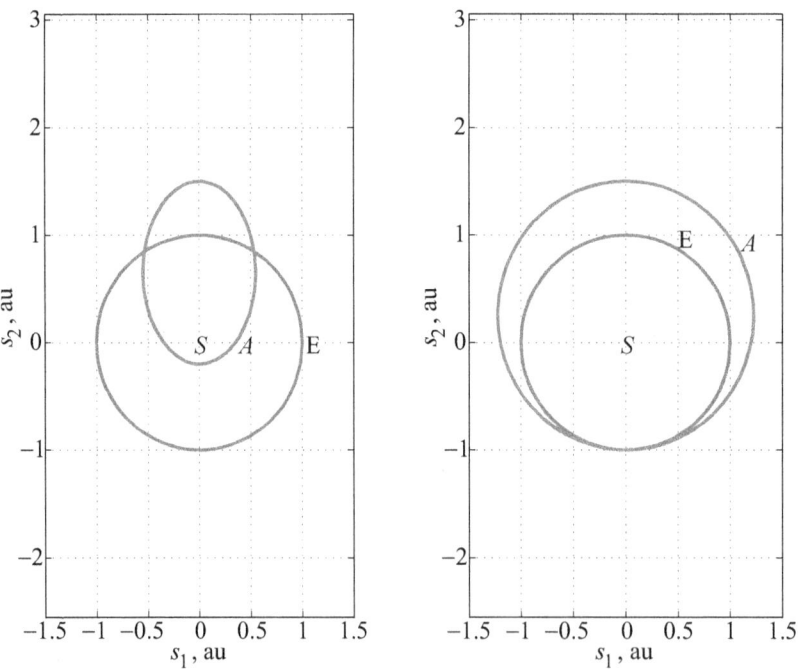

Figure 30. Examples of exterior asteroid orbits.

about the Sun S, and an asteroid A is shown in a coplanar elliptical orbit. The negative signs of eccentric anomalies E and E_k required for preperihelial collision with LPCs (see the discussion of warning time) apply also to exterior asteroids; however, collisions of interior asteroids must always take place after perihelion, therefore the sign changes are unnecessary.

The focus of the present analysis is upon asteroids discovered less than 1 orbital period prior to collision. Preliminary orbit determination for interior and exterior asteroids is performed in much the same way as it is in connection with LPCs, with the following details regarding the seven-step procedure given in connection with erroneous predicted miss distance.

In the first step, the earliest opportunity for observations t_0 is specified as 60 days prior to the time of collision t_k, and the initial value of true anomaly $\nu(t_0)$ is calculated accordingly. The reason for basing $\nu(t_0)$ on time to collision rather than a specified value of r has to do with the observation times chosen in the third step. Four observations permit the determination of orbits inclined 0° or 180° to the ecliptic; we wish to make the fourth and final observation at least 2 weeks prior to collision in order to provide some amount of warning. Hence, the times t_1, t_2, t_3, and t_4 of the observations are taken to be 11, 22, 33, and 44 days after t_0, leaving 16 days between the final observation and collision. A single observatory in a circular heliocentric orbit of radius 1 au in the ecliptic plane is used to make the observations, and the initial true longitude L_0 takes on values of 0°, 90°, 180°, and 270°.

The uniformly distributed measurement errors associated with the fourth step are bounded in magnitude by a telescope resolution of $\rho = 0.1$ arcsec, and once again are constructed with a pseudo-random number algorithm. Average values $\bar{\varepsilon}$ of erroneous predicted miss distances for 100 sets of measurements are shown.

In connection with the fifth step, the position and velocity at t_1 are to be determined by the method of least squares. It so happens that the algorithm is quite sensitive to the initial estimate $\{\hat{x}\}$ of position and velocity; therefore, values corresponding to day 11 are used for exterior asteroids, and to day 12 for interior asteroids. The MATLAB® algorithm failed to converge for a number of measurement sets in a handful of cases; nevertheless, a value of ε is computed from the unconverged estimates. The sensitivity to the initial guess, and the lack of convergence, may be due to changes in an asteroid position that are significantly larger than changes in a comet orbit over the same interval of time.

Orbit determination for interior asteroids is discussed first, and exterior asteroids follow.

Interior Asteroid Orbits

Classical orbital elements of interior asteroids are constructed with $r_a = 1.0$ au, $r_p = 0.2, 0.4, 0.6$, and 0.8 au, $i = -40°, -30°, -20°, \ldots, 40°$, and $\Omega = 0°, 45°, 90°, 135°$, and $180°$. For each of these 180 sets of orbital elements, the argument of perihelion ω is determined to be 180°, according to equation (5). The initial true longitude L_0 of the observatory takes on four values, each differing by 90°, resulting in a total of $4 \times 180 = 720$ cases.

The results from this analysis are presented in figure 31 as a function of asteroid orbital elements i and r_p; each data point represents $\bar{\varepsilon}$ for one case, and a family of 20 cases (in which Ω and L_0 vary) form one stack of points. At the lower values of r_p one can see an inverse relationship of $\bar{\varepsilon}$ to i, which is not unexpected because orbits that are coplanar with the observatory require a minimum of four observations to be determined. Although not as evident at this scale, the relationship exists at the higher values of r_p as well. At $i = 0°$ and $r_p = 0.6$ au one notices two outlying values of $\bar{\varepsilon}$, approximately 0.45 lunar distance. These two cases are distinguished from each other only by the value of L_0 (and of Ω, which is

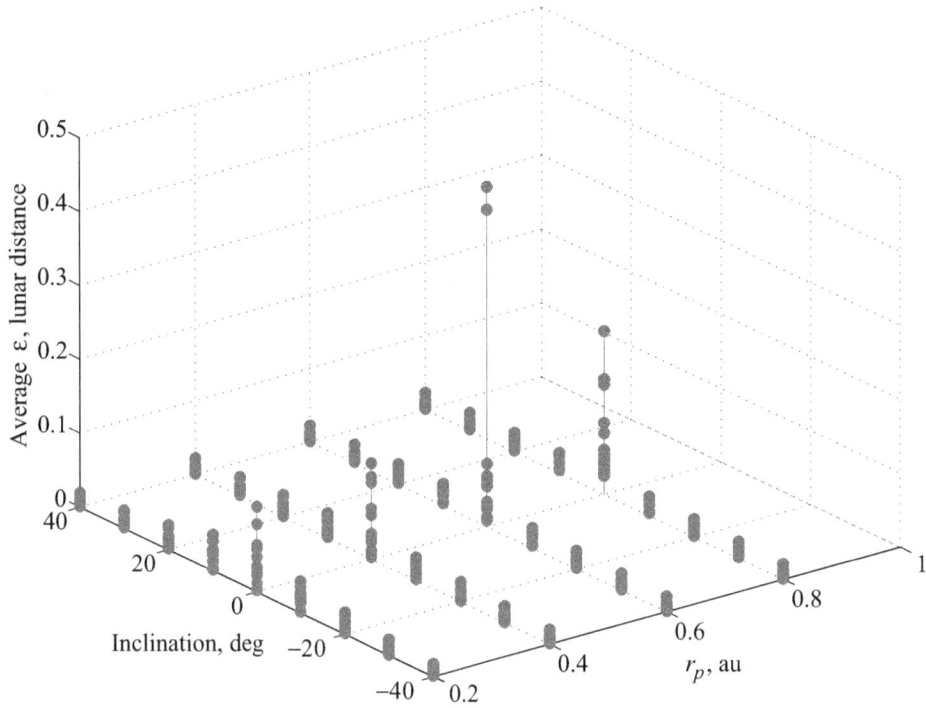

Figure 31. Interior near-Earth asteroids.

undefined for $i = 0$). For a third case in this family the least squares algorithm was unable to converge on a solution for 38 of the 100 measurement sets, and the associated data point is not shown. Of the 719 results displayed, 697 values of $\bar{\varepsilon}$ are less than 0.05 lunar distance, or about 3 Earth radii. When one holds i constant and plots the results in two dimensions, $\bar{\varepsilon}$ is seen to vary proportionally to r_p when $i = \pm 40°$ and inversely to r_p when $i = \pm 10°$. When displayed as functions of orbital elements other than i and r_p, no remarkable trends are visible in the results.

In eight other cases a certain number of measurement sets did not produce convergence. These cases have common values of $r_a = 1$ au, $r_p = 0.2$ au, and $\Omega - L_0 = 45°$. The nine cases that suffered from convergence problems are listed in table 4, together with their orbital parameters, initial true longitude of the observatory, and the number of trials out of 100 that did not converge.

Table 4. Interior Asteroid Orbits With Unconverged Trials

Case	r_a, au	r_p, au	i, deg	Ω, deg	ω, deg	ν_0, deg	L_0, deg	Trials
115	1.0	0.6	0	180	180	123.72	180	38
362	1.0	0.2	−40	45	180	132.52	0	85
367	1.0	0.2	−30	45	180	132.52	0	3
397	1.0	0.2	30	45	180	132.52	0	4
402	1.0	0.2	40	45	180	132.52	0	86
544	1.0	0.2	−40	135	180	132.52	90	1
549	1.0	0.2	−30	135	180	132.52	90	10
579	1.0	0.2	30	135	180	132.52	90	9
584	1.0	0.2	40	135	180	132.52	90	7

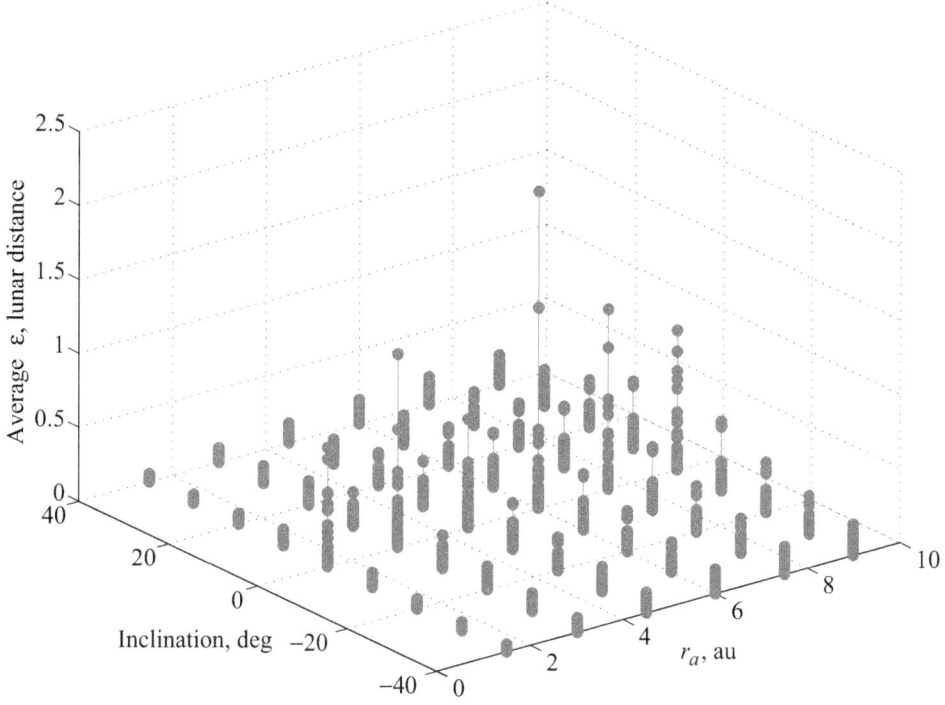

Figure 32. Accuracy of exterior NEA orbits, as a function of i and r_a.

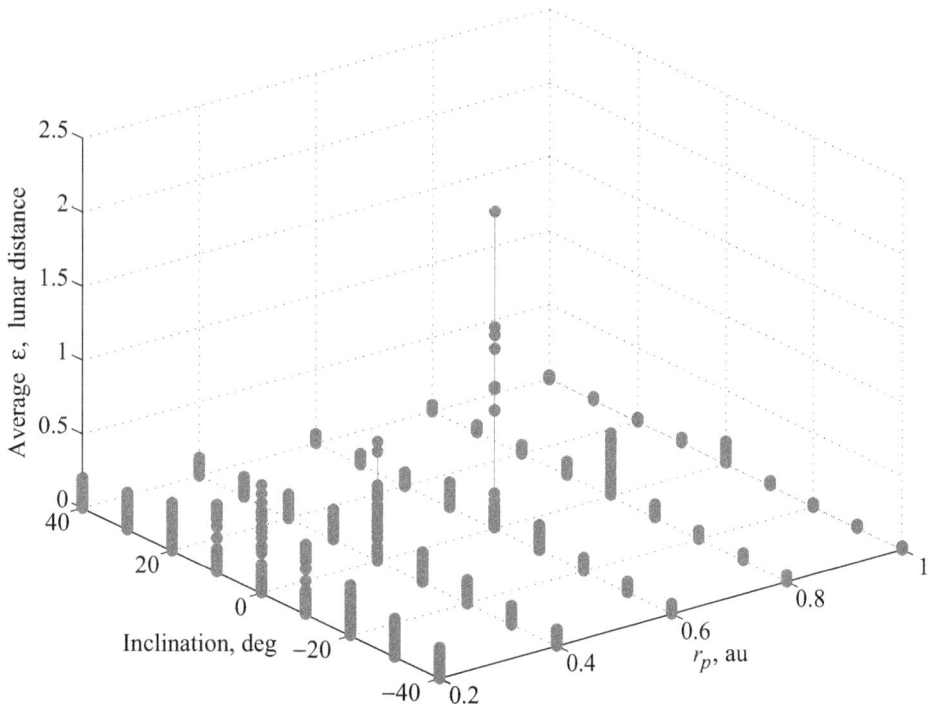

Figure 33. Accuracy of exterior NEA orbits, as a function of i and r_p.

Exterior Asteroid Orbits

For exterior asteroid orbits, 1350 sets of classical orbital elements are constructed with r_a = 1.5, 3.0, 4.5, ..., 9.0 au, r_p = 0.2, 0.4, 0.6, ..., 1.0 au, i = –40°, –30°, –20°, ..., 40°, and Ω = 0°, 45°, 90°, 135°, and 180°. For each set, the argument of perihelion ω is computed according to equation (5). L_0 takes on the same four values as in the interior cases; thus, a total of 4 × 1350 = 5400 cases for exterior asteroid orbits are examined.

The average values $\bar{\varepsilon}$ of erroneous predicted miss distance are shown in figures 32 and 33 as functions of i, and r_a and r_p, respectively. Relatively poor orbit determination is exhibited once again for asteroids traveling in orbits that have low inclination with respect to the observatory, when only four observations are obtained. Accuracy generally decreases for smaller values of r_p and for larger values of r_a.

Of the 5400 cases examined, 5389 yielded $\bar{\varepsilon}$ less than 1 lunar distance. In two of the remaining cases $\bar{\varepsilon}$ was reported to be approximately 4.8 lunar distances; however, the least squares algorithm was unable to converge on a solution for 15 and 38 of the 100 measurement sets, and the associated data points are not shown. The largest value of $\bar{\varepsilon}$ shown in figures 32 and 33 is 2.14 lunar distances, and the smallest is 0.00125 lunar distance, or 0.08 of an Earth radius. In seven other cases involving an object whose orbit is coplanar with that of the observatory (i = 0), a certain number of measurement sets did not produce convergence, as reported in table 5.

Table 5. Exterior Asteroid Orbits With Unconverged Trials

Case	r_a, au	r_p, au	i, deg	Ω, deg	ω, deg	v_0, deg	L_0, deg	Trials
564	4.5	0.6	0	135	85.59	–120.36	180	15
2991	3.0	0.4	0	0	112.62	–140.51	0	49
3082	3.0	0.8	0	45	62.96	–106.90	0	5
3307	4.5	0.8	0	45	59.10	–102.53	0	32
3532	6.0	0.8	0	45	57.42	–100.64	0	30
3757	7.5	0.8	0	45	56.48	–99.58	0	24
3982	9.0	0.8	0	45	55.88	–98.91	0	19
4479	3.0	1.0	0	135	0	–62.90	0	1
4612	4.5	0.6	0	45	85.59	–120.37	0	38

Although the analysis for NEAs has not been extended to include multiple observatories or improved telescope resolution, it is likely that configurations of two or more observatories, together with multiple observations, will improve preliminary orbit determination for asteroids, as has been demonstrated in connection with LPCs.

Concluding Remarks

An examination has been made of the effects of several factors upon determination of orbits of comets and asteroids on a collision course with Earth, including the time interval between successive observations, the distance at which the object is detected, the resolution of the telescope, the orbital parameters of the object, the number of observations, and, in a rather limited manner, placement in the solar system of one or more observatories. The primary factor is seen to be the length of the data arc. As mentioned at the outset, the analysis is based on a crucial assumption that optical measurements can in fact be obtained from a telescope; the validity of this assumption depends upon several factors that have not yet been addressed, among them the quality of the instrument, the extent of solar illumination of the object (solar phase angle), the optical reflectivity (albedo) of the object, occultation by the Sun or other bodies, and interference of light from sources such as the stars in the galactic plane, zodiacal dust, and the Sun.

Provided the measurements are available, the following conclusions are made based upon a study of many hypothetical objects, each of which has been given an orbit resulting in a collision with Earth at a specified time and a specified position. As used here, accuracy is expressed in terms of erroneous predicted miss distance, the difference between an object's determined position and the specified position, evaluated at the specified time of collision. A single observatory in a circular heliocentric orbit of radius 1 au in the ecliptic plane, possessing an optical resolution of 0.1 arcsec, provides good preliminary orbit determination accurate to less than 8 lunar distances for long-period comets (LPCs) observed over a 66-day arc at a heliocentric distance of 6.5 au. Multiple observations (approximately 100, spaced 1 day apart) improve the accuracy to less than 14 Earth radii, and a second observatory phased 180° from the first improves the accuracy further still to less than 1 Earth radius. Of the multiobservatory configurations examined here, two observatories of equal resolution in circular heliocentric orbits of radius 1 au in the ecliptic plane, phased 180° apart, provide the best balance of cost and accuracy in orbit determination for LPCs. Thirty to forty observations taken over a 98-day period appear to give results nearly as good as 99 observations taken 1 day apart. For near-Earth asteroids observed over an interval between 60 and 16 days before collision, the accuracy is better than 2 lunar distances for the exterior variety and 0.5 lunar distance for interior asteroids.

The results reported here give a preliminary indication of what is required to predict impending collisions accurately. We must keep in mind that such predictions must be reliable enough to support a decision to expend considerable resources in attempting to alter the orbit of a dangerous body; very accurate knowledge of the orbit is probably essential if orbital modification is to steer the object away from a collision, rather than toward one. Moreover, predictions must have a degree of integrity that virtually eliminates false alarms and unfounded panic. The results indicate that it may be possible to make reliable forecasts with two observatories whose angular resolution is on the order of 0.1 arcsec, or with a single observatory whose resolution is better by 1 to 3 orders of magnitude. Additional study of these two alternatives will have to weigh the advantages in geometry and redundancy of multiple observatories against the expense of putting them in place and maintaining them; a single observatory near Earth could be easier to maintain, but less likely to be in the best position for obtaining measurements.

Outgassing does not perturb a comet's orbit appreciably until it reaches a heliocentric distance of approximately 2.5 au; the perturbation in position is less than the Earth's radius by the time the comet reaches 1 au. Analysis of dangerous LPCs performed with Kalman filters reveals several interesting results. With a mix of weekly and daily optical observations from one or two observatories, a credible warning of collision, in the absence of outgassing, can be expected at least a year in advance if the object is spotted when it is about 6.5 au from the Sun. If a comet is discovered at a heliocentric distance of 5 au, then supplementing daily angular measurements whose resolution is 0.01 arcsec, with range measurements accurate to 1000 km and range-rate measurements good to 1 m/s, is beneficial when dealing with a short data arc of 15 to 30 days, but doing so offers little advantage when the arc is 80 or 90 days long. The assumption of linearity upon which the sequential filters depend is warranted, even in the presence of outgassing, when daily angular measurements with a resolution of 0.01 arcsec are available for 3 months. These sorts of measurements yield extremely accurate orbit determination, even with only 2 months of observations; the addition of range data can reduce the necessary data arc to 1.5 months, and only 1 month is needed when range-rate is included.

Future work should investigate the effects of perturbations that have been ignored here, including the gravitational attraction of Jupiter, Saturn, and other planets, solar radiation pressure, and relativistic effects. The advantages of placing observatories in heliocentric orbits other than those studied here deserves careful consideration. In addition, the consequences of missed observations (due to the reasons stated earlier) should be examined.

References

1. Bate, R. R.; Mueller, D. D.; and White, J. E.: *Fundamentals of Astrodynamics*, Dover Publications, Inc., New York, 1971.

2. Battin, R. H.: *An Introduction to the Mathematics and Methods of Astrodynamics*, AIAA, New York, 1987.

3. Chodas, P. W.; and Yeomans, D. K.: Orbit Determination and Estimation of Impact Probability for Near Earth Objects. *Proceedings of the 22nd Annual AAS Rocky Mountain Guidance and Control Conference*, Breckenridge, CO, Feb. 3–7, 1999 (A99-34626 09-12), San Diego, CA, Univelt, Inc. (Advances in the Astronautical Sciences, vol. 101), 1999, pp. 21–40.

4. Chodas, P. W., et al.: Automated Detection of Potentially Hazardous Near-Earth Encounters. AAS 01-461, AAS/AIAA Astrodynamics Specialists Conference, Quebec City, Quebec, Canada, July 30–Aug. 2, 2001.

5. Tapley, B. D.; Schutz, B. E.; and Born, G. H.: *Statistical Orbit Determination*, Academic Press, May 2004.

6. Grace, A.: *Optimization Toolbox User's Guide*, The MathWorks, Inc., Natick, MA, 1994. See also <http://www.mathworks.com/access/helpdesk/help/toolbox/optim/lsqnonlin.shtml>

7. Marsden, B. G.; Sekanina, Z.; and Yeomans, D. K.: Comets and Nongravitational Forces. V., *The Astronomical Journal*, vol. 78, no. 2, March 1973, pp. 211–225.

8. Ekelund, J. E.; and Yeomans, D. K.: A Program for the Accurate Generation of Ephemerides for Halley's Comet. AAS 85-350, *Astrodynamics 1985*: *Proceedings of the AAS/AIAA Astrodynamics Conference*, American Astronautical Society, San Diego, CA, 1986, pp. 795–808.

9. Gelb, A., ed.: *Applied Optimal Estimation*, M.I.T. Press, Cambridge, MA, 1974.

Detection Element Concepts—Initial Design[4]

ROBERT H. KOONS
Swales Aerospace, Inc.

JAMES C. BREMER
Swales Aerospace, Inc.

Introduction

The initial design for a basic Comet/Asteroid Protection System (CAPS) detection element uses 3.2-m medium-resolution visible light infrared telescopes to function as whole sky surveyors and 3.2-m high-resolution visible light telescopes to serve as target trackers and collectors for interferometry. The two major types of detector telescopes, survey scope and tracking scope, are discussed and the survey scope physical layout is explained. There is a brief discussion of using the tracking scopes with a collector for interferometry and the additional data that could be acquired using interferometry. CAPS telescope technology development is discussed as well as the support foundations required for the scopes in either a lunar-based or space-based configuration.

Systems Requirements and Design Assumptions

The systems requirements in table 1 were assigned to the detection system. The design assumptions in Table 2 were made and a series of trades were performed to develop an initial set of design parameters for the optical/infrared (IR) telescopes. The quantum and optical efficiencies in table 2 were selected based on best laboratory performance today. It was felt a signal-to-noise ratio of 6 at the detector would be required to initially find a target, but once a target was located, the tracking scope could acquire and follow the target with a signal-to-noise ratio greater than 1.

Table 1. Detector Systems Requirement Table

Target: long-period comet (LPC)	Detection 7 au from Earth and 6 au from Sun Diameter 1000 m Albedo 0.02 Target rate 0.012 arcsec/s Visual magnitude +27
Target: near-Earth asteroid (NEA)	Detection 0.2 au from Earth 1 au from Sun Diameter 50 m Albedo 0.02 Target rate 0.480 arcsec/s Visual magnitude +23
Telescope	Full sky coverage Minimize number of telescopes 0.01-arcsec angular resolution scan mode 0.001-arcsec angular resolution track mode

---

[4]Chapter nomenclature available in chapter notes, p. 217.

Table 2. CAPS Telescope Design Assumptions

Telescope type	Derived systems requirements
Survey scope	Four survey telescopes to cover whole sky in 30 days 90-s integration time between scans 5-percent overlap of scans Quantum efficiency of detectors 80 percent Optical efficiency 80 percent Minimum signal-to-noise ratio 6
Tracking scope	Tracking telescopes as needed, number TBD Assume same focal plane as survey scope 90-s integration time between measurements Quantum efficiency of detectors 80 percent Optical efficiency 80 percent Minimum signal-to-noise ratio 1

Derived Requirements

The detection system was initially sized by assuming that there would be four scan platforms providing whole sky coverage in a 30-day period. After assuming a 5-percent overlap of the scan field, a 90-s dwell time, and a 15-s slew time, a field of view (FOV) of 0.83° for each survey telescope was selected and formed the basis for analysis. Initially it was felt that a 0.1-arcsec angular resolution would be acceptable for track determination, leading to the 32k × 32k focal charged-injected device (CID) plane selected for this analysis. Later studies demonstrated that a higher resolution was necessary, creating the need for a tracking scope with an accuracy of 0.01 arcsec. The technique of blur centroiding was suggested to improve the accuracy of the survey telescope to 0.01 arcsec using the same 32k × 32k focal plane. On the basis of those design assumptions two different types of telescopes were engineered to perform the mission and are described subsequently.

During this analysis phase, the effects of target size on detection location and target signal strength as a function of illumination angle were examined using the survey scope. Figure 1 shows the distance at which a target is initially sighted (from the Sun) as a function of target size, assuming a minimum signal-to-noise ratio of 6 for detection and best opposition. Figure 1 effectively bounds the earliest opportunity for detection for the survey scope. Figure 2 demonstrates the effect of solar illumination angle on target signal to noise for a long-period comet (LPC) at 7 au. Figure 2 demonstrates that a target can be tracked over a wide range of illumination angles after acquisition.

CAPS Optical/Infrared Telescope Concept

The tracking and survey telescopes were further refined. Additional analysis was performed on the survey telescope to demonstrate blur centroiding and the advantages of different scanning techniques. An IR detection system was added to the survey telescope to provide for earlier identification of small NEAs. An optical analysis of the survey telescope provided for more detailed design of the optical path including optical hardware, mirror sizes, and optical path lengths. The tracking telescope was modified to include an interferometry option.

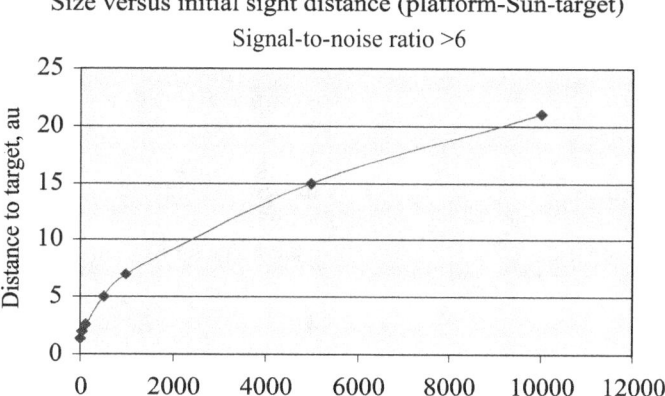

Figure 1. Initial sighting as a function of target size for CAPS survey telescope.

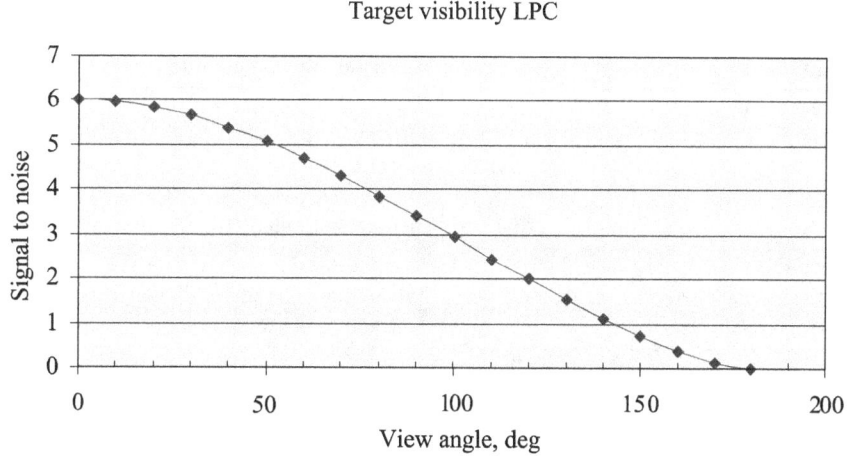

Figure 2. Target signal to noise as a function of view/illumination angle.

Survey Telescope

The survey mode for the Comet/Asteroid Protection System (CAPS) telescope identifies candidate targets and determines their angular positions to an accuracy of 0.01 arcsec. Survey telescope design and performance are summarized in table 3. Survey telescope hardware and operational requirements identified in the study are included in tables 4 and 5. Figure 3 shows the detailed layout of the survey telescope optical path and the sizes of the mirrors. A dichroic beam splitter was added to the optical path to permit concurrent IR and visual light detection of targets. A filter wheel should be added after the dichroic beam splitter to provide for a visual light spectral analysis of the target.

Performance. More detailed analysis of the survey telescope design with blur centroiding incorporated demonstrated that the signal-to-noise ratio for each scan can be reduced to 3.6 and still provide positive indication of a target. Blur centroiding also permits target acquisition to within three pixels of an object,

Table 3. Survey Telescope Design Summary

Performance	
Area of coverage	0.85 sterrad every 30 days
Angular resolution	0.01 deg
Field of view	0.83 deg^2
Integration time/slew time	6 frames at 15 s/frame/15 s slew
Overlap per grid point	5 percent
Minimum signal to noise	3.6 per frame for LPC
Optics	
Mirror diameter	3.2 m
Surface quality	Micro roughness <2 Å
Effective focal length (EFL)	16.5-m acquisition
Includes	Dichroic beam splitter allowing for two focal planes to view same field
Detector array (primary)	32k × 32k Si CID
Resolution	0.1 arcsec/pixel, blur centroided to 0.01 arcsec/pixel
Wavelength	0.4 µm to 1.0 µm
Pitch	8 µm
Temperature	230 K ± 0.1 K
Image quality	Diffraction limited 0.5 µm
Detector array (secondary)	4k × 4k Si:GAs CID
Wavelength	8 µm to 18 µm
Pitch	64 µm
Temperature	10 K (TBD) ± 0.001 K (TBD)
Image quality	Diffraction limited 10 µm
Innovative feature	
Allowance for subpixel angular resolution	Provided by blur centroiding algorithm plus Kalman filtering of moving blur

Table 4. Survey Telescope Hardware Requirements

All-reflective, unobscured telescope with 3.2-m aperture
Off-axis telescope with no central obscuration
Apodization to facilitate detection of LPCs in vicinity of stars and provide a multipixel blur for centroiding yielding subpixel measurement accuracy
Accessible focal plane for field stop
Accessible pupil for apodization masks and Lyot stops
Baffle tube length 6.4 to 9.6 m
Baffle vanes with knife edges radii <25 µm
Mirror surface roughness <2 Å
Dichroic beamsplitter
32k × 32k Si FPA with 8-µm pitch, QE > 0.8, CTE > 0.999999, read noise <4 electrons, dark current <0.003 electrons/pixel/s, well capacity >100 000 electrons, <1 dead pixel/100 000 operative pixels
4k × 4k thermal IR array
16.5-m EFL and 0.91 × 0.91 degree FOV (for 0.1 arcsec/NIR pixels)
Able to slew 0.91° and settle in <15 s
Filter wheel for reflected spectra analysis

Table 5. Survey Telescope Operational Requirements

Detection in window with guardband signal subtracted to suppress stray light, dark current, read noise, etc.
Summation of signals in 3- × 3-pixel window and subtraction of 9/16 the sum of the surrounding 16 pixels
Frame integration time limited to 15 s for LPCs and 0.375 s for NEAs
Detection in 4 of 6 frames required to declare a potential target
Ratios of intensities in 2- × 2-pixel array analyzed to determine centroids of potential target detections to <0.01 arcsec
Analysis of shape and intensity of energy distribution to reject charged particle events and detections of the centroid of multiple objects
Analysis of potential target trajectories and rejection of those with physically impossible trajectories
Refinement of trajectories of true targets
Correlation of true targets with track files from prior observations and with observations from other platforms
Simultaneous observations in two or more thermal IR bands
Use of visible data to identify frames where IR contribution target dominates background from stars, etc.
Estimation of target cross section and temperature from thermal IR measurements and reflected spectra

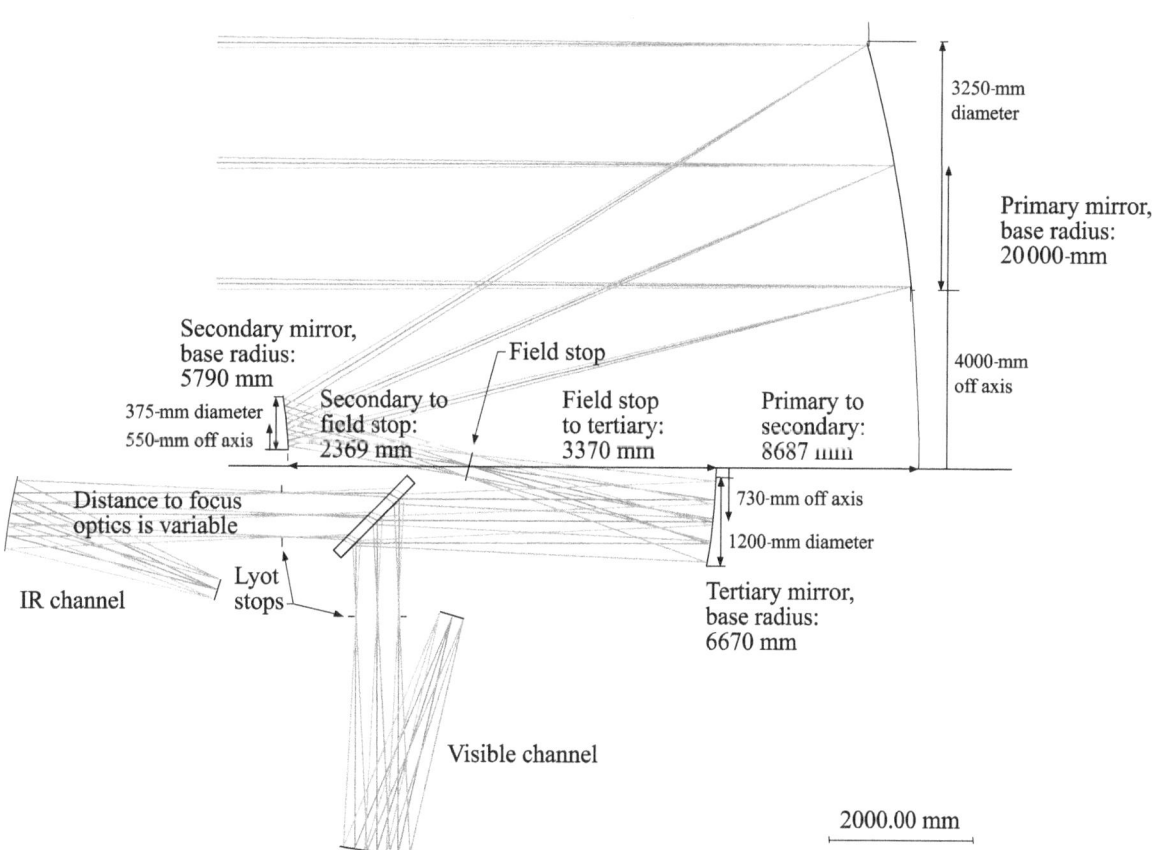

Figure 3. CAPS telescope optical layout survey telescope.

which is up to seven times brighter in visual magnitude than the target object, so targets can be picked out near bright stars.

The survey telescope is Rayleigh limited at ≈0.039 arcsec. This means that the survey telescope cannot distinguish between objects of similar magnitude if they are less than 0.039 s apart. The blur centroid for two such objects would tend to be dumbbell-shaped rather than ellipsoidal. The survey telescope will thus tend to track the midpoint of two objects that are closer than 0.039 arcsec.

Blur centroiding techniques. The centroid measurement algorithm determines the center of a point spread function (PSF) within a fraction of a pixel dimension from the ratio of energy in four pixels centered around a vertex of a focal plane array (FPA). The following description of blur centroiding refers to figure 4. The PSF on the left is centered on a vertex, so all four pixels have the same signal. Because the sum of the top two pixels equals the sum of the bottom two pixels, the vertical coordinate of the displacement of the centroid from the vertex is zero. Because the sum of the right two pixels equals the sum of the left two pixels, the horizontal displacement of the centroid from the vertex is also zero. The central PSF is centered horizontally, but the sum of the two bottom pixels exceeds the sum of the two top pixels, so the centroid lies below the vertex. A formula, based on the known angular distribution of the PSF, determines the vertical coordinate from the normalized difference. The PSF on the right lies above and to the right of the vertex, so the sum of the two top pixels exceeds the sum of the two bottom pixels and the sum of the two right pixels exceeds the sum of the two left pixels. The same approach is used to compute the horizontal and vertical components of the displacement of the PSF from the vertex. This centroiding technique is used routinely by star cameras and in astrometric measurements.

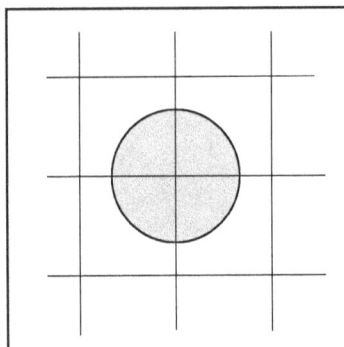 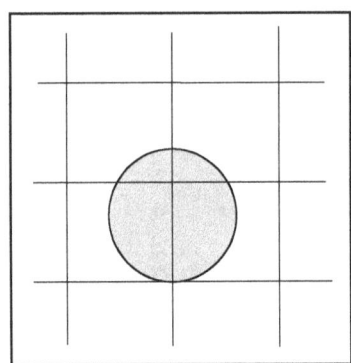 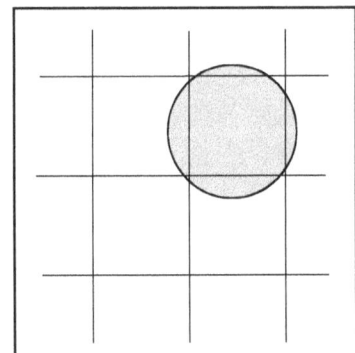

Figure 4. Centroiding example.

Errors in the centroid measurement algorithm can occur due to multiple sources within the Rayleigh limit. The centroid measurement algorithm can determine the centroid of a blur circle that has a diameter of 0.2 to 0.3 arcsec to an accuracy of 0.01 arcsec (assuming adequate signal levels). If two or more point sources are too close together (<3 pixels = 0.3 arcsec), the algorithm will not resolve them as independent objects but will measure the centroid of the sum of their signals. The CAPS survey approach observes the moving comet or asteroid in a series of frames. It is possible for the algorithm to reject those spurious measurements in which the desired signal is mixed with that of a background source using one or more of the following techniques: identification of the background source from a star catalog, identification of the background source from previous and subsequent frames, rejection of one frame out of several because the signal level is much higher (also rejects spurious detections produced by charged particles), and Kalman filtering to optimize the trajectory by identifying and rejecting an outlying position measurement.

Sorting. The survey telescope must distinguish between targets and the stellar background. Static objects can be quickly identified during the scan process and quickly eliminated. Any moving object likely to enter the telescope FOV will need to have an ephemeris created and propagated for when that object enters the FOV. The first time a target is identified it will have to be tagged for identification as a target in subsequent scans. That target will also need to have an ephemeris propagated for subsequent reacquisition by the tracking telescopes. The scanning telescope IR detector may serve as a target discriminator for NEAs. A filter wheel for the visible band may make additional target tagging possible for NEAs and LPCs.

Survey visual band. The 32k × 32k silicon (Si) CID detector provides angular resolution in the visual mode to 0.01 arcsec. A field stop is provided after the secondary mirror to shut light and heat out of the detector optical path. A Lyot stop is provided to shut down the visible light path if required.

Survey IR band. The 4k × 4k Si:GAs CID detector provides angular resolution in the IR mode to 0.08 arcsec. The IR detector has a larger pitch due to the longer IR wavelengths. The IR band can serve as a target discriminator by rapidly identifying the higher IR signature of NEAs. Additional studies should examine the system costs of the IR detector and its value in establishing target classification.

Tracking Telescope

The CAPS tracking telescope reacquires targets identified in the survey mode, tracks a target, and determines its angular positions to an accuracy of 0.001 arcsec. Tracking telescope design and performance are summarized in table 6. Survey telescope hardware and operational requirements identified in the study are included in tables 7 and 8. The optical layout of the tracking scope would include the hardware from the survey scope (fig. 4) minus the dichroic beam splitter, IR Lyot stop, IR focusing mirror, and IR detector. The mirror focal lengths would also need to be changed.

Table 6. Tracking Telescope Design Summary

Performance	
Angular resolution	0.001 deg
Field of view	0.083 deg^2
Integration time/slew time	90 s/frame
Minimum signal to noise	>1 per frame for LPC
Optics	
Mirror diameter	3.2 m
Surface quality	Micro roughness <2 Å
Effective focal length (EFL)	16.5-m acquisition
Detector array (primary)	32k × 32k Si CCD or CID
Resolution	0.01 arcsec/pixel, blur centroided to 0.001 arcsec/pixel
Wavelength	0.4 μm to 1.0 μm
Pitch	8 μm
Temperature	230 K ± 0.1 K
Image quality	Diffraction limited 0.5 μm
Innovative feature	
Interferometer	Two scopes plus a collector can serve as an interferometer

Table 7. Tracking Telescope Hardware Requirements

All-reflective, unobscured telescope with 3.2-m aperture
Off-axis telescope with no central obscuration
Apodization to facilitate detection of LPCs in vicinity of stars and provide a multipixel blur for centroiding yielding subpixel measurement accuracy
Accessible focal plane for field stop
Accessible pupil for apodization masks and Lyot stops
Baffle tube length 6.4 to 9.6 m
Baffle vanes with knife edges radii <25 μm
Mirror surface roughness <2 Å
32k × 32k Si FPA with 8 μm pitch, QE > 0.8, CTE > 0.999999, read noise <4 electrons, dark current <0.003 electrons/pixel/s, well capacity >100 000 electrons, <1 dead pixel/100 000 operative pixels
165-m EFL and 0.091 × 0.091 degree FOV (for 0.01 arcsec visible/NIR pixels)
Slew and settle times TBD
Filter wheel for reflected spectra analysis
Breakout mirrors for interferometry mission

Table 8. Tracking Telescope Operational Requirements

Detection in window with guardband signal subtracted to suppress stray light, dark current, read noise, etc.
Summation of signals in 3 × 3 pixel window and subtraction of 9/16 the sum of the surrounding 16 pixels for target and guide stars only
Frame integration time 90 s or greater for LPCs, less for NEAs
Target identified in acquisition phase, telescope centers on target
Ratios of intensities in TBD pixel array analyzed to determine centroids of known guide stars detections to <0.001 arcsec
Analysis of shape and intensity of energy distribution to reject charged particle events and detections of the centroid of multiple objects
Analysis of potential target trajectories and rejection of those with physically impossible trajectories
Refinement of trajectories of true targets
Correlation of true targets with track files from prior observations and with observations from other platforms
Estimation of target cross section and color from visible band measurements only

Performance. The tracking mode uses the 3.2-m diameter that has a narrower FOV and longer focal length than the survey telescope mirror. The tracking telescope is required to follow a target while maintaining that target on a particular set of pixels on the detector array. The dwell times for the tracking telescope are on the order of 90 s/frame.

Tracker visual band. Initially 10X optical magnification of the survey telescope focal array was suggested for the tracking telescope. Subsequent analysis demonstrated the need for a telescope optimized for tracking. The tracking telescope uses the same visual light focal plane as the survey telescope.

The tracking telescope is also Rayleigh limited at ≈0.039 arcsec. This means that the tracking telescope cannot distinguish between objects of similar magnitude if they are less than 0.039 s apart. The blur centroid for two such objects would tend to be even more dumbbell shaped when magnified by the tracking telescope.

Tracker IR Band. The initial design for the tracking telescope has no IR band.

Tracker Interferometry Mode

The CAPS telescope has provisions for being used for interferometry while in the tracking mode. Two or more tracker scopes serve as collectors to send collimated light to a centrally located combiner satellite or building. The resulting interferometer provides greater resolution of the target object for more detailed size and composition information.

Improving the accuracy of the orbit determination for a near-Earth object (NEO) could be achieved by obtaining better resolution in the observations. In order to achieve better resolution of the asteroids and comets, one could increase the aperture diameter of the telescope. However, it is not pragmatic to increase the diameter to large values. Optical interferometry is another way the resolution can be improved by increasing the baseline distance between two or more telescopes forming interference patterns.

Space-based interferometry requires that three "grand technological challenges" be met and overcome: (1) nanometer-level control and stabilization of optical element positions, (2) subnanometer-level sensing of optical element relative positions over meters of separation distance, and (3) overall instrument complexity and the implications for interferometer integration and test and autonomous on-orbit operation.

Interferometry linkage. To provide interferometry capability, the CAPS telescope design would have to be modified. Additional mirrors would have to be inserted into the collimated light beam (after the tertiary mirror) to break the beam out for interferometry processing. The collimated light beam would then have to be sent to a central location (equally distant from two tracking telescopes) where a fairly simple combiner array would be located. A series of mirrors would be necessary to rotate the collimated light source from each of the scopes so the images from each telescope could be overlaid to form the standing wave pattern necessary for interferometry.

Precision interferometry is extremely difficult on Earth due to the absorption of the collimated beam by the surrounding atmosphere. A space-based system provides the opportunity for very long baseline interferometry. Preliminary analysis has shown that the practical limit for telescopic interferometry is $\approx$+24 magnitude, given the need to establish a standing wave pattern. Interferometry could be performed on the CAPS 50-m diameter NEA immediately after establishing a track (target 1 au from Earth) and could be established on a CAPS 1000-m diameter LPC when the target is 3 au from Earth.

Interferometry collection. The interferometry processing on a target object would occur as follows:

1. A target would be identified as a candidate for interferometry.

2. Two target scopes with the same combiner would have to acquire the target while maintaining equal distance from that combiner.

3. Tracker scopes establish a constant track of the target.

4. Image rotation from each of the target scopes is sent to the image alignment mirrors.

Interferometry advantages. Interferometry is used to do subpixel resolution to identify local energy maxima within a pixel. The advantages of interferometry are increased resolution of the object, more accurate determination of size (object is no longer Rayleigh limited), and earlier detection of closely spaced multiple targets.

Use of interferometry to increase angular resolution. It is suggested that an investigation be performed to determine if interferometry can be used to increase the angular resolution accuracy of the CAPS telescope. Such a method might make use of a highly accurate star catalog (<0.0001 arcsec) and long dwell times (>90 s) on a target with extremely accurate angular rate sensors (error/dwell period <0.0001 arcsec). If angular accuracy of the telescope can be increased by another order of magnitude, even greater accuracy in orbit determination and thus earlier and more accurate notice of impacting trajectories may be possible.

Comparison With Current Earth-Based Systems and Space-Based Designs

The CAPS system is revolutionary when compared with current ground- or space-based systems. The CAPS detector telescopes are optimized for LPC and NEA detection as well as angular resolution for orbit determination. In particular, the CAPS telescope makes use of blur centroiding to get subpixel angular resolution, and has no central obscuration of the main mirror.

CAPS versus Earth-based systems. The CAPS telescope is compared with two ground-based systems in table 9. Due to atmospheric diffraction, blur centroiding is not possible with ground-based systems; thus, CAPS has a higher angular resolution. Although CAPS has a smaller coverage area, CAPS is potentially available 24 hr/day as opposed to 6 hr/day for a typical ground-based telescope. At least two (and probably three or four) ground-based systems will be needed to obtain whole sky coverage. Atmospheric absorption and background tend to eliminate the possibility of the use of IR detection. Atmospheric absorption also effectively limits the ground-based systems from seeing magnitude 27 LPCs.

Table 9. Comparison of CAPS Telescope Versus Ground-Based Systems

Feature	Individual CAPS telescope (survey/track)	Current state-of-the-art ground-based	Future ground-based system: large-aperture synoptic survey telescope (LSST)
Mirror diameter, m	3.2	1.0	8.4[a]–6.9, effective
Angular measurement resolution (visible), arcsec	0.01/0.001	1.96 (est.)[b]	0.3 (est.)[b]
Optical resolution limit (visible), arcsec	0.05	0.15	0.02
Field of view, deg^2	0.83/0.083	1.96	6.76
Visible array	32k × 32k 8 μ pitch	2560 × 1960 24 μ pitch	30k × 30k 10 μ pitch
IR array 4k × 4k	64 μ pitch	None	None
Target limiting magnitude	27	22	24
Integration time, sec	90	100	30
Coverage, deg^2/hr[c]	32	71	410

[a]With central obscuration.
[b]Assumes no blur centroiding.
[c]Exposure time only.

CAPS versus near term space-based systems. The Hubble Space Telescope (HST) and James Webb Space Telescope (JWST) are similar to CAPS in that they are primarily visible light, space-based telescopes; however, neither of these telescopes have been optimized for asteroid or comet detection and angular resolution (ref. 1). Both HST and JWST have central obscurations that make blur centroiding impossible. HST and JWST have lower resolution focal planes (800k × 800k for HST and ≈10 000k × 10 000k for JWST) and therefore have lower angular resolution. An IR array is projected for the JWST, but current designs utilize an expendable coolant system with a systems life of ≈18 months. HST requires very long dwell times to detect faint (magnitude 20 and higher) objects. While both HST and JWST are very capable visual light telescopes, they will not meet the CAPS angular resolution requirement.

Advanced Technologies for Optical/Infrared Telescope

Advanced technologies that can be applied to the CAPS detection system are discussed below.

Detectors. By the 2015 to 2020 time frame, charge-coupled device (CCD) and charge-injected device (CID) detectors appear to provide the solutions that meet CAPS sensor requirements. Both technologies are developing at a rate leading to near-term production of CAPS compatible sensors that will provide the required performance. CCD technology is currently one or two generations ahead of CID technology; however, the CID's ability to eliminate bloom coupled with its higher sampling rate accumulator makes CIDs the preferred sensor technology for future investment.. Commercial industry is currently performing an enormous amount of work to enhance both technologies: higher pixel density, increased number of pixels, improved quantum efficiencies, increased response at visible and IR bandwidths, and improving response characteristics at higher temperatures.

As a result of CAPS trade studies and analyses, the baseline CAPS sensor detector requirements are listed in table 10.

Table 10. Baseline CAPS Sensor Detector Requirements

Requirement	Principal FPA	Auxiliary FPA
Detector type	CID	CCD (CID is available)
Array size	32k × 32k	4k × 4k
Pitch	8 μm	64 μm
Material	Si	Si:GAs
Wavelength	0.4 to 1.0 μm	8 to 18 μm

CCD detectors. A CCD is a metal-oxide semiconductor (MOS) or solid-state integrated circuit that stores electrons produced by incident photons as discrete packets of charge in potential wells maintained by an electric field. A CCD comprises photosites, typically arranged in an X-Y matrix of rows and columns. Each photosite, in turn, comprises a photodiode and an adjacent charge holding region, which is shielded from light. The photodiode converts light (photons) into charge (electrons). The number of electrons collected is proportional to the light intensity. Typically, light is collected over the entire imager simultaneously and then transferred to the adjacent charge transfer cells within the columns.

Next, the charge is read out. Each row of data is moved to a separate horizontal charge transfer register. Charge packets for each row are read out serially and sensed by a charge-to-voltage conversion and amplifier section. This architecture produces a low-noise, high-performance imager.

The quantum efficiency of a sensor indicates the percentage of photons (light) reaching the detector that are detected and converted. Silicon CCDs currently have quantum efficiencies (in the lab) on the order of 60 to 80 percent in most of the visible light region. Changing the material of the substrate changes the response of the CCD to vary with wavelength, thus allowing response to be tuned to a particular wavelength.

CID detectors. CID structure, principle of operation, and readout techniques are fundamentally different from that of CCDs. Each pixel in a CID array is individually addressed during readout. Scanning routines are implemented via electronic switching of row and column electrodes that are fabricated in a thin, clear polysilicon matrix over the surface of the array. While CCDs transfer the collected charge out of the pixels during readout (thus erasing the image stored on the sensor), CIDs do not transfer any charge from site to site in the array.

CID readout is accomplished by transferring or shifting the collected charge packet within an individually addressed pixel and by sensing displacement values across the electrodes at the site. Readout is nondestructive because charge remains intact in the pixel after the signal level has been determined. To empty the pixel for new frame integration, row and column electrodes are biased to inject the charge packet into the substrate collector.

CIDs have advantages over other detector technologies because CIDS do not transfer any charge. The charge from light falling on a CID pixel stays on that pixel and pixels are individually addressable. CIDs are highly resistant to space-based radiation and can withstand over 100 times the lethal dose of radiation as that of CCDs. CIDs are randomly addressable, thereby allowing for real time detection and cancellation of noise induced by cosmic radiation. CIDs do not bloom, so as a pixel saturates, the charge is not spilled to adjacent pixels, thus a bright object will not overwhelm its surrounding area.

CAPS visible light detector. CID technology was selected for the CAPS telescope for visible light detection. The baseline 32k × 32k Si sensor with 8-μm pixel has slightly greater quantum efficiency than the CCDs now available commercially, but various CCD manufacturers have demonstrated quantum efficiencies greater than 90 percent (ref. 2). In addition to the nonblooming and radiation hardening demonstrated by CIDs, the ability to randomly access the data in an array adds the capability to vary the dwell times for individual pixels (or groups of pixels), provides the opportunity to eliminate data from individual pixels during the read cycle, and finishes the ability to oversample to prevent saturation of brightly illuminated pixels. The visible light CAPS detector will require some additional development, but current commercial technology is progressing at a rate which will provide detectors similar in size and pitch to the CAPS detector within the next 5 years (ref. 3).

CAPS infrared detector. CID technology was also selected for CAPS telescope IR detection. The baseline 4k × 4k Si:GAs sensor with 64-μm pixel also has greater quantum efficiency than the IR CCDs available commercially, but 80-percent efficiency seems doable in the CAPS time frame. The main area for development of the IR detector is achieving a detector that will not have to be cooled to 10 K. Spacecraft operational temperature (–15 to 60 °C) IR CID detectors would eliminate the need for active cooling. While there are IR CIDs, development seems to be at least two technology generations behind development for visible light CIDs. Quantification of the higher detector operating temperatures awaits a decision on the background environment encountered at the selected basing location.

Mirrors. Four advanced mirror technologies are under development that may be applied to CAPS. The technologies are: segmented thin silica glass mounted on a glass isochor honeycomb with actuators, thin segmented super polished metallic mirrors with actuators, thin solid metallic mirrors with actuators, and lightweight composite mirrors with actuators. These technologies should be available in the 2005 to 2010 time frame according to current NASA technology development plans. At this time none of these technologies stands out as the best selection for CAPS. The following section discusses these technologies in more detail.

Kodak is developing a thin silica glass on a glass honeycomb mirror for the Advanced Mirror System Demonstrator (AMSD) Project (ref. 1). Thin glass is first fused to both sides of a glass honeycomb substructure. The combined glass structure is then placed on an inverse mirror-shaped mandrel and heated until it takes the shape of the mandrel. After polishing, the mirror surface is coated with gold. Mirror weights for this process will be on the order of 15 kg/m^2 (without actuators) as opposed to the HST telescope weight of 180 kg/m^2 with a solid glass mirror. The Kodak mirror proposed for AMSD consists of 19 segments for a 4.3-m range telescope. A TBD sensor/actuator set will be used to provide and maintain alignment for each of the mirror segments.

Danbury (Goodrich Optical and Space Systems, formerly Raytheon, formerly Perkin Elmer) has taken a similar approach for AMSD using an isochor glass sandwich only. Danbury will make the mirror of nine segments: eight petal-shaped mirrors around a hexagonal core (ref. 4). The Danbury mirror would fold four petals forward and four petals rearward for launch. Danbury would also use a TBD sensor/actuator set to maintain the alignment of the mirror. Mirror weights for this process will be on the order of 15 kg/m^2.

Ball Aerospace is examining a 36-segment thin beryllium (Be) metal mirror with an actuator set similar to that used by Kodak and Danbury for the AMSD mirror (ref. 5). The Be mirror has the advantage of increased rigidity, but Be does not have the same isochoric properties as glass; the crystalline structure of Be tends to expand in one direction more than others. Large solid blanks of Be are also unavailable, thus resulting in the increased number of mirror segments required. Be can be super polished, therefore eliminating the need for a gold mirror surface. Mirror weights for this process will be on the order of 15 kg/m^2.

NASA Goddard Space Flight Center has proposed a single blank superpolished aluminum (Al) disk mirror for future space telescopes (ref. 6). The thin Al mirror would be relatively lightweight and use known manufacturing processes to form the shape of the mirror. However, Al has a crystalline structure that limits the polished surface flatness to about 5 angstrom (Å). A 2-Å mirror is needed for CAPS. Further metallurgical development with Al would be needed to provide a CAPS quality mirror surface. Al has a relatively large thermal expansion coefficient. Heater control can maintain the mirror shape at a constant temperature (effectively eliminating the mirror usefulness in the IR band), or the mirror control actuators could compensate known thermal distortions of the optical shape of the mirror. Mirror weights for this process will be on the order of 15 kg/m^2.

The Jet Propulsion Laboratory and Composite Optics, Inc., have proposed a composite disk mirror with a gold reflective finish for future space telescopes (ref. 7). Using composites results in a very lightweight mirror, and manufacturing techniques are similar to those used for aircraft structures. Currently, composites have a polished surface roughness finish greater than Al, but additional materials studies may lead to a polished surface roughness closer to that of glass. Further composite development would be needed to provide a CAPS-quality mirror surface. Provided care is taken in assembly, composites are relatively insensitive to thermal distortion. Mirror weights for this process will be on the order of 10 kg/m^2.

Each mirror discussed has the capability for some type of active cooling to be added for sensing in the IR bands. Active cooling impacts the shape of the mirror, so additional adaptive actuators are required to maintain mirror shape at low temperatures. Mirror technology is an area that requires further development prior to CAPS deployment.

Mirror controller. CAPS requires some type of active mirror control for the advanced mirrors selected. Active shaping in the secondary and tertiary mirrors will maintain focus under varying thermal conditions in addition to providing calibration focusing after assembly and focal calibration from time to time. Active mirror control is currently accomplished on some of the larger ground-based telescopes by using a flexible secondary mirror and a set of actuators at 2- to 6-in. intervals to bend the mirror to the required shape. The thin metal or composite mirrors discussed above would be ideal candidates for active mirror control. Current mirror control designs require a relatively low computational capacity (small personal computer) and fairly simple sensors and actuators to shape the mirror. These technologies have a relatively high weight-to-mirror ratio (one to five times the weight of the mirror being shaped). This area is a candidate for further technology improvement (thin bonded piezoelectric actuators and advanced mirror shape detection techniques).

Active focusing. The HST is focused by staring at a known object (such as a star or a galaxy) and manually adjusting the focal actuators through a series of ground-based inputs. This process may take hours or days depending on the degree of focal adjustment necessary since the last calibration. The possibility of using an active focusing technique—where a known light pattern may be projected onto the mirror surfaces (possibly using lasers) and used to rapidly focus and calibrate the telescope optical paths in conjunction with the active mirror control discussed previously—should be examined. This technology would minimize the requirement for man in the calibration loop and provide for more frequent and accurate optical path maintenance. Such technology might also provide the capability to dramatically change the focus of thin mirrors, resulting in the capability to magnify objects of interest within the FOV without the use of refractive optics.

Telescope cooling (IR). The IR optical path will have to be cooled to 10 K and controlled to 0.001 K to get the performance required with current CCD/CID technology. This requires that the telescope barrel and baffles; primary, secondary, and tertiary mirrors; dichroic beam splitter; IR focusing mirror; and IR focal plane be cooled. The original JWST had a similar cooling requirement and some development work has been performed on cooling.

Shading/baffling technology. Depending on the final selection of telescope locations, CAPS will have to be baffled and/or shielded from the Sun. CAPS studies have demonstrated that the CAPS scan telescope can look to within 90° of the Sun for magnitude +27 LPCs and 45° of the Sun for magnitude +22.7 NEAs given a two to three times mirror diameter baffle system and a mirror smoothness of 2 Å or less. To look closer to the Sun, some form of shading technology must be developed. This shading technology should block the main disk of the Sun and also block the solar corona for at least two solar diameters to get the solar background down to ≈magnitude +2.

Various methods of shading are being investigated for the JWST including a large deployable shade attached to the telescope (ref. 8). Such technology would need to be deployed for any spaced-based CAPS telescope. Any space-based option should provide for a cover over the telescope aperture that can be closed when the background light exceeds a certain threshold. The cover will serve to protect the mirrors and focal planes from excess radiation in the event of loss of attitude control. A properly designed cover could also serve as an orientable shade to shield the telescope from nonsolar background light (such as that from the Earth or Moon) by orienting the shade normal to the source of the background light.

A surface-based option would require a fairly standard telescope-housing dome for stray surface light rejection. It may be possible to provide a deployable Sun shade for the surface-based system so that the telescope can look within 90° of the Sun during daylight, provided the background surface radiation is also minimized. Shielding and shading are an integral part of the telescope design and should be examined further.

Computational processing requirements. The computational processing requirements for LPC and NEA detection should not be underestimated. The sequence for turning the inputs on the focal planes into defined targets is listed below.

1. Store the data from each pixel.

2. Perform a grouping analysis to identify clumps of data as individual objects.

3. Subtract the known star field (such as stars or galaxies) from the scene (the background blur from each object should be subtracted).

4. Identify known nontargets and subtract the blurs created by them (planets, moons, main belt asteroids, known nonimpacting NEOs).

5. From the remaining scene, centroid on each of the blurs and obtain position information.

6. Identify known objects and provide position and velocity data to propagation algorithm.

7. Remaining objects are unknown targets; establish tracking files for each of the new objects and propagate objects to establish preliminary track files.

8. Identify any known objects that should have been in the scene and were not; tag these objects as lost and continue to propagate them according to their last known trajectories.

9. Identify objects with trajectories that have a TBD probability of impacting the Earth.

Steps 1 through 9 occur every 15 s when scanning for LPCs and every 0.375 s when scanning for NEAs.

Advanced image processing techniques may provide a way to reduce the computational power by applying an algorithm to the focal plane image that subtracts out the stationary light objects at step 2, eliminating the need for a priori knowledge of the star field and reducing computational load. Computational processing load is an area for further investigation.

Scan Methodology

This section is a general discussion of how the detection field is moved (or not) during a CAPS scan of a particular area of sky. Each of the scan methodologies discussed has advantages or disadvantages that will need to be tested during CAPS systems studies in the next phase. The scan strategy selected may affect the computational load, sensing accuracy, telescope position, and velocity control accuracy, as well as the more obvious sky coverage with time.

Survey. There are two types of scan strategies for the survey mode: continuous motion where the telescope moves at a very slow rate along the sky field collecting images every 15 s and a stare mode where the telescope remains inertially fixed and the image is refreshed every 90 s.

Continuous scan. During the continuous scan mode the background images are shifted a known number of pixels during each scan cycle. The image from the previous cycle is correlated with the image from the current cycle and any differences between the images are identified as possible targets. The possible targets are then correlated with the targets from previous cycles, and if the possible target is moving along a physically possible target track a real target is declared. The real target motion is provided to the tracking algorithms and the target track updated. This method has the following advantages: target image is continually moved across the focal plane reducing susceptibility to bad pixels and cosmic background radiation, and settle times are minimized creating a faster sky scan. The primary disadvantage is that extremely accurate attitude rate sensing is needed to maintain constant (or at least known) scan speed. Correlation of moving background scene may be more difficult than correlation with a static background.

The CAPS telescope analysis was performed assuming a continuous scan.

Stare scan. During a stare scan the telescope is held inertially fixed in space and images are collected every 15 s for a 90-s time span; the telescope is then slewed to a new portion of the sky (0.83° away for CAPS) and a new image taken. Targets are identified by the streaks created as they move across the static star field. The centroid for each target image is taken and then propagated in time to establish a target track. This method has the advantage that the stellar background is static (no need to correct for motion of the star field), moving objects are readily apparent (streaks), and attitude rate accuracy is minimized. Known guide stars can easily be selected for attitude determination while staring and serve as hold points (the image of the star is held on a particular set of pixels) during each stare. The disadvantages of this method are: susceptibility to error due to cosmic radiation, increased sensitivity to bad pixels (a bad pixel will appear in the same point in every image), and a settling period time is required before each scan.

Track. Once a target has been declared in the scan mode, a track mode may be used to increase target position accuracy by an order of magnitude. This track mode requires that either an accurate initial target track be established so that the telescope can acquire the target hours or days later, or the tracking telescope must be pointed near the target object and the target track acquired while the scan mode is staring at a target. Two types of track modes have been identified: track and follow, and a high-resolution stare. Both modes require a telescope with a higher angular resolution than that used in the survey mode.

Track and follow. The first method for acquiring higher resolution accuracy is the track and follow mode wherein the tracking telescope maintains the target at a fixed point in the tracking focal array and known fixed objects (guide stars) are correlated to the target while the telescope moves through the star field. This method would require a fairly complex correlation algorithm along with Kalman filtering of the position rate data to provide accurate target position location. The advantages of this method are: target signal strength is maximized reducing chances of target loss and known bright stars may be used for relative position and rate determination. Disadvantages are: highly accurate rate measurements are required and telescopes must be dedicated for a period of time to each target object.

High-resolution stare. An alternative method for tracking the object is to center on the predicted point of the target motion and remain inertially fixed like the stare scan in the survey mode. The target is then tracked as it moves across the focal plane. This method requires that the target smear be bright enough to be detected and that the resolution of the tracking scope be increased. The advantages of this method are: the target object moves relative to the static background, similar algorithms to the scan mode may be employed for angular position determination, accurate motion of the telescope is not required (it is fixed), and simple optical magnification of the survey scope may provide the resolutions needed provided the target is bright enough.

Of the two tracking modes discussed, high-resolution stare was the mode selected for this analysis.

Basing Concepts

Two basing concepts were considered: space-based and lunar surface-based (or other planetary bodies without an atmosphere). Modifications to the basic design are necessary depending on the basing strategy selected.

Space-based concept. Space basing for the detection system is defined as any configuration not located within the Earth's atmosphere and not based on the surface of any extraterrestrial body. Some typical space-based locations would be orbit around the Earth, Earth-Moon libation points, in Earth's solar orbit but phased some distance away from the Earth, and any other orbit inside the solar system. Assembly, servicing, attitude control and pointing, and thermal environment are identified as issues unique to space basing. A space-based subsystem breakdown identifying major components required is included in table 11.

Table 11. CAPS Space-Based Subsystems Required

Guidance navigation and control	Reaction wheels (CMGs), attitude rate sensors, angular rate sensors, accelerometers, attitude determination sensor, momentum unloading devices, navigation sensors (if performed onboard), navigation algorithm (if maintained by ground base), propellant and tanks, propellant plumbing, reaction engines
Power and electrical	Power generation device (solar arrays, RTGs), batteries, wire harnesses and cables, power converters
Thermal	Sensors, cooling devices (thermionic, heat pipes, coolant pumps, coolant storage tanks), passive devices
Mechanical	Structure to hold mirrors, baffles, door and mechanism for interferometer breakout, door and mechanism for telescope cover, shielding
Command and data handling	Antennas, communications devices (RF/laser), flight computers, component controllers, component data converters
Payload	Includes mirrors, Lyot stops, focal planes, beam splitters, baffles, mirror controllers, calibration devices
Payload processing	Onboard processing devices, onboard processing algorithms

Note: Collector satellite for interferometry requires all of the above plus a combiner mirror minus all of the payload optics except a focal plane.

Figure 5 shows a concept for the CAPS space-based detector telescopes. Note the stray light baffles in the main tube of the telescope to minimize stray light. The mirror is offset relative to the optical bore sight, eliminating any central obscurations.

Assembly. The telescope as currently designed is somewhat larger than the shuttle cargo bay and would therefore require assembly in space. The space-based design could probably be repackaged so that no more than one shuttle launch is required. The telescope could be split into a barrel with a main mirror

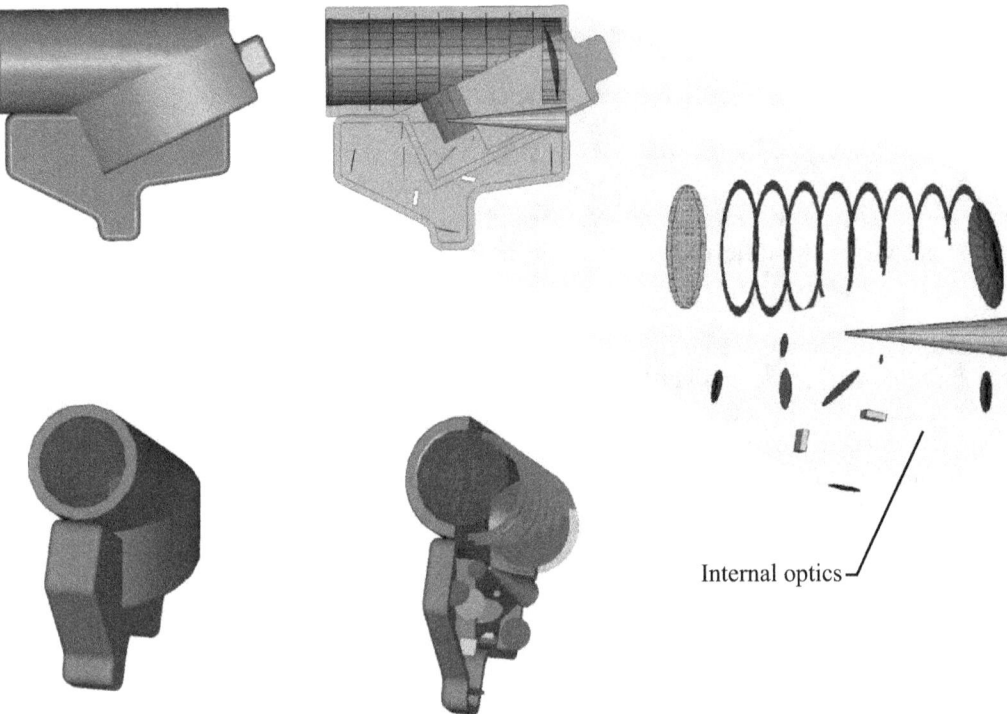

Figure 5. CAPS space-based detector concept (including internal optics).

section and a secondary optics with subsystems section. These two sections could be combined to form one self-contained telescope/satellite that could be transported to the basing location using a direct transfer or by using an in-space orbital transfer system. The loads encountered by the space-based system would be shuttle launch loads, assembly loads, and transfer orbit loads. While on-orbit assembly is minimized, this scenario also provides the ability to check out the completed telescope-satellite combination prior to moving it to its permanent location.

Servicing issues. The servicing issues for the CAPS telescope are similar to those encountered by the HST or those to be encountered by the JWST (ref. 1). Consumables (liquid hydrogen for cooling) could be replenished in space with the appropriate telescope/satellite design. The telescope could be modular like the HST so that as components break or need upgrading the telescope could be repaired in Earth orbit or at a basing location by human or telerobotic means. The systems trades involved with selecting the appropriate level of modularity were beyond the scope of this study.

Attitude control and pointing issues. The telescope attitude will have to be maintained to better than 0.001 arcsec, the telescope rates controlled to better than 0.001 arcsec/s, and jitter controlled to 0.01 arcsec^2 to achieve telescope angular resolution of 0.001 arcsec. The star catalog (including at least two guide stars for each 0.89°) should be known to at least 0.0005 arcsec. The star catalog should be corrected for stellar proper motion to meet the above requirements. The telescope is small enough that reaction wheels could probably be used for attitude control. Control moment gyros with vernier reaction wheels might be required for the tracking mode for widely spaced targets, but a target spacing systems analysis should be performed to identify nominal slew angular distances and rates for the track mode.

Thermal issues. A detailed thermal analysis should be performed on the selected CAPS basing concept to identify thermal background environments for sizing the thermal control system for the CAPS detector (satellite or ground base). The CAPS IR CIDs require cooling the IR focal plane to 10 K ± 0.001 K as well as cooling the hardware in the optical path. At those temperatures active cooling needs to be provided in the form of consumables (liquid hydrogen) or an as yet to be developed cooling technology. The energy needs of this cooling system are to be determined.

Subsystems. The subsystems required for the CAPS telescope-satellite are briefly listed in table 11. The subsystems are similar for both the tracking and survey telescopes with the main difference between the two scopes being the narrower FOV of the tracker scope and different mirror focal lengths.

Lunar-based concept. Lunar basing for the detection system is defined as any configuration located on the Moon. Details of a manned versus unmanned lunar basing concept have not been performed, so this section confines itself to identifying the changes required in a lunar-based telescope from a space-based telescope. Assembly, gimbaling, housing, and thermal environment are identified as issues unique to lunar basing. A lunar-based subsystem breakdown identifying major components required is included in table 12.

Table 12. CAPS Space-Based Subsystems Required

Pointing and control	Gimbals, gimbal rate sensor, position sensors (if performed on-board), position algorithm (if maintained by ground base)
Power and electrical	Power generation device (solar arrays/collectors, RTGs), batteries, wire harnesses and cables, power converters
Rail or track system	Some type of transport system to keep the tracking telescopes equally distant from the collector
Thermal	Sensors, cooling devices (thermionic, heat pipes, coolant pumps, coolant storage tanks), passive devices
Mechanical	Telescope dome/housing, structure to hold mirrors, door and mechanism for interferometer breakout, door and mechanism for dome cover, shielding
Command and data handling	Antennas, communications devices (RF/laser), dome computers, component controllers, component data converters
Telescope	Includes mirrors, Lyot stops, focal planes, beam splitters, baffles, mirror controllers, calibration devices
Payload processing	Local processing devices, local processing algorithms

Note: Collector satellite for interferometry requires all of the above plus a combiner mirror minus all of the payload optics except a focal plane.

One possible configuration for the lunar-based design is shown in figure 6. This concept has the space-based telescope mounted on a monorail-type platform above the lunar surface. The track keeps the telescope away from the dust near the lunar surface. This concept uses the same telescope housing as designed for the space-based system mounted on a gimbaled platform. Details of the track system are shown in figure 7. The track could be used for power and data transfer from a remote central facility.

Assembly. The entire telescope would need to be repackaged for a lunar-based system. No decision has been made as to where or how a lunar-based system would be assembled. Keeping the telescope optical train free of lunar dust during assembly would be a particular challenge. Lunar assembly techniques need to be developed.

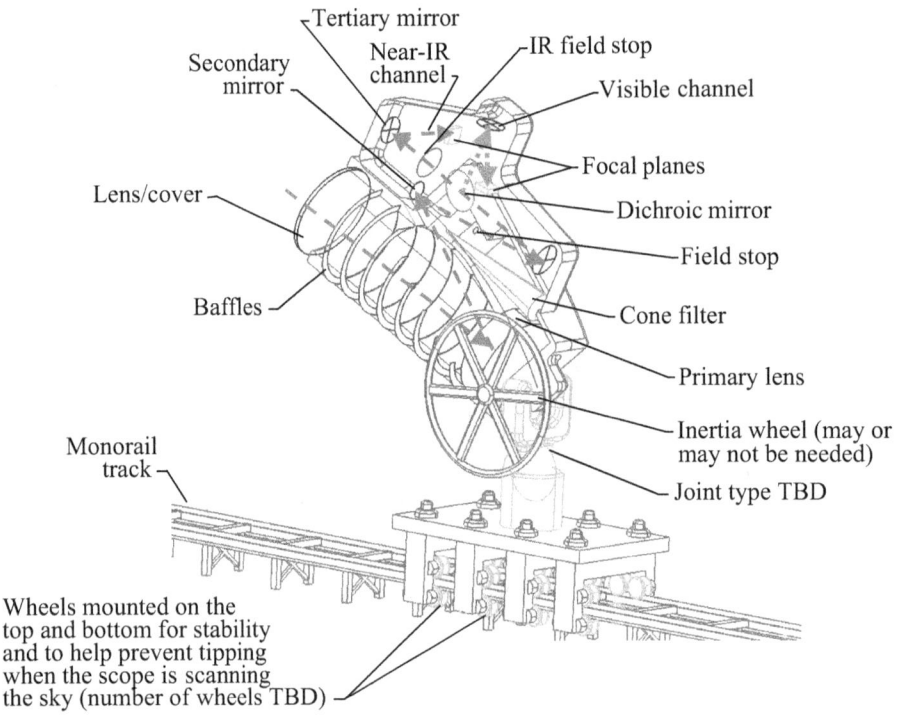

Figure 6. Surface-based CAPS telescope concept (minus housing dome).

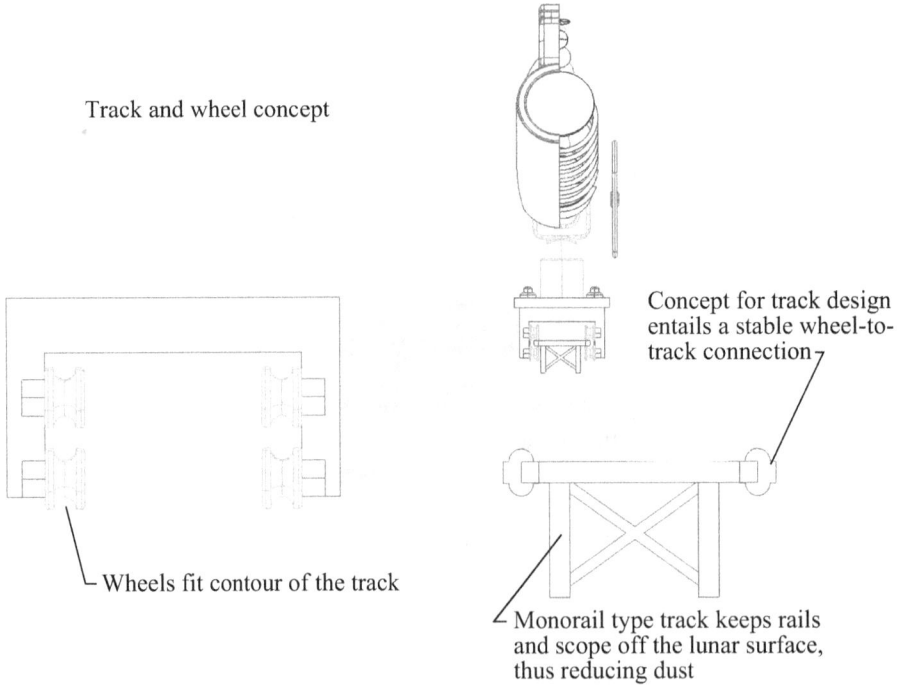

Figure 7. Lunar surface-based CAPS track details.

Gimbaling issues. Surface basing creates the need for gimbals for pointing. The CAPS telescope will require a three-axis gimbal to provide for interferometry, as it is anticipated that entire images (32k × 32k) from the tracking telescopes will have to be overlaid in the collector telescope for interferometry. A three-axis gimbal provides some flexibility in pointing the telescope, minimizing keyholing effects when tracking. The gimbals should track to an accuracy of better than 0.001 arcsec/s. Technology development areas that focus on the extremely long-life gimbal/bearing systems would be extremely beneficial for a surface-based CAPS detection element to minimize servicing and replacement of components.

Housing issues. Some form of dust free environment must be provided for the telescope assembly on the Moon. A dome similar to those constructed for Earth-based telescopes would keep dust out of the telescope and provide shelter for thermal protection during the lunar day. The dome needs an additional cutout for an interferometry light path to the combiner from the tracker/collector telescopes. The dome may even eliminate the need for baffles. Some of the telescope support systems (computers, communications, electronics, thermal coolers) would no longer need to be housed near the telescope mirrors but could be placed elsewhere in the structure, or located in a remote facility.

Thermal issues. The lunar-based system requires a background thermal analysis similar to the space-based systems. The large day/night thermal variations may be a particular problem. The lunar surface will also radiate additional heat toward the telescope that must be removed for IR target detection. Greater amounts of power from large surface-based arrays or nuclear reactors may be available with the lunar basing. The thermal systems trades are an area for further investigation.

Subsystems required. The subsystems required for a lunar-based telescope are listed in table 12.

Telescope Design Observations

As a result of the analysis of the CAPS optical detection system, the design observations in table 13 were noted.

Table 13. Telescope Design Observations

It is possible to design a four-mirror, all reflective, unobscured telescope with an accessible field stop between its secondary and tertiary mirrors, a collimated output beam from its tertiary mirror, an accessible pupil in collimated space, and nearly diffraction-limited performance in its focal plane (some refractive elements may be desirable).
It is possible to detect mV = +22.7 NEAs within 45° of the Sun, provided that the primary mirror has a surface microroughness of 2 Å or less and that a baffle tube of reasonable length (2 to 3 X mirror diameter) is used.
It is possible to detect LPCs to mV = +27 and NEAs to mV = +22.7 using a system with the baseline parameters.
Greater sensitivity to LPCs is due to their lower angular velocities, requiring a longer frame integration time, and their near-opposition to the Sun, reducing stray light levels.
A higher level of apodization is optimal for viewing LPCs because the star field is relatively brighter, and apodization allows the LPC to be detected in close proximity to a brighter star.
The detection algorithm sums the signals in each 3×3 pixel and subtracts an equivalent signal averaged over a surrounding guardband. This approach is able to suppress the effects of uniform background signals due to stray light, dark current, read noise, etc. A target moving in an unknown direction can be detected with respect to the inertial star field if the target's motion during a frame is <2 pixels.
A high probability of detecting a true target and a manageable number of preliminary false alarms can be achieved by requiring that multiple detections lie on a physically realistic trajectory (four out of six in this analysis). Determining the centroids of the detections and refining the trajectory estimate rejects virtually all of the preliminary false alarms.
The multiple detection approach rejects false alarms due to charged particle events even though a single frame may appear to contain a target with a high signal-to-noise ratio. The centroid of the point spread function of a target can be determined within 1/10 pixel (0.01 arcsec for the survey telescope) if only one source contributes to the energy distribution. It is possible to reject observations of the centroid of multiple targets by several means: energy distributions that are "dumbbell" shaped rather than circular, energy signal levels significantly larger than those in previous and subsequent frames, and energy distributions too close to a much brighter object.

References

1. *James Web Space Telescope*, (NASA NGST web page). <http://www.ngst.nasa.gov/Hardware/text/Kodak.htm> Accessed October 2003.

2. *CCD Quantum Efficiency*, (MIT web page). <http://space.mit.edu/ACIS/cal_report/node83.html> Accessed October 2003.

3. *Calypso HR CAM* and *CCD Quantum Efficiency*, (Calypso Web page). <http://www.calypso.org/~neill/cameras/hrcam/HRCamQE.html> Accessed October 2003.

4. *The James Web Space Telescope—Hubble's Scientific and Technology Successor*, (NASA NGST web page). <http://ngst.gsfc.nasa.gov/project/text/JWST_HST_successor.pdf> Accessed October 2003.

5. *James Web Space Telescope*, (NASA NGST web page). <http://ngst.gsfc.nasa.gov/Hardware/text/Ball.htm> Accessed October 2003.

6. *Aluminum Super Polish Technology*, (RTI International). <http://www.rti.org/page.cfm?objectid=786D7F10-C6A8-46EA-AEA79AE7FCB4C238> Accessed October 2003.

7. *Composite Optics, Inc. Successfully Completes the World's Largest, Lightest Weight, Composite Mirror*, (Composite Optics web page). <http://www.coi-world.com/first.htm> Accessed October 2003.

8. *James Web Space Telescope*, (NASA NGST web page). <http://www.ngst.nasa.gov/Hardware> Accessed October 2003.

Near-Earth Object Astrometric Interferometry[5]

MARTIN R. WERNER
The George Washington University

Introduction

Using astrometric interferometry on near-Earth objects (NEOs) poses many interesting and difficult challenges. Poor reflectance properties and potentially no significant active emissions lead to NEOs having intrinsically low visual magnitudes. Using worst case estimates for signal reflection properties leads to NEOs having visual magnitudes of 27 and higher. Today the most sensitive interferometers in operation have limiting magnitudes of 20 or less. The main reason for this limit is due to the atmosphere, where turbulence affects the light coming from the target, limiting the sensitivity of the interferometer. In this analysis, the interferometer designs assume no atmosphere, meaning they would be placed at a location somewhere in space. Interferometer configurations and operational uncertainties are looked at in order to parameterize the requirements necessary to achieve measurements of low visual magnitude NEOs. This analysis provides a preliminary estimate of what will be required in order to take high resolution measurements of these objects using interferometry techniques.

Theory

Interferometry is the measurement of interference patterns produced by combining the light from two or more telescopes that observe the same source. The theory behind interferometry is actually quite simple: when two coherent beams of light are overlapped, the waves will interfere with each other, creating an interference pattern. If the two waves are in phase, they will combine constructively and produce an intensity that is more than twice the combination of the two separate source intensities. When the two waves are 180° out of phase, they will cancel destructively, reducing the intensity to zero. This fluctuation in the intensity as the phase is changed is measured by the interferometer and used to produce a position measurement.

Setup and Measurements

Figure 1 shows the basic setup of an interferometer. The vector between the two telescopes is known as the baseline **B**. The vector from the centerline of the interferometer to the source being observed is called the position vector **S**.

Because the target being observed does not usually lie on the perpendicular axis of the interferometer, the light waves from the source will reach one telescope before the other. The distance the wave must travel to reach the second telescope after reaching the first is the delay distance or D as shown in figure 1. The value of D for an interferometer is given by

$$D = \hat{\mathbf{s}} \bullet \mathbf{B} + C \tag{1}$$

where $\hat{\mathbf{s}}$ is the unit position vector, **B** is the baseline vector, and C is a calibration term. Both $\hat{\mathbf{s}}$ and **B** are three-dimensional vectors. The calibration term should remain constant for a given interferometer, and for simplification it is assumed to be zero. This equation can then be written in the form

$$D = |\mathbf{B}| \cos \theta \tag{2}$$

---

[5]Chapter nomenclature available in chapter notes, p. 217.

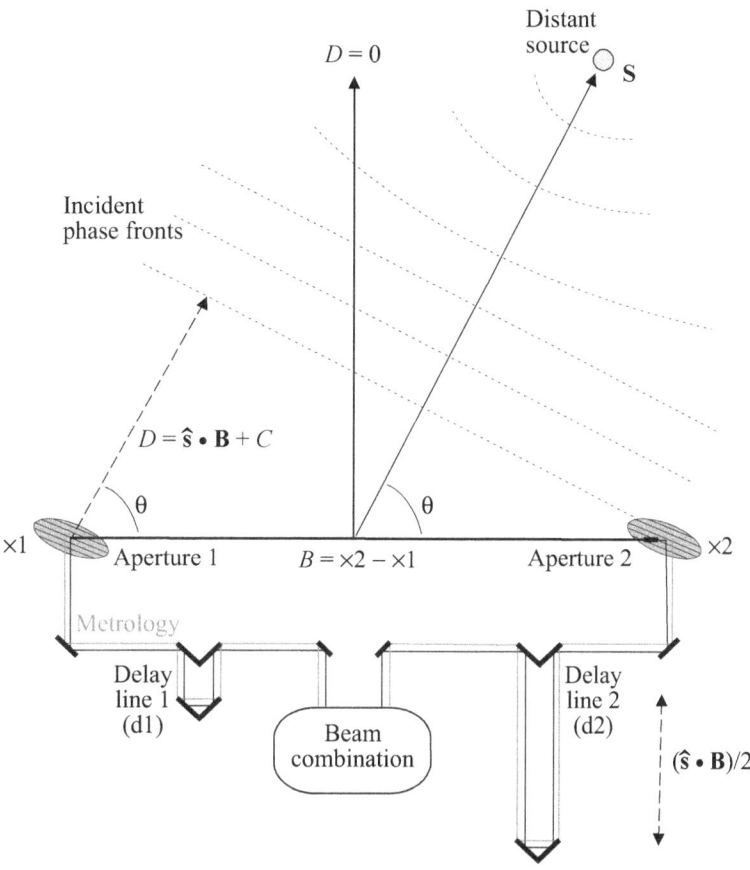

Figure 1. Basic interferometry layout.

where θ is the angle between the $\hat{\mathbf{s}}$ and **B** vectors. The goal is to measure the baseline distance and the delay position and then solve for the target's angular position θ. The delay position is found by moving the internal delay lines until the internal delay distance equals D. As the delay line is varied, the interferometer measures an intensity pattern, which will look similar to figure 2. The internal delay is equal to D at the point of maximum fringe intensity. When this point is reached, the value of the internal delay is then measured and used to solve for θ. To find the position of the star in inertial space, the baseline vector orientation with respect to inertial space must be found. For ground-based interferometers, baseline orientation is usually done before the science measurement by first taking measurements of reference stars that have well-known positions in inertial space. This can be done reliably on the ground because the baseline is relatively stable over long time periods. For a space-based interferometer, such as the one planned for the Space Interferometry Mission (SIM) in reference 1, two guide interferometers can be used that are tied to the science interferometer at the picometer level. These guide interferometers will lock on to reference starts and measure the spacecraft's inertial position at the same time as the science measurement. This is necessary because there is no stable surface to which the interferometer can be attached, and the baseline will fluctuate over the period of science observations. A given baseline orientation will only give a one-dimensional measurement of θ. To get the actual two-dimensional target position (i.e., latitude and longitude), two measurements will need to be taken using orthogonal baselines.

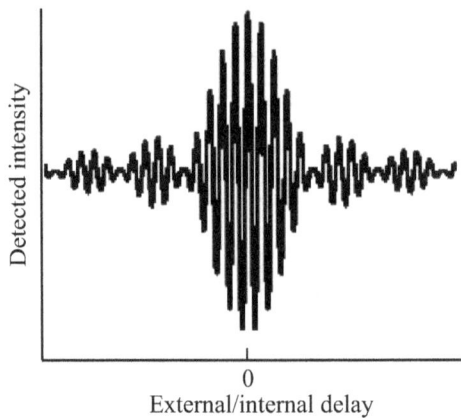

Figure 2. Interferometric intensity pattern.

Resolution

The preliminary requirement for a Comet/Asteroid Protection System (CAPS) astrometric interferometer is to achieve angular position accuracy (σ_θ) of 100 microarcseconds (µas), which can also be referred to as the resolution. Therefore, an error taken in the angular measurement will be off by no more than 100 µas. To achieve this resolution, the delay position and baseline distance must be measured with an accuracy of at least 1 nm (ref. 2). Distance measurements can be done using laser metrology systems that are already being used in ground-based interferometers. Measuring the delay position will not be the problem. The problem is getting the internal delay mirror in the correct place to produce the maximum fringe intensity because this value of delay corresponds to the actual angle. It is necessary to establish how accurately the position of the central fringe can be determined. Accuracy that the interferometer can achieve in detecting the central fringe can be estimated using

$$\sigma_\theta \approx \frac{\lambda/B}{\text{SNR}} \qquad (3)$$

where λ is the average wavelength in micrometers, B is the baseline in meters, and SNR is the signal-to-noise ratio. The top part of the equation gives the width of the fringes that are produced. The smaller the width of the fringe, the easier it is to pick out from an adjacent fringe. Increasing the baseline decreases the fringe width, improving the resolution of the interferometer. Furthermore, an interferometer has the ability to improve on this resolution depending on the SNR. As the SNR is increased, the interferometer can better pinpoint the exact center of the fringe, resulting in a higher accuracy measurement. This improvement in accuracy is analogous to blur centroiding with a single aperture telescope where the centroid position of the star can be measured to much less than the size of the airy disk (ref. 3). It must be noted that this equation is only an approximation and will ultimately depend on the detection system and tracking algorithms used.

Fringe Visibility

Fringe visibility (V) is the apparent contrast between light and dark areas of the fringe pattern and is defined by

$$V = \frac{I_{\text{MAX}} - I_{\text{MIN}}}{I_{\text{MAX}} + I_{\text{MIN}}} \qquad (4)$$

where V is a dimensionless value contained on the interval [0,1], I_{MAX} is the maximum fringe intensity, and I_{MIN} is the minimum fringe intensity (ref. 4). There are many factors that will affect the fringe visibility. With strict interferometer design, construction, and operation, most of the factors can be taken out. However, one factor that must be considered for observations of near-Earth objects (NEOs) is the visibility loss caused by overresolving the target. Ideally, the target object should be a point source; however, an NEO in the solar system is potentially close enough such that it is no longer a point source, but instead is a uniform disk. When looking at a uniform disk, each point on the disk will create its own fringe pattern that will overlap the other fringe patterns. As the baseline gets longer, the more the fringes overlap, blurring out the ideal fringe pattern. Overresolving reduces the fringe visibility, reducing the amplitude of the intensity peaks, making the fringes harder to pick out from one another above the noise and ultimately reducing the measurement accuracy. This visibility effect is given by

$$V = \frac{2J_1(\pi B \theta_{ud}/\lambda)}{(\pi B \theta_{ud}/\lambda)} \quad (5)$$

where θ_{ud} is the width of the uniform disk measured at the detector in radians (rad), and J_1 is the first order Bessel function (ref. 5). Referred to as the "target width" throughout the rest of this analysis, θ_{ud} is calculated using

$$\theta_{ud} = \tan^{-1}\left(\frac{d}{x_{det}}\right) \quad (6)$$

where d is the diameter of the target and x_{det} is its distance from the detector (assuming that $x_{det} \gg d$). The target width will affect fringe visibility and place limitations on the length of baseline that can be used.

Dwell Time

Dwell time (T) is the amount of time that can be spent at a single delay position, before uncertainty overwhelms the measurement, and is defined by

$$T \leq \frac{c/\delta v}{\sigma \dot{D}} \quad (7)$$

where c is the speed of light, δv is the bandwidth, and $\sigma \dot{D}$ is the delay rate uncertainty (ref. 6). The maximum dwell time depends on uncertainties in the motions of the interferometer and of the target, which affect the delay rate uncertainty. Uncertainties in the delay rate will cause the fringes to move on the detector, so the dwell time must be short enough that the intensity measurement can be taken before the desired fringe moves off the detector.

Photon Rate

To successfully take a measurement of an NEO, a certain amount of signal in a given amount of time is needed. There are many factors that affect the signal that reaches the detector. The goal is to determine how many photons per second are being collected by the interferometer. In this analysis, the photon rate (R) at the detector is found using

$$R = R_0 \cdot 2.512^{-M} \cdot \frac{\pi}{4} D_T \cdot \eta \cdot \delta v \quad (8)$$

where R_0 is the zero visual magnitude photon rate, M is the visual magnitude of the NEO, D_T is the diameter of the collecting aperture, η is the throughput efficiency of the interferometer, and $\delta\nu$ is the instrument bandwidth.

Signal-to-Noise Ratio

SNR, by definition, is the ratio of total signal received to the noise. The SNR is estimated using

$$\text{SNR} \approx \frac{R \cdot T \cdot V}{\sqrt{R \cdot T + 4\frac{\nu}{\delta\nu}r^2}} \qquad (9)$$

The numerator of equation (9) is the total amount of signal received by the interferometer and is affected by the photon rate, dwell time, and visibility. The denominator is the total amount of noise, which includes photon noise and read noise. Possible noise caused by background sources was excluded from this analysis and needs to be studied and accounted for in future analysis. Background noise, if present, will lower the SNR for a given measurement, increasing the requirement on telescope diameter or dwell time to complete a successful measurement. Photon noise is due to the nature of light. The actual rate of photons received from the target object will fluctuate over time, giving a natural fluctuation in the intensity that affects the measurement of the fringes. The standard deviation of the photon noise is given by $\sqrt{N}$, where N is the total number of photons collected in a given time. Read noise is a property of the detector and is due to random energy being picked up by the detector and recorded as photons during each pixel readout. The read noise has a standard deviation (r), which results in a variance in each read of r^2. For each fringe measurement, the detector should read out four times. The total number of fringes in the fringe pattern is approximated by taking the average wavelength over the bandwidth of the interferometer (ref. 1). Multiplying these three terms together gives the total read noise. Both the photon noise and read noise are added together, and the square root of that number gives the total noise for each measurement. Detectors are being developed that suggest subelectron read noise errors will be possible (ref. 7), essentially eliminating the effects of read noise. However, read noise is still included in this analysis to place an upper bound on the results.

Assumptions

Many different factors can be looked at for interferometry calculations. For simplification, multiple assumptions are made about the interferometer setup in this analysis. This section includes the assumptions used for the calculations that are performed.

Observing Conditions and Location

The main assumption of a CAPS-based interferometer is that it is placed in a location absent of atmosphere. The atmosphere is the largest source of error that limits Earth-based interferometers. Turbulence in the atmosphere places great restrictions on the signal and maximum dwell time that can be achieved. The limitations on dwell time alone would make the telescope size requirements impractical for a system designed to view such low magnitude objects. Two possibilities for an interferometer location could be either in a heliocentric orbit around the Sun or on the Moon. The advantages and disadvantages are briefly discussed for both these locations subsequently; however, for this analysis, no specific location is defined. The only thing specific about the detector location is that it is positioned at a distance of 1 astronomical unit (au) from the Sun.

Heliocentric orbit. At a location 1 au from the Sun, the detector could be placed in an Earth leading or trailing orbit. An advantage of this orbit is that the interferometer could theoretically be aimed in any direction. However, this also can be a disadvantage because of the fuel requirements and vibrations introduced during the maneuver. Keeping a stable platform is also a problem at this location. All vibrations must be actively controlled or passively damped. Thermal issues are a concern as well because the spacecraft will always be in sunlight.

Moon. Placing an interferometer on the Moon would be very similar to an Earth-based interferometer. Having the hardware fixed to the lunar surface keeps the baseline very stable. Additionally, thermal variations could be minimized during lunar night or by locating the interferometer in a crater that offers significant shadowing. Disadvantages of lunar-based interferometers include the problem of lunar dust degrading the optics, and observations are constrained by the Moon's rotation rate and the detector's geographic location.

Control and Measurement Accuracy

The interferometer is assumed to have metrology systems that achieve 1 nm accuracy in delay and baseline distance measurements. The only error that is assumed to be present in the CAPS interferometer is the delay rate uncertainty. Also, the only fringe visibility loss present is that due to the baseline length and target width parameter discussed previously. All other visibility losses are assumed to be negligible and not included in the calculations.

Interferometer Configuration

This analysis assumes that the interferometer is using a pupil plane configuration with a single pixel charge-coupled device (CCD) type detector. The throughput efficiency is assumed to be 80 percent and includes losses due to imperfections in the optics and the quantum efficiency of the detector. This value of throughput efficiency is very high, and superior design of the interferometer and improvements in detectors are needed to produce a throughput efficiency of this value. Advancements in the next 25 years should allow this value to be obtained and makes it a reasonable estimate for this simulation. The interferometer is assumed to have a bandwidth from 500 to 1000 nm (3.0×10^{14} to 6.0×10^{14} Hz), which results in an average wavelength of 750 nm (4.0×10^{14} Hz). The read noise of the detector is assumed to be 3 electrons per read.

Analysis Cases

Four cases were analyzed consisting of an impacting long-period comet trajectory with four different initial observation distances. The CAPS detection scenario defines that interferometry, or another method that can achieve 100 µas resolution, will be used once the impacting orbit has been refined such that the mean erroneous predicted miss distance ε is less than 1 lunar distance. Therefore, the object will have traveled some distance closer to the Sun between the time when it is first observed and when ε is found to be 1 lunar distance. This new distance is referred to as the "watch distance." Once the NEO reaches the watch distance, the orbit determination using the single 10-milliarcsecond (mas) resolution telescope will have an estimated trajectory of the object. The trajectory will have uncertainties in its angular rate. Values for the watch time and angular rate uncertainty were provided by the orbit determination analysis, and averages were taken of these values and listed in table 1. The four analysis cases will be defined by their respective watch distances. Figure 3 shows a plot of the angular rate uncertainties versus watch distance for each analysis case. The farther an object is away from the Sun, the higher its angular rate error will be when it reaches the watch distance.

Table 1. Analysis Case Initial Rate Uncertainty

Start distance, au	Watch distance, au	Rate uncertainty, rad/s	
		Longitude	Latitude
5	4.69	2.42745×10^{-13}	9.15576×10^{-13}
7	6.73	4.18491×10^{-13}	1.01544×10^{-12}
9	8.78	1.40205×10^{-12}	1.65268×10^{-12}
11	10.79	1.80448×10^{-12}	1.92439×10^{-12}

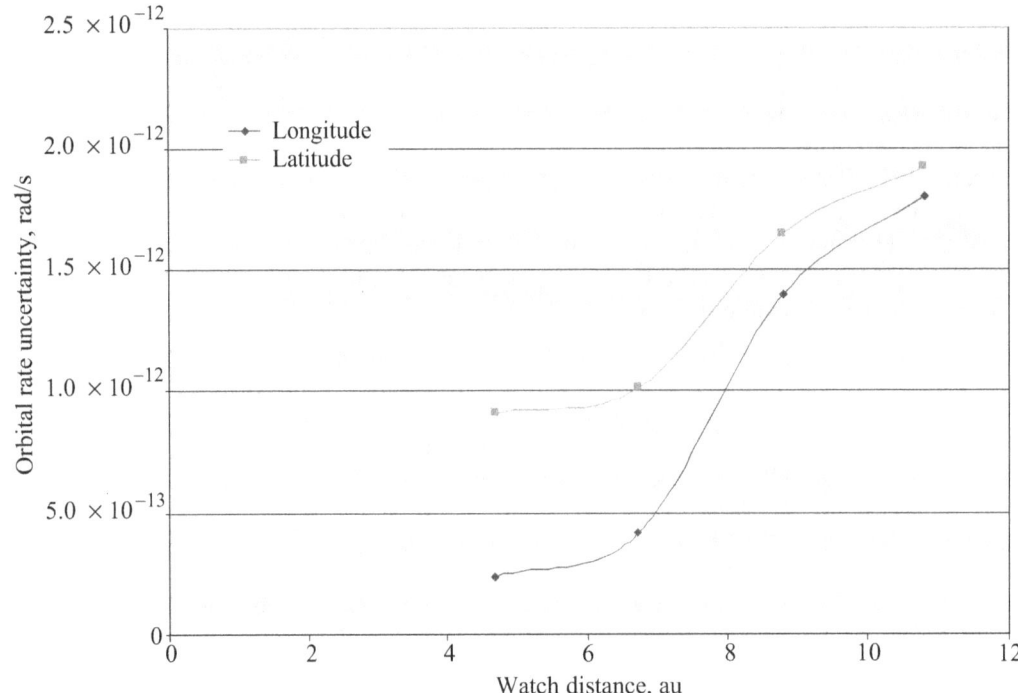

Figure 3. Average orbital rate uncertainty versus watch distance.

NEO Properties

For all four analysis cases, the NEO is assumed to be a 1-km diameter, spherical object. The value chosen for the albedo is 0.02. This value is estimated as the lowest albedo that will be encountered and is used for worst case analysis. If the detection system is designed to observe objects with this albedo at a given distance and diameter, it will be able to detect the entire range of objects at this size and distance.

Default Values

Table 2 includes the default parameter values and constants used in this analysis. These values are used for all calculations unless otherwise specified.

Table 2. Default Parameters and Constants for Analysis

Parameter	Value	Units
Target diameter	1000	meters
Target albedo	0.02	
Desired resolution	0.0001	arcsec
Average wavelength	750	nm
	4.0×10^{14}	Hz
Instrument bandwidth	500–1000	nm
	3.0×10^{14}	Hz
Throughput efficiency	0.8	
Read noise (std. dev.)	3	electrons
Speed of light	2.99792458×10^8	m/s
Magnitude zero photon rate	1.0×10^{-4}	photons/s/m^2/Hz

Signal Properties of NEOs

To obtain visual magnitude estimates, an algorithm by Basil H. Rowe was used (ref. 8). The algorithm was modified to output the visual magnitude given an NEO's diameter. As previously mentioned, the detector distance from the Sun is assumed to be 1 au. The value used for the photometric slope was 0.15. Using the modified algorithm, a visual magnitude contour plot was created that shows the values of visual magnitude as a function of distance and direction from the Sun and the detector (fig. 4). The blue circles represent the watch distance for each analysis case. Therefore, depending on the detector location, the NEO can have any visual magnitude that the blue lines pass through.

To determine the actual range of visual magnitudes that will be expected, the Sun angle θ_S is introduced to specify where the NEO is with respect to the Sun and detector. Shown in figure 5 are the geometry of the Sun, the detector, and the definition of the Sun angle. A Sun angle of 180° is defined as that point when the detector is located directly between the Sun and the NEO; whereas, a Sun angle of 0° is when the detector is located on the opposite side of the Sun. Due to symmetry, it is not necessary to go higher than 180°.

Using this configuration, a plot of Sun angle versus visual magnitude was made and is shown in figure 6. The brightest visual magnitude is when the Sun angle is 180°. The Sun angle that gives the dimmest visual magnitude changes with the object distance from the Sun. As the object gets farther away, the Sun angle that gives the worst visual magnitude increases toward 90°. For the four analysis cases, the values of the Sun angle that produce the lowest visual magnitude range from 48° to 55° and are listed on figure 6.

Best and worst case scenarios are defined for each analysis case using the brightest and dimmest visual magnitudes and their respective Sun angles. These values are summarized in table 3. As a general rule-of-thumb, for each analysis case, as the distance is increased by 2 au, the visual magnitude drops by approximately 1 magnitude.

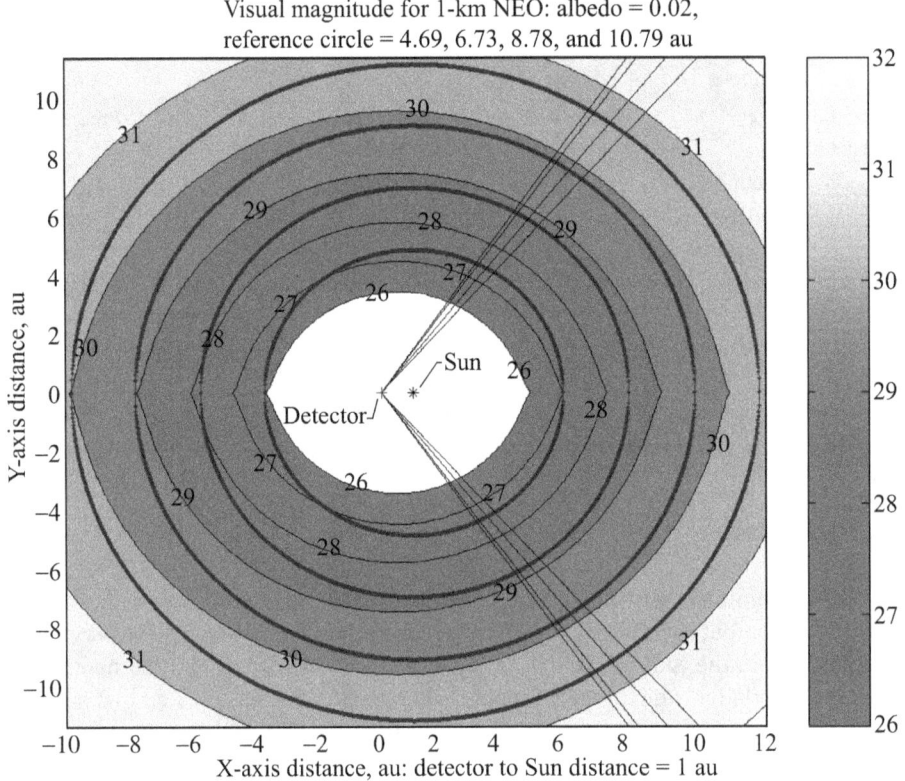

Figure 4. Visual magnitude contour plot for 1-km-diameter NEO.

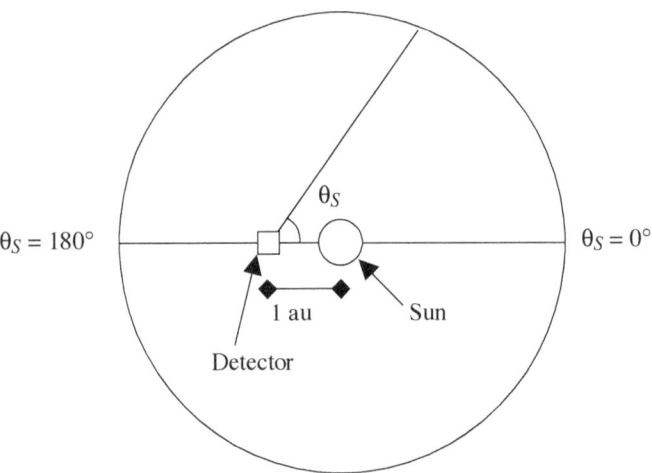

Figure 5. Sun angle geometry.

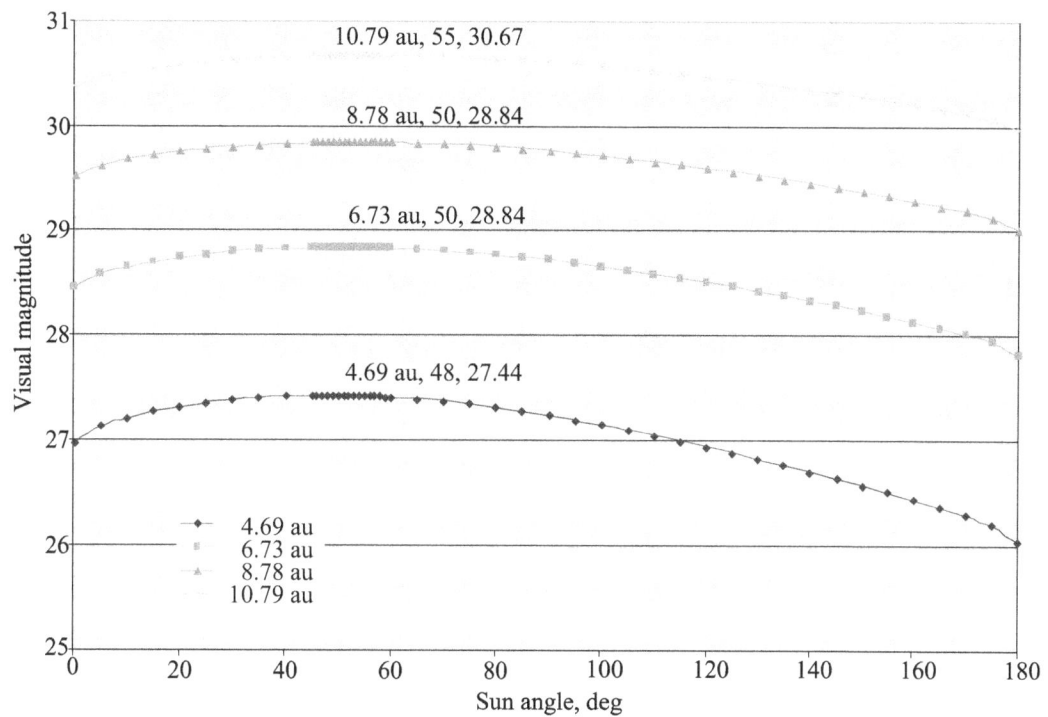

Figure 6. Visual magnitude versus Sun angle.

Table 3. Visual Magnitude Ranges Using Worst Case Albedo

Watch distance, au	4.69	6.73	8.78	10.79
Best scenario visual magnitude	26.04	27.78	29.02	29.97
Worst scenario visual magnitude/Sun angle, deg	27.44/48	28.84/51	29.85/54	30.67/55

Observing Considerations

Before final analysis can be performed, value ranges and limitations for different interferometer parameters and configurations must be looked at.

SNR, Baseline Length, and Visibility

From equation (3), in order to achieve the desired resolution of 100 µas, there needs to be a certain SNR for each baseline in order to achieve this accuracy. Using the mean wavelength value of 750 nm, baseline length versus the required SNR to achieve a resolution of 100 µas is plotted in figure 7.

Figure 7 shows that as the baseline is increased, the SNR required to achieve the desired resolution gets smaller. From this plot alone it would appear that using an extremely large baseline would be ideal for keeping the necessary photon rate to a minimum; however, there is a limitation to the maximum

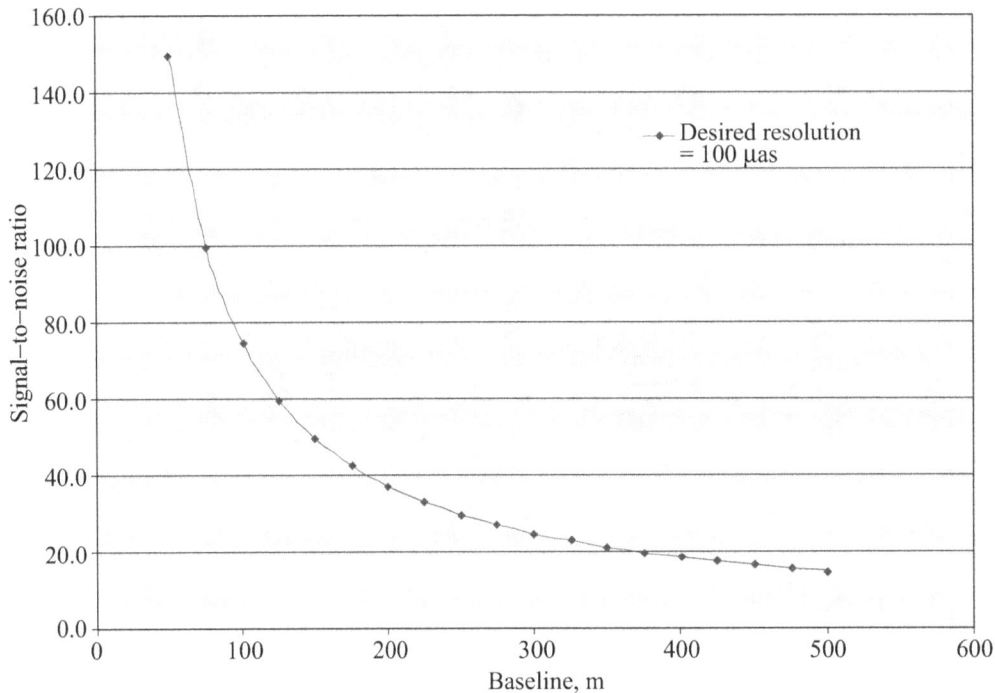

Figure 7. SNR requirement for given baseline length.

baseline that can be used for observing the desired targets. Increasing the baseline overresolves the target object, causing the fringe visibility to drop. If the baseline is increased too far, the fringe visibility will drop to zero and no interference fringes will be produced. Using equation (5) and the value the target diameter associated with the best scenario for each of the four cases, a plot of fringe visibility versus baseline length is made and shown in figure 8.

In figure 8, as the target width increases, increasing the baseline length causes the fringe visibility to drop off faster. This means for larger width objects, a smaller baseline must be used to produce the same visibility as a smaller width object with a longer baseline. In other words, the farther the object is away from the detector, the smaller its diameter will be, allowing a longer baseline to produce the same visibility.

Simulation Results

Increasing the baseline reduces the SNR required to achieve 100-μas accuracy, but it also decreases the fringe visibility, which in turn reduces the amount of signal. Therefore, a comparison of these two properties was considered to see if an optimum baseline exists that balances the required SNR and fringe visibility. This was done by taking equation (9) and solving for R, then substituting in equations (3) and (5) for the SNR and V, respectively. The required photon rate to achieve 100-μas measurements is graphed versus baseline length and is shown in figure 9.

Figure 9 clearly shows that for each case there is an optimal baseline length giving the lowest photon rate requirement. Detailed analysis at these optimal points results in a constant value of the fringe visibility for each case of approximately 0.63205. Because V is constant, the only factor that affects the optimal

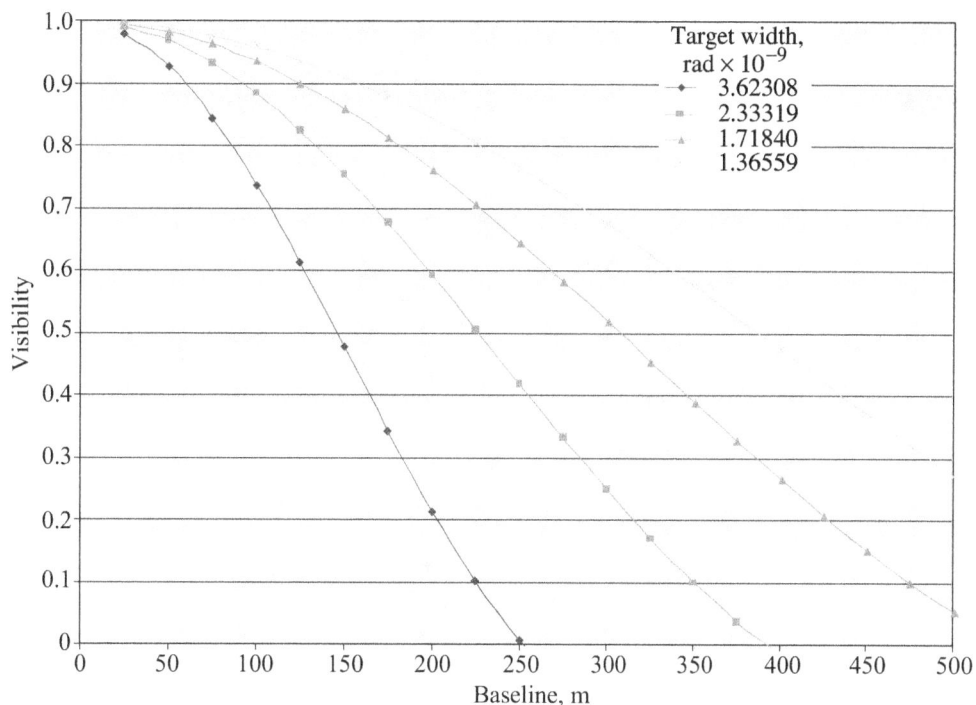

Figure 8. Effect of baseline on fringe visibility.

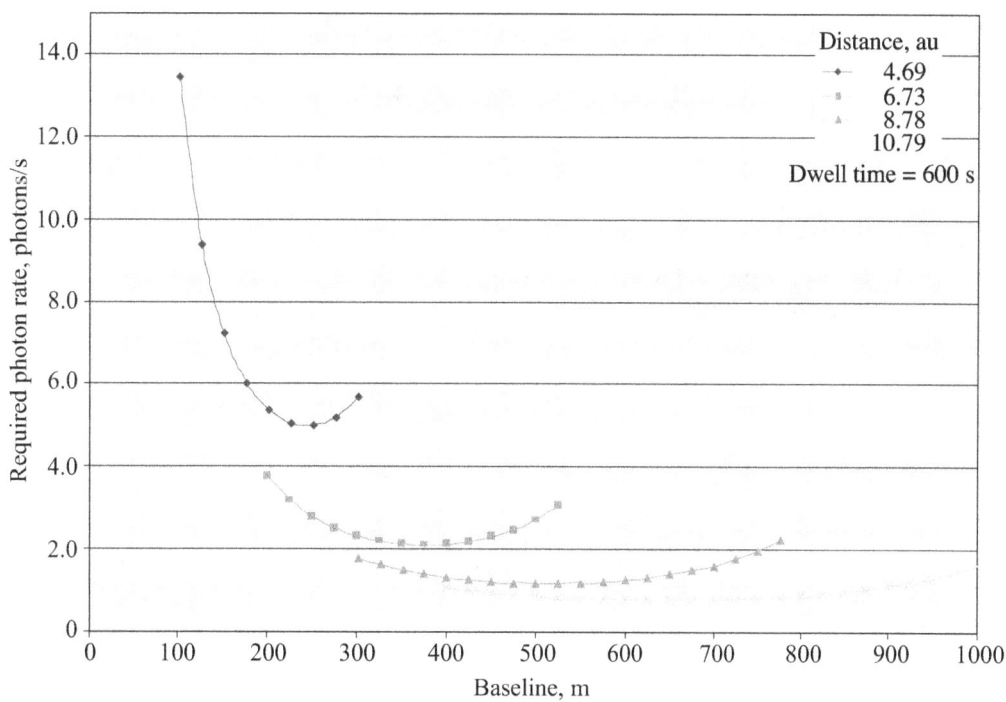

Figure 9. Effect of baseline on required photon rate.

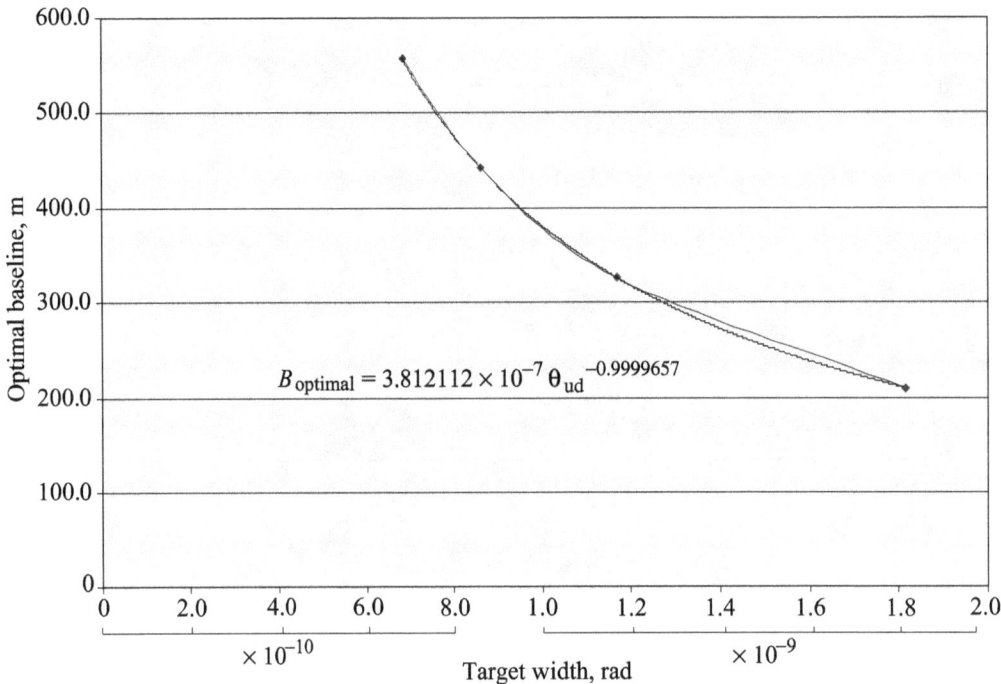

Figure 10. Optimal baseline versus target width.

baseline length is the object's diameter and distance from the detector, which changes the target width at the telescope. Therefore, a plot of the optimal baseline versus target width was made and shown in figure 10. A power series line of best fit was performed on the data points and is given by

$$B_{\text{optimal}} = 3.812112 \times 10^{-7} \theta_{ud}^{-0.9999657} \tag{10}$$

Equation (10) can be used to estimate the optimal baseline for any object diameter and distance from the detector.

For each analysis case, upper and lower bounds are set based on the best and worst case visual magnitudes. Using the detector distances at the best and worst scenario positions of each analysis case, the optimal baseline is found and listed in table 4.

Table 4. Optimal Baselines: Best and Worst Scenarios

Watch distance, au	Best case			Worst case		
	Distance from detector, au	Target width, rad	Optimal baseline, m	Distance from detector, au	Target width, rad	Optimal baseline, m
4.69	3.69	1.81×10^{-9}	210.3	5.30	1.26×10^{-9}	302.0
6.73	5.73	1.17×10^{-9}	326.5	7.36	9.14×10^{-10}	416.6
8.78	7.78	8.59×10^{-10}	443.4	9.42	7.17×10^{-10}	531.1
10.79	9.79	6.83×10^{-10}	557.9	11.43	5.90×10^{-10}	645.7

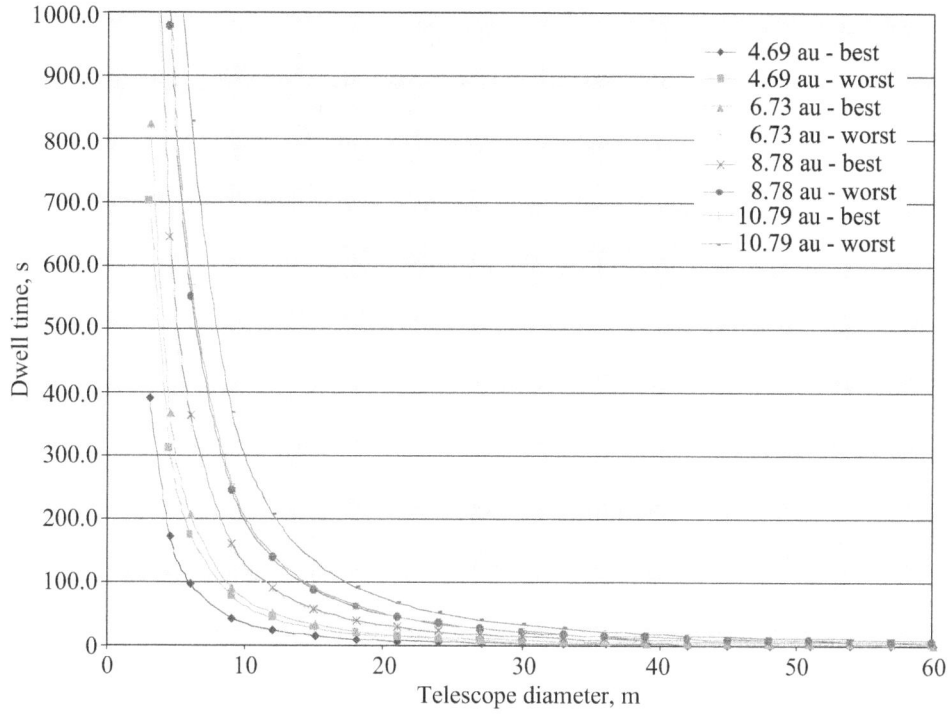

Figure 11. Dwell time and telescope size using optimal baselines.

As the optimal baseline gets longer, the SNR requirement to achieve the desired resolution goes down. Using these optimal baselines, a comparison between telescope diameter and the required dwell time needed to achieve the necessary SNR was made for each analysis case and is shown in figure 11. A study of figure 11 shows that as the telescope diameter is increased, the dwell time to achieve the necessary SNR goes down. A telescope diameter of approximately 5.5 m is sufficient to keep the dwell time requirements below 1000 s.

Telescope size will ultimately be determined by the maximum allowable dwell time that can be achieved. Referring back to equation (7), the bandwidth and speed of light will be constant. The parameter that can vary is the delay rate uncertainty. Delay rate uncertainty is caused by two effects, the angular rate error in the orbit of the NEO and the uncertainty in the motion of the interferometer. The range of possible delay rate uncertainties and the corresponding maximum allowable dwell time are listed in table 5.

If the interferometer was perfectly stable, the angular rate uncertainty would be the only factor affecting the delay rate uncertainty. The delay rate uncertainty can be found by multiplying the angular rate uncertainty by the length of half the baseline. Using the angular rate uncertainty of the worst case orbit (10.79 au = 1.92439×10^{-12} rad/s) leads to a delay rate uncertainty of approximately 1.0×10^{-9} m/s, which leads to a maximum dwell time of approximately 1000 s. Therefore, the effects of the angular motion of the NEO can be a factor when taking measurements. It is unknown how this angular uncertainty compares with that found in a space-based interferometer baseline. Once the SIM is launched in a few years, reliable data on space platform performance will be available. Until then only estimates can be made. For a lunar-based interferometer, uncertainty rates from existing Earth-based interferometers can

Table 5. Maximum Dwell Time Given Delay Rate Uncertainty

Delay rate uncertainty, m/s	Maximum dwell time, s
1.00×10^{-13}	9993082
5.00×10^{-13}	1998616
1.00×10^{-12}	999308
5.00×10^{-12}	199862
1.00×10^{-11}	99931
5.00×10^{-11}	19986
1.00×10^{-10}	9993
5.00×10^{-10}	1999
1.00×10^{-9}	999
5.00×10^{-9}	200
1.00×10^{-8}	100
5.00×10^{-8}	20
1.00×10^{-7}	10
5.00×10^{-7}	2

Table 6. Telescope Diameter Required for Given Dwell Time (Optimal Baseline)

Dwell time, s	Watch distance, au							
	4.69		6.73		8.78		10.79	
	Best, m	Worst, m	Best, m	Worst, m	Best, m	Worst, m	Best, m	Worst, m
2	42.0	56.2	61.0	78.8	80.8	100.3	101.4	122.8
10	18.8	25.1	27.3	35.3	36.1	44.8	45.3	54.9
20	13.3	17.8	19.3	24.9	25.6	31.7	32.1	38.8
100	5.9	7.9	8.6	11.1	11.4	14.2	14.3	17.4
200	4.2	5.6	6.1	7.9	8.1	10.0	10.1	12.3
999	1.9	2.5	2.7	3.5	3.6	4.5	4.5	5.5

be used as a close approximation. Because precise values of baseline angular uncertainties are not yet known, estimates for the possible delay rates that might be achieved in space are assumed to be the six largest values of delay rate uncertainty taken from table 5. Incorporated in table 6 are these values and the telescope diameter required to enable the desired astrometric accuracy in the given dwell time. Analysis of the data in table 6 shows that the dwell time plays a significant factor in the final design of an interferometer. If the delay rate uncertainty cannot be kept to less than 5.00×10^{-9} m/s, the required telescope sizes become impractical for a space-based system. Ideally, the telescope size used will be as large as possible but it will ultimately depend on the cost and size limitations and the technology available to put them into space. An ideal value to achieve in delay rate uncertainty would be around 1.0×10^{-9} m/s. This value would allow a telescope with a diameter of 6 m to take measurements in all four cases regardless of Sun angle.

To investigate the effects of using a fixed baseline, the data in figure 12 show the same results using a baseline of 150 m for each analysis case. The data in figure 12 demonstrate that using a fixed baseline

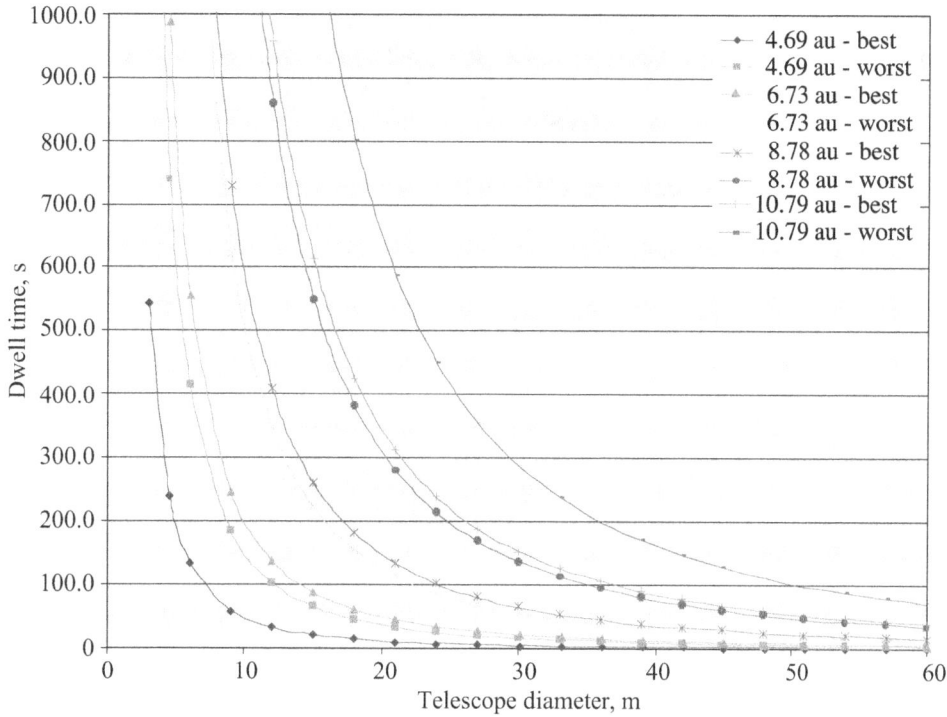

Figure 12. Dwell time and telescope size using constant baseline of 150 m.

will increase the necessary dwell time significantly, for a given telescope diameter, over a variable baseline that can be optimized. To highlight the differences in necessary dwell time, a 6-m-diameter telescope is chosen, and the respective dwell time for each case is listed in table 7. For the 150-m fixed baseline, a telescope almost three times larger ($\approx$16 m) would be needed to keep the dwell time below 1000 s. The percent difference in dwell time between the optimal and fixed baseline is the same for each case regardless of the telescope diameter used.

Even with this limited set of analysis cases, there is a significant difference when an optimal baseline is used compared with a fixed baseline. The fixed baseline will be optimized for only a small range of

Table 7. Dwell Time Comparison Using 6-m-Diameter Telescope

Sun distance/case, au	Optimal baseline	150-m baseline	Difference, percent
	Dwell time, s	Dwell time, s	
4.69 - best	97.82	135.95	39.0
4.69 - worst	175.45	415.83	137.0
6.73 - best	206.53	555.89	169.2
6.73 - worst	345.26	1399.55	305.4
8.78 - best	362.7	1637.92	351.6
8.78 - worst	550.99	3438.11	524.0
10.79 - best	570.67	3825.93	570.4
10.79 - worst	828.77	7202.54	769.1

objects. For example, the fixed 150-m baseline is near the optimized baseline length for the 4.69 au range, giving only a small percent difference in dwell time. However, as the object gets farther away, the dwell time required to take the measurement using the fixed 150-m baseline increases rapidly compared with using the optimal baseline. To take a measurement of the 10.79-au object, the necessary dwell time increases by almost 800 percent. If the fixed baseline was longer, to give the best performance in the range of 10.79 au, the close objects would be overresolved and could not be measured at all. Therefore, a changeable baseline is recommended for a CAPS interferometry system. If a fixed baseline is used, it will either restrict the viewing of NEOs closer than a certain distance or place tremendous requirements on the telescope size or control accuracy in order to measure those objects farther away.

Adaptability for Small NEOs Near the Interferometer

If a system is designed that can take measurements of the four analysis cases, how will it perform on smaller sized objects? In figure 13, a visual magnitude contour plot for a 50-m NEO at distances of 0.8 and 1.2 au from the Sun is shown.

In table 8, the properties of the 1.2-au case are listed; the 0.8-au case is kept separate because the Sun angle definition breaks down once the NEO is less than 1 au from the Sun. Evaluating the numbers in table 8 shows that the CAPS interferometer would have no problem observing the 50-m object. For the best case, the visual magnitude is almost 3 magnitudes brighter than the best 4.69-au analysis case. Also, the optimal baseline is slightly larger. Observing the 50-m NEO at this location should actually be much

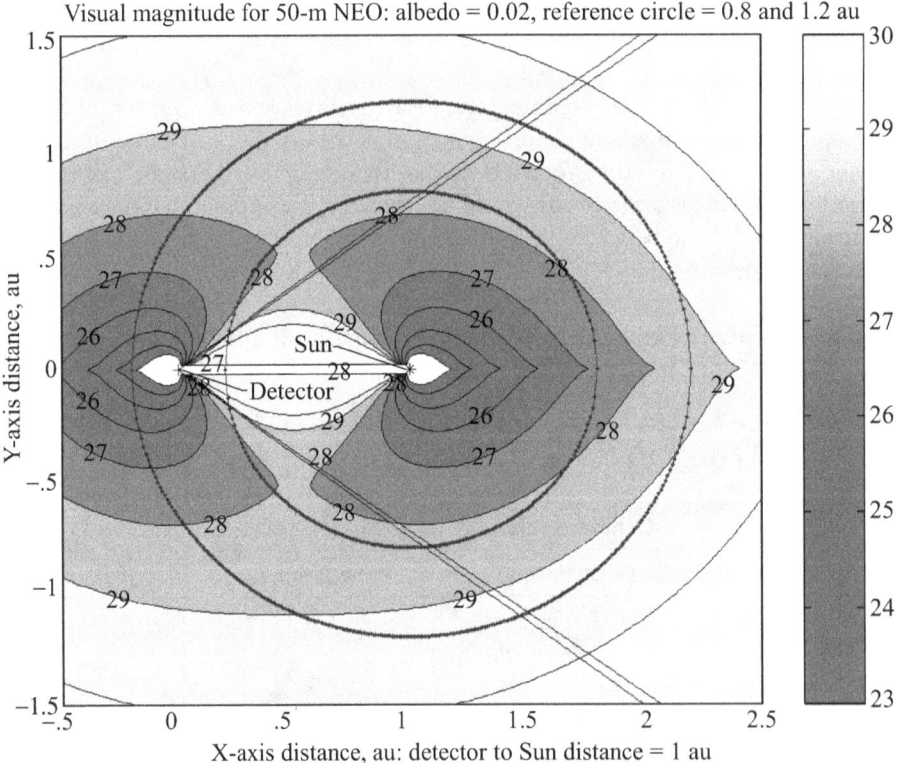

Figure 13. Visual magnitude contour for 50-m-diameter NEO.

Table 8. Detection Properties of 50-m NEO With 1.2-au Watch Distance

	Visual magnitude	Sun angle, deg	Distance from detector, au	Target width, rad	Optimal baseline, m
Best case	23.25	180	0.2	1.67115×10^{-9}	228.0
Worst case	29.37	37	1.837	1.81943×10^{-10}	2093.6

easier than any of the analysis cases. At the worst case position, the visual magnitude is comparable with that of the best 10.79-au analysis case. However, the optimal baseline is substantially larger, which would reduce the necessary photon rate to achieve the desired accuracy. Even if the baseline could only be extended to the distance of the longest analysis case, 645.7 m, the interferometer should still receive adequate signal to take the 100-µas measurement. If the NEO is 0.8 au from the Sun, the definition of the Sun angle is no longer valid because there are two possible locations of the NEO for each Sun angle. For most of the locations in the orbit, the visual magnitude varies between 27 and 29. When the NEO starts to come between the detector and the Sun, the visual magnitude increases rapidly, essentially going to infinity when the NEO is directly between the two.

Assuming that the interferometer can view the NEO during other parts of its orbit, the interferometer should be able to take a measurement for the majority of these locations using the same reasons defined for the 1.2-au object. The actual range of NEOs that can be observed in orbits less that 1 au from the Sun will ultimately depend on the pointing restrictions of the interferometer toward the Sun. One problem might arise in viewing these 50-m objects: their initial input errors might be considerably higher than the 1-km objects due to the fact that they are moving a lot faster relative to the detector. This factor may lead to higher angular rate uncertainties and smaller maximum dwell times.

Concluding Remarks

Acquiring space-based astrometric measurements of faint near-Earth objects (NEOs) using interferometry is a formidable task. Stable interferometers with very accurate measuring systems are needed to produce the delay rate uncertainty that will lead to feasible telescope sizes. Using the desired value of 1.0×10^{-9} m/s for the delay rate uncertainty still requires telescopes approximately 6 m in diameter. Six-meter telescopes would be a good starting point for a space-based interferometer because they, as well as larger diameter telescopes, have already been built for ground-based interferometers. The next challenge is figuring out how to deploy and operate interferometers in space. Advances in technology should allow this to be possible in the foreseeable future, and even larger telescopes might be available 25 years from now. All the results in this analysis assumed the worst case for most estimates, especially the NEO albedo. Realistically, most of the NEOs that will be encountered will have a higher value of visual magnitude than used in this analysis. If the system is designed with a variable baseline, the range of objects that can be observed will be quite large.

Future Work

All calculations performed in this analysis are only first order approximations. For a more detailed analysis, specific interferometer control systems, detectors, and scanning and tracking algorithms should be used to generate more robust equations. Also, many possible sources of error were assumed to be zero for this analysis. Real life values for these errors need to be determined and used, especially values for background noise. The effects of changing the default values (like using a larger bandwidth) of the interferometer should be studied. Location will play a big role in the errors produced by the interferometer. Specific locations need to be determined and estimated errors need to be calculated at those locations.

There are some space-based interferometry technology demonstrator missions, including the Space Interferometry Mission, planned for the near future that should provide more information regarding this problem.

References

1. Halverson, P. G., et al.: Progress Towards Picometer Accuracy Laser Metrology for the Space Interferometry Mission. *International Conference of Space Optics*, ISCO 2000, Dec. 5–7, 2000, Toulouse, France.

2. Boden, A.: Interferometric Narrow-Angle Astrometry: Data Analysis. Slides from the 2001 Michelson Interferometry Summer School, Lowell Observatory, Flagstaff, AZ, May 21–25, 2001.

3. Lay, O.: Jet Propulsion Laboratory, Pasadena, CA. (Personal communication 2002–2003.)

4. Boden, A.: Elementary Theory of Interferometry. *Principles of Long Baseline Stellar Interferometry*, P. Lawson, ed., 1999, pp. 9–28.

5. Traub, W.: Beam Combination and Fringe Measurement. *Principles of Long Baseline Stellar Interferometry*, P. Lawson, ed., 1999, pp. 31–58.

6. Lay, O.: Separated Spacecraft Interferometry. Presentation given at the 2002 Michelson Interferometry Summer School, Harvard-Smithsonian Center for Astrophysics, Cambridge, MA, June 24–28, 2002.

7. Mackay, C. D.; Tubbs, R. N.; Bell, R.; Burt, D.; and Moody, I.: Sub-Electron Read Noise at MHz Pixel Rates. SPIE 4306 Conference Proceedings, Jan. 2001, pp. 289–298.

8. Rowe, B. H.: Astronomical Computing. *Sky & Telescope*, R. W. Sinnott, ed., June 1993, pp. 83–85.

Mission Functionality for Deflecting Earth-Crossing Asteroids/Comets[6]

SANG-YOUNG PARK
Swales Aerospace

DANIEL D. MAZANEK
NASA Langley Research Center

Introduction

An infrequent but significant hazard to life and property due to impacting asteroids and comets exists. Earth-approaching asteroids and comets also represent a significant resource for commercial exploitation, space exploration, and scientific research. The impact problem and those planetary bodies that could be a threat have been discussed in great depth in a wide range of publications. A popular planetary defense method is the deflection of asteroids and comets on a collision orbit with the Earth by changing their orbital velocities. It is fundamental to estimate required changes in the orbital velocities of dangerous objects to avoid a collision.

A detailed optimization problem is formulated to calculate optimal impulses for deflecting Earth-crossing asteroids/comets, using nonlinear programming. The constrained optimization problem is based on a three-dimensional patched conic method to include the Earth's gravitational effects and the asteroid/comet's orbital inclination. The magnitudes and impulse angles of optimal impulse change in velocity (ΔV) are accurately computed at various points on the asteroid/comet's orbit to provide a given target separation distance. Based on these change in velocity analyses, we can establish the approximate cost and build strategies to prevent possible catastrophe due to the objects.

Background

The optimal ΔV and deflection strategy are dependent on the size and the orbital elements of the asteroid/comet, as well as the amount of warning time. Assuming a linear approximation between orbital energy and velocity increment, a velocity increment of about 1 cm/s is required to deflect an asteroid by a distance equal to 1 Earth radius for the order of a decade ahead of an impact (ref. 1). Using Keplerian motion and perpendicular impulse direction, both kinetic-energy deflection and nuclear-explosive deflection are treated in reference 2. A method in reference 3 shows the instantaneous correction of asteroid velocity as a result of a spacecraft's collision with an asteroid at perihelion only. In reference 4 the use of a nuclear explosive is discussed and it is concluded that a nuclear device having a yield of about 1 Mt may be required to deflect an object with a 0.3-km radius at a distance of 1 astronomical unit (au). Direct spacecraft impacts on Earth-crossing objects (ECOs) can provide kinetic energy to suitably disrupt up to about 0.1-km stony asteroids and 0.3-km ice comets (ref. 5). One clear conclusion from these simplified analyses is that early detection gives longer reaction time, and asteroid/comet interceptions far from Earth are much easier and more desirable than interceptions near Earth because small deflections far away will produce a greater miss distance at the Earth. It is important to consider interception several orbital periods in advance as well as interceptions close to the Earth because both of these situations could be encountered depending on the size and orbital characteristics of an asteroid/comet.

Recently, the astrodynamical optimization problem was formulated for Earth-crossing asteroids (ECAs) and presented accurate impulse angles as well as impulse magnitude according to impulse time

---

[6]Chapter nomenclature available in chapter notes, p. 217.

based on a two-dimensional, two-body analysis (ref. 6). The minimum impulse change in velocity (ΔV) requirement is not a monotonically decreasing function of warning time; rather, there is a finer structure associated with the orbital period of the colliding asteroid. Thus, the "optimal time" for application of ΔV is the earliest possible perihelion for warning times greater than 1 orbital period (ref. 6). However, this research has simplified the problem by assuming two-body orbital mechanics between the Sun and an ECA. This assumption neglects perturbations due to the Earth's gravity. While these perturbations may not be present until the terminal phase of the impact scenario, they affect both long and short warning time analyses. Further research refines the heliocentric two-body analysis presented in reference 6 by including the gravitational effect of the Earth (ref. 7). The minimum ΔV is increased due to gravitational effects of the Earth, and effects on the minimum ΔV are sensitive to orbital elements of a target asteroid because the impact parameter is dependent upon those orbital elements. Gravitational effects of Earth have the strongest influence on minimum ΔV calculations for deflecting ECAs in nearly circular heliocentric orbits around the Earth. Hence, reference 7 concludes that the problem of deflecting ECAs should include the gravitational effects of Earth; the results in reference 7 followed from two-dimensional analysis. A near-optimal analysis for three-dimensional deflecting problems is described by using a system transition matrix (ref. 8). The study in reference 8 does not include Earth's gravitational effects and is a first-order approximation because of the nature of the system transition matrix used. To have insights in deflecting Earth-approaching asteroids and comets with nonzero inclination, this paper expands the works of reference 7 by formulating an optimal deflection problem in three-dimension that also includes the Earth's gravitational effects. The analysis of the impact deflection problem in this paper is based on a three-dimensional patched conic method. The analysis centers on how impulses applied to an asteroid at various points on the asteroid's orbit affect the outcome when there is a presumption of collision otherwise. The analysis tool presented can be utilized in determining an accurate estimate for optimizing the time and position of intercepting the asteroid and comet for impact deflection.

Deflection or orbit modification methods will be dependent on the size and composition of asteroids/comets and the amount of warning time available. There are several deflection schemes such as the propulsive mode, kinetic deflection, nuclear detonation, and laser ablation. A study of spacecraft vertical landing on an asteroid indicates the possibility of giving a significant boost to an asteroid by means of a buried explosion of practical yield (refs. 9 and 10) and a high-thrust or low-thrust engine. Because every kilogram of propellant landed on an asteroid requires thousands of times more mass to lift it from the Earth and deliver it to the object, landing propellant on the asteroid in order to provide working mass is extremely inefficient (ref. 11). However, this strategy can easily control motion of an asteroid/comet because it has many opportunities to choose the direction and time of impulse application (ref. 3). Kinetic deflection and nuclear detonation demand much less "cost" than the propulsive mode and, consequently, require much less initial mass for deflection missions (ref. 12). A high-thrust, laser ablation technique should be seriously considered for use in deflecting or disrupting threatening celestial objects because these schemes appear both practical and possible. In this paper, we briefly discuss the potential of each strategy to deflect ECOs in conjunction with several future spacecraft concepts.

Three-Dimensional Optimality Problem for ΔV Analysis

Given an ECO and an established Earth collision, the problem is to minimize the impulse change in velocity (ΔV) required to deflect the ECO in such a manner as to miss the Earth by a minimum target miss distance (see fig. 1). The performance index is defined by the magnitude of the required deflection velocity:

$$J(\mathbf{u}) = \sqrt{\Delta V_T^2 + \Delta V_N^2 + \Delta V_W^2} \tag{1}$$

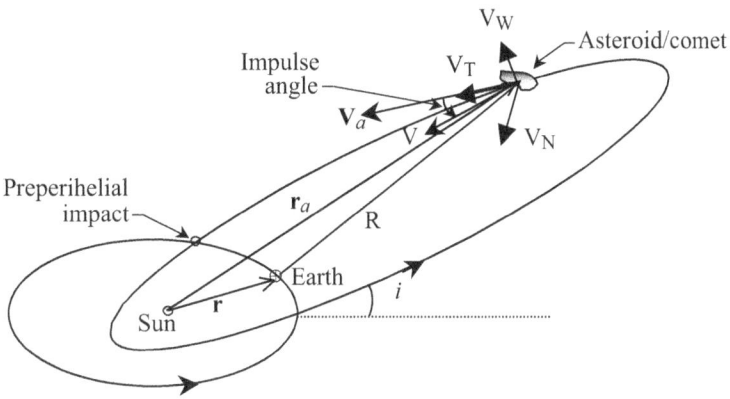

Figure 1. Geometry of Earth-crossing asteroid/comet (not to scale) with TNW coordinate system.

where ΔV_T is the velocity increment aligned with the ECO velocity, ΔV_N is the velocity increment normal to ΔV_T in the ECO orbital plane, and ΔV_W is the velocity increment normal to the ECO orbital plane. The impulse controls, **u**, are ΔV_T, ΔV_N, and ΔV_W, which effectively give the magnitude $\left(|\Delta V| = \sqrt{\Delta V_T^2 + \Delta V_N^2 + \Delta V_W^2} \right)$ and angle of the deflection velocity. The geometry of an ECA or comet with the "TNW" coordinate system is shown in figure 1. The calculation is always done to move the ECO's trajectory from crossing the Earth's orbit at the Earth's center. This problem is subject to the heliocentric two-body equation outside the Earth's sphere-of-influence (SOI) and geocentric two-body equations inside the SOI. Thus, the analysis of the impact deflection problem is based on a three-dimensional patched conic method. It is assumed that an ECO is influenced by the gravitational field of the Earth only when it is within the Earth's SOI. Beyond the SOI, the object is considered to be affected only by the Sun's gravitation. In the case of an impact scenario, the ECO begins in an elliptical orbit about the Sun. Once within the Earth's SOI, the object's motion is described by two-body orbit equations for a hyperbolic orbit about the Earth. The radius of the Earth's SOI is about 9.31×10^5 km, or 0.00621 au. In figure 2, the approaching distance and a hyperbolic trajectory (ref. 13) are illustrated, and the geometry is still valuable for three-dimensional analysis. Under this framework, the constraints for the optimization problem can be described in terms of the terminal boundary conditions at the time (t_f) when the ECO intersects the Earth's SOI:

$$R - R_{SOI} = 0 \qquad (2)$$

$$b - b_i = 0 \qquad (3)$$

$$\dot{R} < 0 \qquad (4)$$

where R is the distance between Earth and the ECO, R_{SOI} is the radius of Earth's SOI, $\dot{R}$ is the time derivative of R, b is the approach distance of the ECO, and b_i is an impact parameter of the Earth. The constraints above mean that the approach distance (b) of the approaching ECO must be equal to the impact parameter (b_i) at the edge of the Earth's SOI in order to deflect the ECO by a minimum target

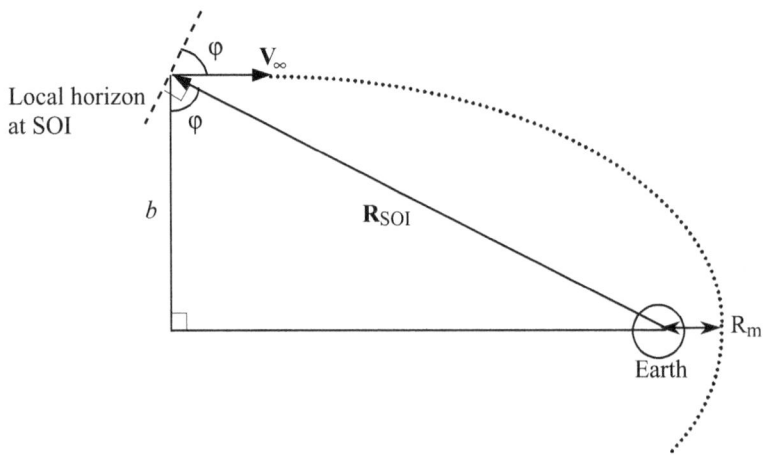

Figure 2. Approach distance, b, at SOI of Earth (not to scale).

miss distance R_m. If the approach distance is less than the impact parameter, the ECO can be inside the target miss distance. When we set $R_m = 1$ Earth radius, and $b < b_i$, the ECO will collide with Earth. Note that if $b = b_i$, there will be a surface graze and the case is not considered as a collision in our analysis. If the approach distance is greater than the impact parameter, the impulse will not be a minimum value, forcing the ECO to miss the Earth by R_m. The variable R is given by

$$R = \sqrt{\mathbf{r}_\oplus \cdot \mathbf{r}_\oplus + \mathbf{r}_a \cdot \mathbf{r}_a - 2\mathbf{r}_\oplus \cdot \mathbf{r}_a} \tag{5}$$

Then, we have

$$\dot{R} = \left[\mathbf{r}_\oplus \cdot \mathbf{V}_\oplus + \mathbf{r}_a \cdot \mathbf{V}_a - \left(\mathbf{r}_\oplus \cdot \mathbf{V}_a + \mathbf{r}_a \cdot \mathbf{V}_\oplus\right)\right]/R \tag{6}$$

where $\mathbf{r}_\oplus$ is the radius vector from the Sun to the Earth, $\mathbf{r}_a$ is the radius vector from the Sun to the ECO, $\mathbf{V}_a$ is the velocity of the ECO with respect to the Sun, and $\mathbf{V}_\oplus$ is the velocity of the Earth with respect to the Sun. In the heliocentric system, $\mathbf{V}_\infty$ is the vector difference at $\mathbf{R}_{SOI}$ between $\mathbf{V}_a$ and $\mathbf{V}_\oplus$, so that

$$\mathbf{V}_\infty = \mathbf{V}_a - \mathbf{V}_\oplus \tag{7}$$

The vector $\mathbf{V}_\infty$ is described by its magnitude V_∞ and an elevation angle, φ. From figure 2, the elevation angle is given by

$$\cos(\varphi + 90°) = \mathbf{V}_\infty \cdot \mathbf{R}_{SOI} / V_\infty R_{SOI} \tag{8}$$

and the approach distance is obtained from

$$b = R_{SOI} \cos \varphi \tag{9}$$

The impact parameter, b_i, yields the expression

$$b_i = R_m \sqrt{1 + \frac{V_{esc}^2}{V_\infty^2}} \tag{10}$$

where

$$V_{esc} = \sqrt{2\mu_\oplus/R_m} \qquad (11)$$

V_{esc} is the escape velocity at R_m, and $\mu_\oplus$ is the gravitational constant of the Earth.

As noted in reference 6, closed-form solutions cannot be obtained for this optimization problem. To solve the heliocentric three-dimensional two-body motion, the method of Lagrangian coefficients (ref. 14) is used for describing $\mathbf{r}(t)$ and $\mathbf{V}(t)$ in terms of initial position vector, $\mathbf{r}(t_0)$, and velocity vector, $\mathbf{V}(t_0)$, in the inertial coordinate frame. At impulse time, $t_{impulse}$, the original orbit of an ECO is perturbed by an impulse; hence,

$$\mathbf{r}(t_{impulse}) = \mathbf{r}_0(t_{impulse}) \qquad (12)$$

$$\mathbf{V}(t_{impulse}) = \mathbf{V}_0(t_{impulse}) + \Delta \mathbf{V}_{XYZ}(t_{impulse}) \qquad (13)$$

$\Delta \mathbf{V}_{XYZ}$ can be expressed in the inertial coordinate system (X,Y,Z) as follows:

$$\Delta \mathbf{V}_{XYZ} = R_3(-\Omega) R_1(-i) R_3(-\omega) R_3(-\alpha) \Delta \mathbf{V}_{TNW} \qquad (14)$$

where $R_3(-\Omega)$, $R_1(-i)$, $R_3(-\omega)$, and $R_3(-\alpha)$ are rotation matrices about the longitude of ascending node (Ω), the inclination (i), the argument of periapsis (ω) for ECO, and the angle (α) between ECO perifocal coordinate and TNW system, respectively. Using $\mathbf{r}(t_{impulse})$ and $\mathbf{V}(t_{impulse})$ as initial conditions for the perturbed orbit, $\mathbf{r}(t_f)$ and $\mathbf{V}(t_f)$ are computed at the time when $R = R_{SOI}$, t_f.

Nonlinear programming (NLP) problems require determining $\mathbf{x} \in R^n$, with $\mathbf{x}$ as the state vector, that minimize the scalar objective function $J(\mathbf{u})$ subject to the equality constraints, $C_i(\mathbf{x}) = 0$, and the inequality constraints, $C_j(\mathbf{x}) \leq 0$. It is assumed that the objective function is continuously differentiable through second order. The parameters needed to obtain the constraint equations can be calculated using the relationship between the orbital elements and the position and velocity vectors. For this deflection problem, the free parameters are ΔV_T, ΔV_N, ΔV_W, and t_f, equations (2) and (3) are the equality constraints, C_1 and C_2, and equation (4) is the inequality constraint, C_3. The parameter t_f describes the time at which the constraints given by equations (2) through (4) are satisfied. The problem of minimizing the ΔV required for deflecting ECOs can now be cast in terms of a standard NLP problem. This formulation is also applicable to any Earth-approaching objects.

Numerical Results of ΔV Analysis

This analysis indicates the optimal impulses applied on an ECO at various points on the object's orbit to provide the miss distance equal to R_m when there is otherwise a presumption of collision at Earth's center. The NLP formulated in the previous section was solved by using the MATLAB™ optimization tool box (ref. 15). In the discussions to follow, we will use the term impulse time to specifically mean either ($t_{impact} - t_{impulse}$) or its absolute value. Here t_{impact} denotes the time at collision. Although the impact time is quite close to t_f, it is not the same because t_f is the time when $R = R_{SOI}$. Also, because t_{impact} and $t_{impulse}$ are not independent quantities, we choose $t_{impact} = 0$. This initialization has the advantage of interpreting $t_{impulse}$ as the time interval prior to impact if no action (i.e., ΔV maneuver) is undertaken. Also, it is apparent that we must have a warning time (i.e., the time interval between detection and collision) greater than the impulse time. The impulse time provides a crude measure of the warning time. The gravitational effects of the Earth are considered by using the three-dimensional optimization problem described previously. The minimum required impulses for the deflecting problem have a targeted distance of 1 Earth radius. The solutions to the deflection problems represent impulse vectors that can be described by the magnitude of the minimum impulse and the optimal impulse angle. The impulse angle is described in the ECO's orbital plane and is defined as the angle from the ECO's original velocity vector to the impulse vector toward the Sun-ECO line, r_a (see fig. 1). As reference 8 indicates, most numerical simulations show that the velocity increment (ΔV_W) normal to the ECO's orbital plane is insignificant and ignored for all cases in these analyses. Thus, the magnitude of impulse consists of only ΔV_T and ΔV_N. For any given impulse time, the problem has two solutions for the optimal impulse angle separated by 180° while keeping the same magnitude of impulse (ref. 6). In this section, only one solution for the optimal angle will be mentioned.

There are two categories of impact scenarios: one (impact before perihelion, or preperihelial impact) is that an impact occurs before the ECO sweeps its perihelion (fig. 1), the other (impact after perihelion, or postperihelial impact) is that an impact occurs after the ECO passes its perihelion. The minimum impulse requirements for the two impact scenarios are similar but slightly different. The differences are caused by the varied geometric positions of the impulse point with respect to the Sun. Each impact scenario has two subscenarios: an impact occurs either at the ascending node or at the descending node of an ECO. These two subscenarios have identical impulse angles and magnitudes because the subscenarios have the same geometric positions of the impulse point. We present results from some ΔV analyses while our method is applicable to any ECO deflection mission.

Near-Earth Asteroids

There are three classes of near-Earth asteroids (NEAs): Atens, Apollos, and Amors. Aten-type asteroids have a semimajor axis smaller than 1 au and an aphelion greater than 0.983 au, while Apollo-type asteroids have a semimajor axis greater than 1 au and a perihelion smaller than 1.017 au. Hence, Apollo-type and Aten-type asteroids can have Earth-crossing orbits. Amors have orbits that lie completely outside Earth's orbit (perihelial distance between 1.017 and 1.3 au) but have the potential to be perturbed into Earth-crossing trajectories.

We consider fictitious Apollo-type asteroids with semimajor axis $a = 1.5$ au, eccentricity $e = 0.5$, inclination $i = 0°$, 20°, 40°, and 60° first. For this example, the asteroids have orbital periods of approximately 1.84 years, 0.75 au perihelial distance, and 2.25 au aphelial distance. Figures 3 and 4 include the magnitudes and angles of the optimal impulses. The impulse time is normalized to the period of the unperturbed asteroid for ease of interpretation. The abscissa represents the time when impulse is applied (as a fraction of the period of the asteroid) prior to collision. It can also be noticed that the separation between Earth and an asteroid, which can be achieved by an impulse, depends strongly on location of the impulse on the orbit as well as direction of the impulse with respect to the orbital velocity. The required minimum ΔV has a cyclic component imposed upon a secular variation that varies inversely with the impulse time. Generally, the minimum ΔV for inclined orbits ($i > 0°$) is slightly less than that for planar orbits ($i = 0°$). The case of preperihelial impact has more fluctuated magnitude variations than the case of postperihelial impact. Figure 4 contains a parallel history of the optimal impulse angle with respect to impulse time. It is apparent that the optimal angles for both planar and inclined cases are almost the same if the impulse takes place at more than half an orbital period of the asteroid before impact. There are fluctuations in the optimal impulse angle when the ΔV occurs less than half an orbital period before impact. The fluctuations are more for an orbit with a higher inclination than for one with a lower inclination. Generally, postperihelial cases have more fluctuations in impulse angle than preperihelial cases.

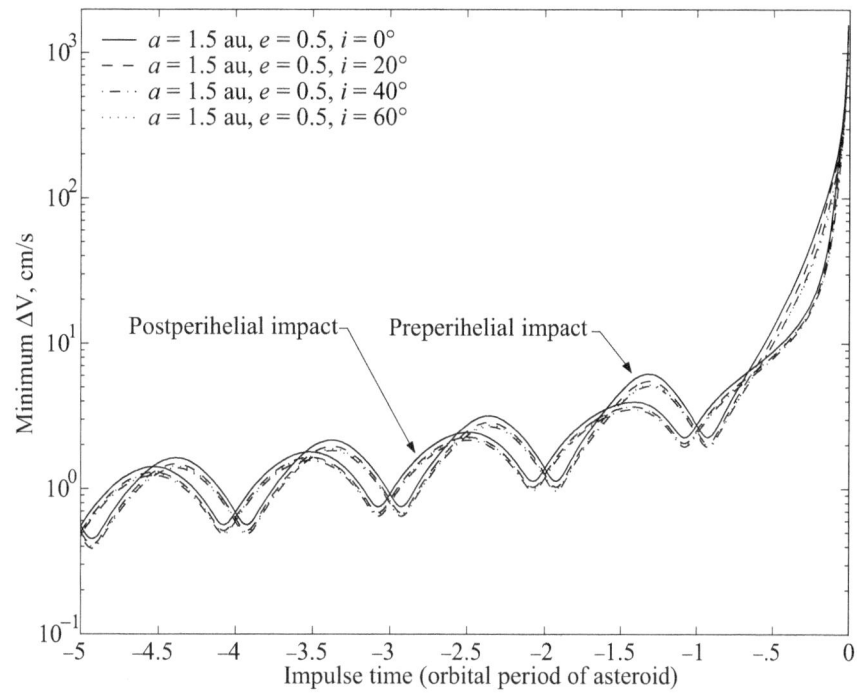

Figure 3. Minimum ΔV: Apollo-type asteroids with $a = 1.5$ au, $e = 0.5$, $i = 0°, 20°, 40°, 60°$.

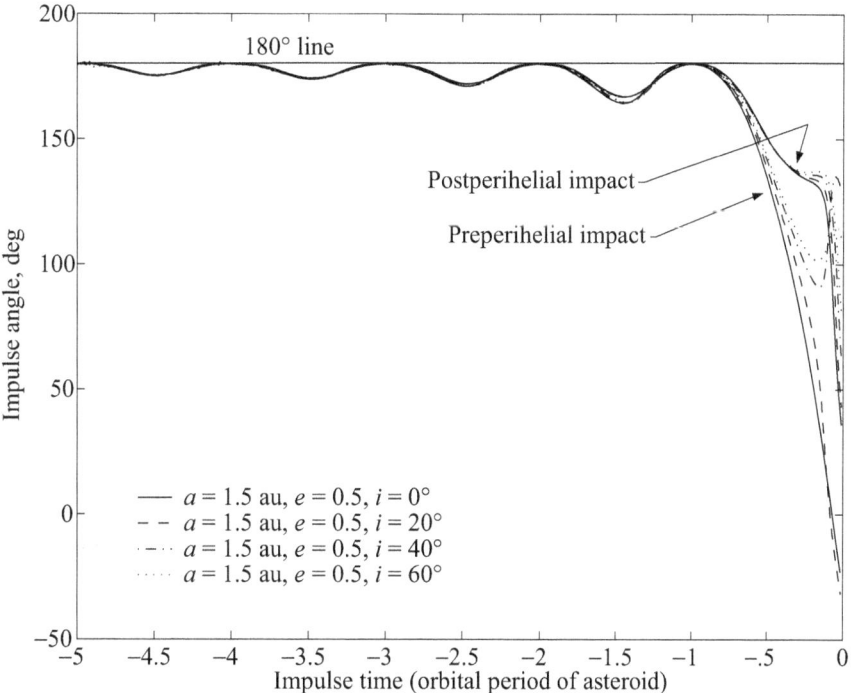

Figure 4. Impulse angle: Apollo-type asteroids with $a = 1.5$ au, $e = 0.5$, $i = 0°, 20°, 40°, 60°$.

Figures 5 and 6 include the magnitudes and angles of the optimal impulses for another set of Apollo-type asteroids whose orbital elements are $a = 2.0$ au (orbital period ≈ 2.83 years), $i = 20°$, and $e = 0.6, 0.7, 0.8$, and 0.9. From these analyses, we know that higher eccentricity yields more fluctuated histories of impulse magnitude and angle. Figures 7 and 8 include the magnitudes and angles of the optimal impulses for a third set of Apollo-type asteroids whose orbital elements are $e = 0.7$, $i = 10°$, and $a = 1.2$ au (orbital period ≈ 1.31 years), 1.6 au (orbital period ≈ 2.02 years), 2.0 au (orbital period ≈ 2.83 years), and 2.4 au (orbital period ≈ 3.72 years). The minimum ΔV decreases with increasing orbital period (i.e., larger semimajor axis) because a longer orbital period yields a longer warning time.

Figures 9 through 14 contain data demonstrating that the same trends are apparent for Aten-type asteroids. In figures 9 and 10, the semimajor axis and the eccentricity are fixed at 0.9 au (orbital period ≈ 0.85 years) and 0.4 au, respectively, and the inclination is varied from 0° to 60°. In figures 11 and 12, the semimajor axis and the inclination are fixed at 0.8 au (orbital period ≈ 0.72 years) and 10°, respectively, and the eccentricity is varied from 0.3 to 0.6. In figures 13 and 14, the eccentricity and the inclination are fixed at 0.6 au and 20°, respectively, and the semimajor axis is varied from 0.65 au (orbital period ≈ 0.52 years) to 0.95 au (orbital period ≈ 0.93 years). In all cases examined, the orbital inclination, i, has a relatively small effect on the minimum ΔV compared with eccentricity e and semimajor axis a (see figs. 7 and 11).

When only a two-body approximation is applied to the deflecting problem, the minimum ΔV is linearly proportional to the miss distance, R_m (ref. 6). For example, in the two-body approximation, deflecting a dangerous celestial body by 10 $R_\oplus$ requires exactly 10 times more ΔV than that needed to deflect the object by 1 $R_\oplus$. When the gravitational effects of Earth are considered, the minimum ΔV is linearly proportional not to the miss distance but to the impact parameter, b_i. Figure 15 includes the minimum ΔV (impact after perihelion) of a fictitious asteroid whose orbital elements are $a = 1.5$ au, $e = 0.5$, and $i = 20°$ as the miss distance R_m varies from 1 $R_\oplus$ to 100 $R_\oplus$. One Earth-Moon distance is approximately 60.27 $R_\oplus$. Analysis of equations (10) and (11) explains that the ΔV requirement for a deflection of $NR_\oplus$ is less than N times the ΔV requirement for 1 $R_\oplus$ deflection. The reason is that V_{esc} is reduced as R_m is increased; hence, b_i is not linearly proportional to R_m. For instance, at $t = -1$ impulse time of the asteroid's orbital period, the ΔV requirement for 1 $R_\oplus$ deflection is about 2.5187 cm/s while the ΔV requirements for 10 $R_\oplus$ and 100 $R_\oplus$ deflection are about 21.6146 cm/s and 212.1871 cm/s, respectively.

Short-Period Comets and Long-Period Comets

Earth-crossing comets are classified into two types: short-period comets (SPCs), or those with an orbital period <200 years, and long-period comets (LPCs), or those defined here as having an orbital period >200 years. In figure 16, the histories of the minimum ΔV (miss distance of Earth radius) for a fictitious SPC whose orbital elements are given by $a = 4.0$ au, $e = 0.85$, and $i = 0°, 20°, 40°$, and $60°$ are shown. For this example, the comets have an orbital period of approximately 8.0 years, perihelial distance of 0.6 au, and aphelial distance of 7.4 au. It is again noted that the ΔV_W component is small enough to be neglected. The effect of inclination on the SPC minimum ΔV is also relatively small, as was observed previously for NEAs. Figure 17 includes the histories of optimal impulse angles corresponding to ΔV in figure 16. As impulse time approaches impact time, the class of postperihelial impact has peaks in optimal impulse angle while the class of preperihelial impact has an optimal impulse angle more sensitive to the inclination. Like other ECOs, the SPC has a slightly larger magnitude of minimum impulse for the case of preperihelial impact than for the case of postperihelial impact. For the SPC, when the impulse time is very close to impact time, the case of preperihelial impact also has more variation in the optimal impulse angle than the case of postperihelial impact as the inclination varies. The data in figures 18 and 19 demonstrate that the same trends are apparent for LPCs, and the same trends may be expected for all

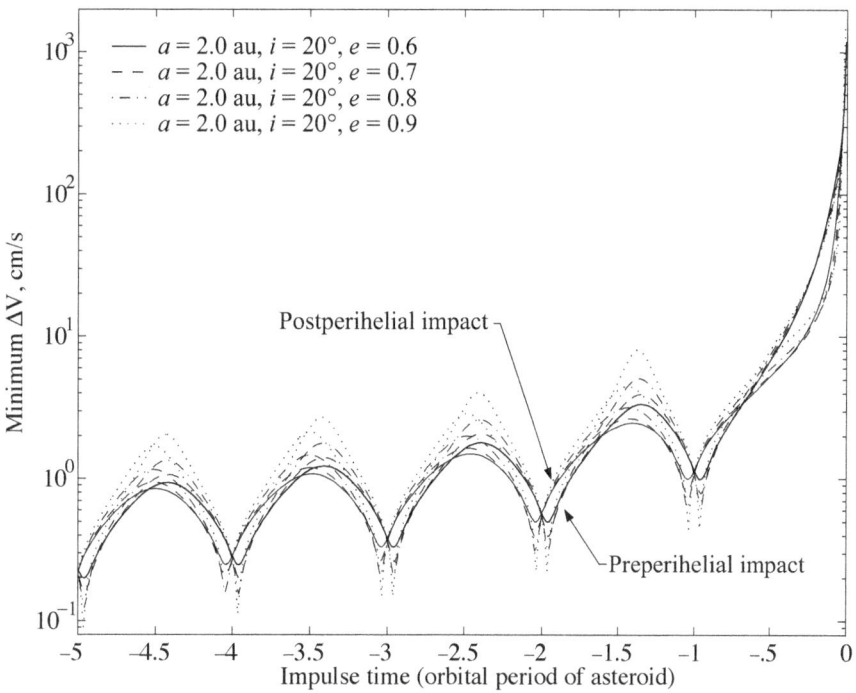

Figure 5. Minimum ΔV: Apollo-type asteroids with $a = 2.0$ au, $i = 20°$, $e = 0.6, 0.7, 0.8, 0.9$.

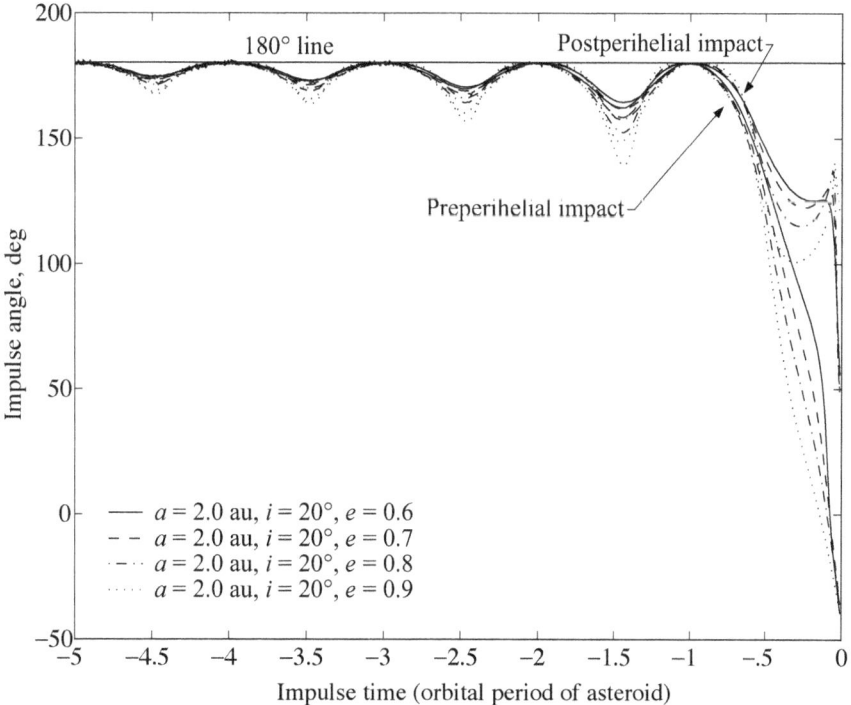

Figure 6. Impulse angle: Apollo-type asteroids with $a = 2.0$ au, $i = 20°$, $e = 0.6, 0.7, 0.8, 0.9$.

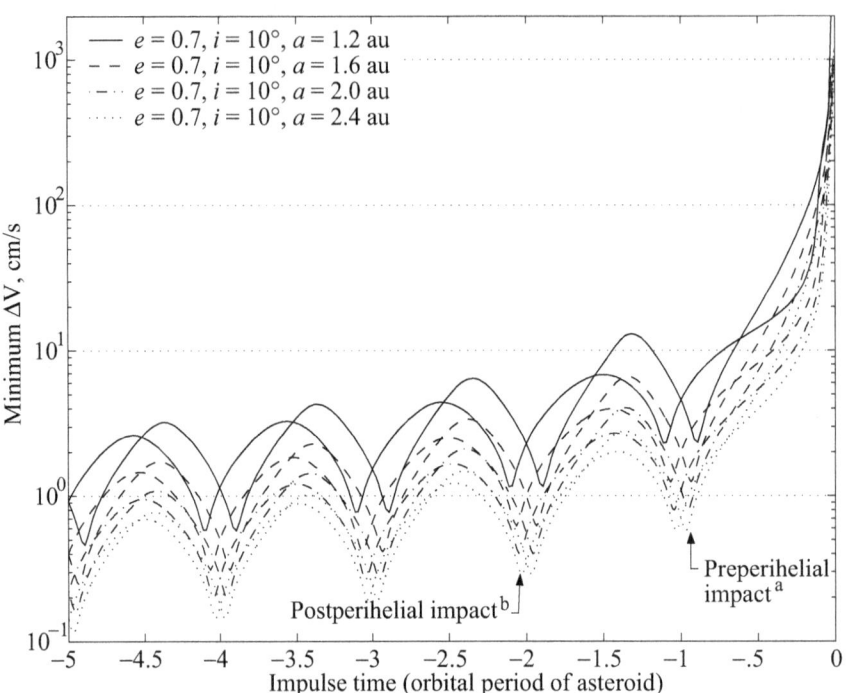

[a] Preperihelial impact: local minimum occurred just before a minus integer impulse time
[b] Postperihelial impact: local minimum occurred just after a minus integer impulse time

Figure 7. Minimum ΔV: Apollo-type asteroids with $e = 0.7$, $i = 10°$, $a = 1.2, 1.6, 2.0, 2.4$ au.

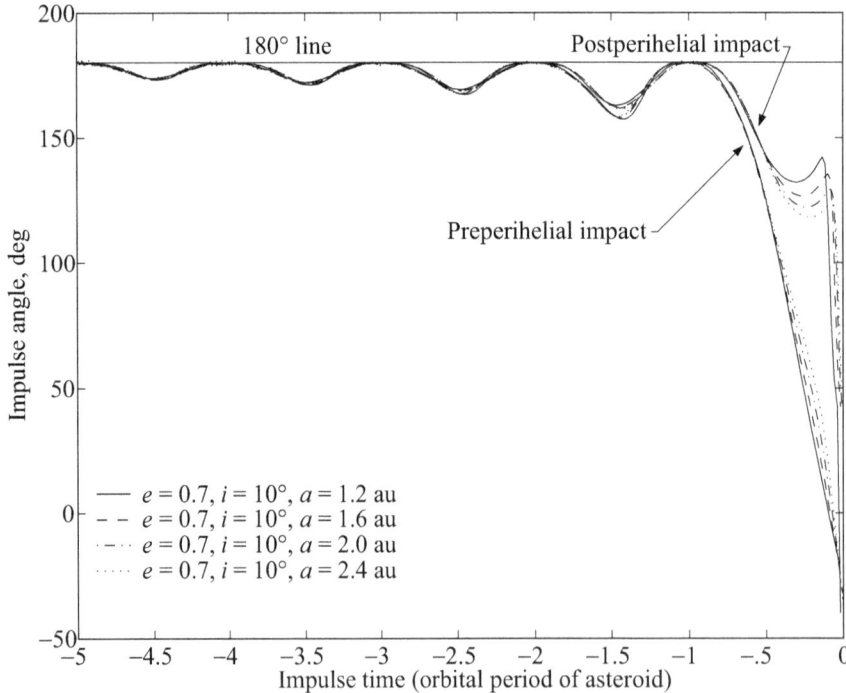

Figure 8. Impulse angle: Apollo-type asteroids with $e = 0.7$, $i = 10°$, $a = 1.2, 1.6, 2.0, 2.4$ au.

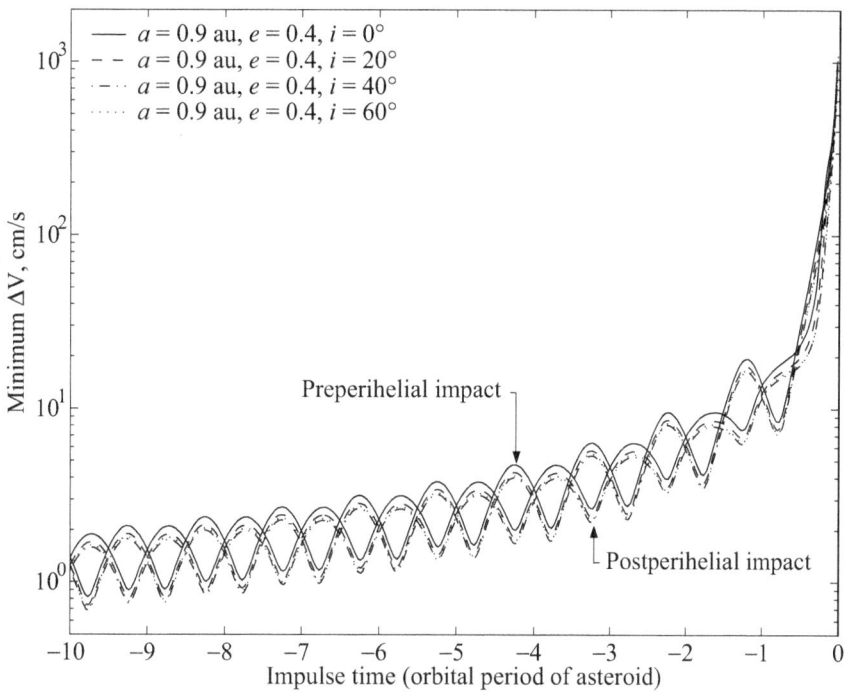

Figure 9. Minimum ΔV: Aten-type asteroids with $a = 0.9$ au, $e = 0.4$, $i = 0°, 20°, 40°, 60°$.

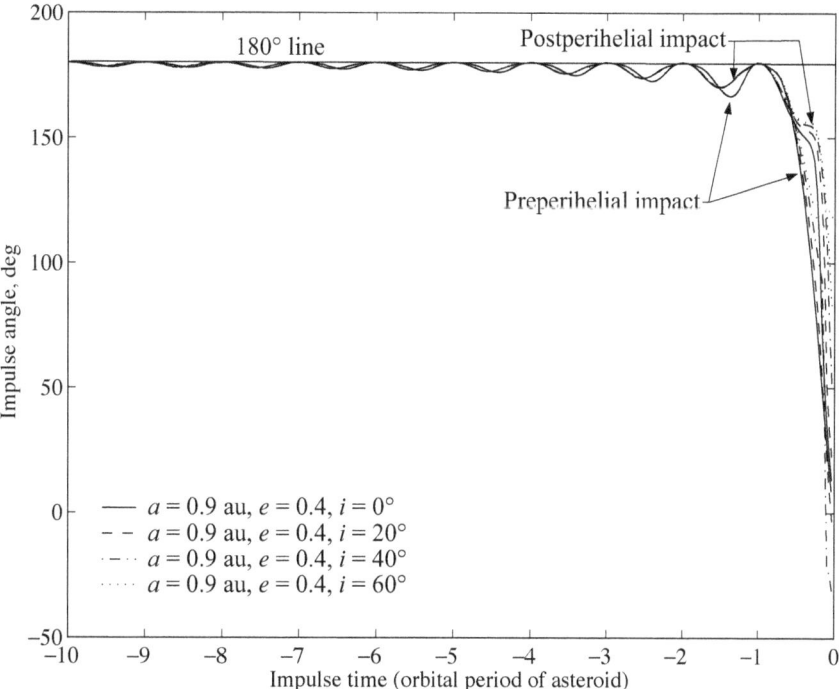

Figure 10. Impulse angle: Aten-type asteroids with $a = 0.9$ au, $e = 0.4$, $i = 0°, 20°, 40°, 60°$.

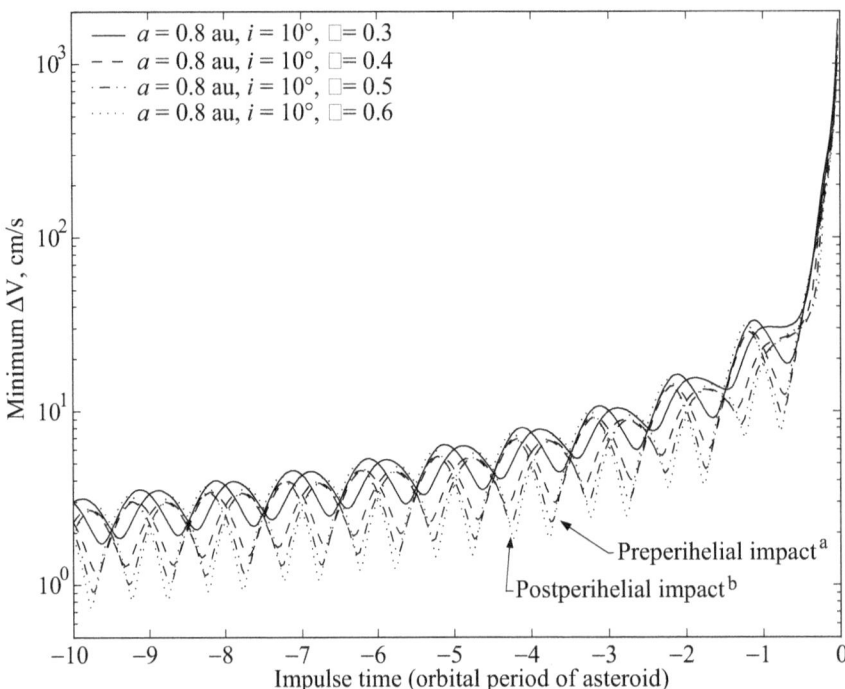

[a] Preperihelial impact: local minimum occurred just before a minus integer impulse time
[b] Postperihelial impact: local minimum occurred just after a minus integer impulse time

Figure 11. Minimum ΔV: Aten-type asteroids with $a = 0.8$ au, $i = 10°$, $e = 0.3, 0.4, 0.5, 0.6$.

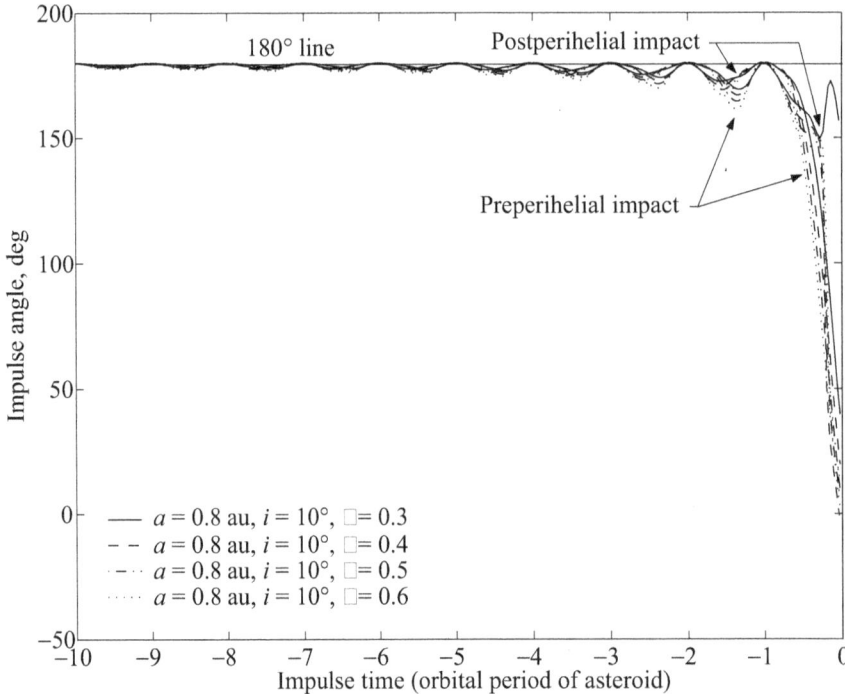

Figure 12. Impulse angle: Aten-type asteroids with $a = 0.8$ au, $i = 10°$, $e = 0.3, 0.4, 0.5, 0.6$.

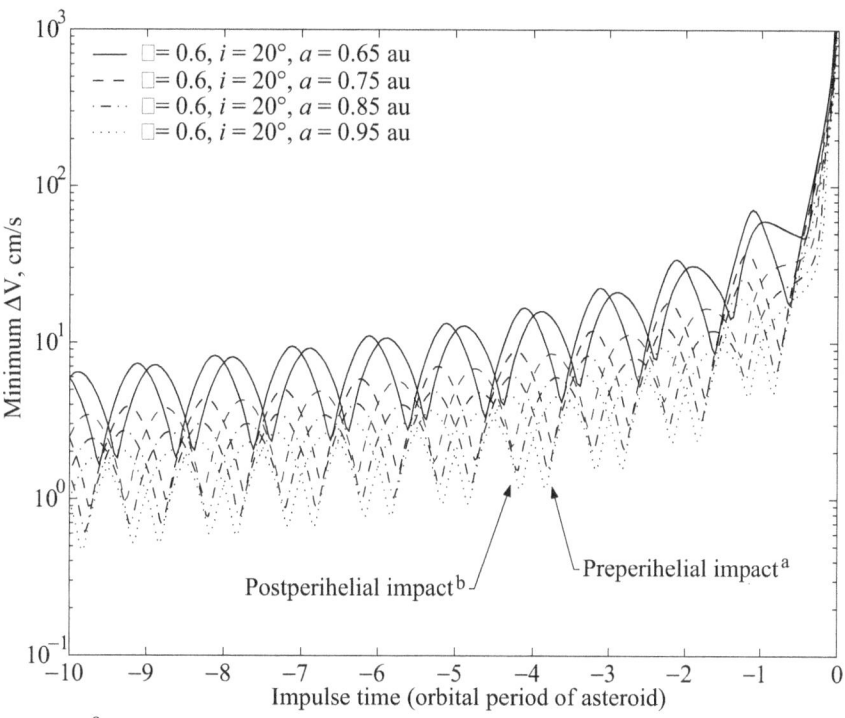

a Preperihelial impact: local minimum occurred just before a minus integer impulse time
b Postperihelial impact: local minimum occurred just after a minus integer impulse time

Figure 13. Minimum ΔV: Aten-type asteroids with $e = 0.6$, $i = 20°$, $a = 0.65, 0.75, 0.85, 0.95$ au.

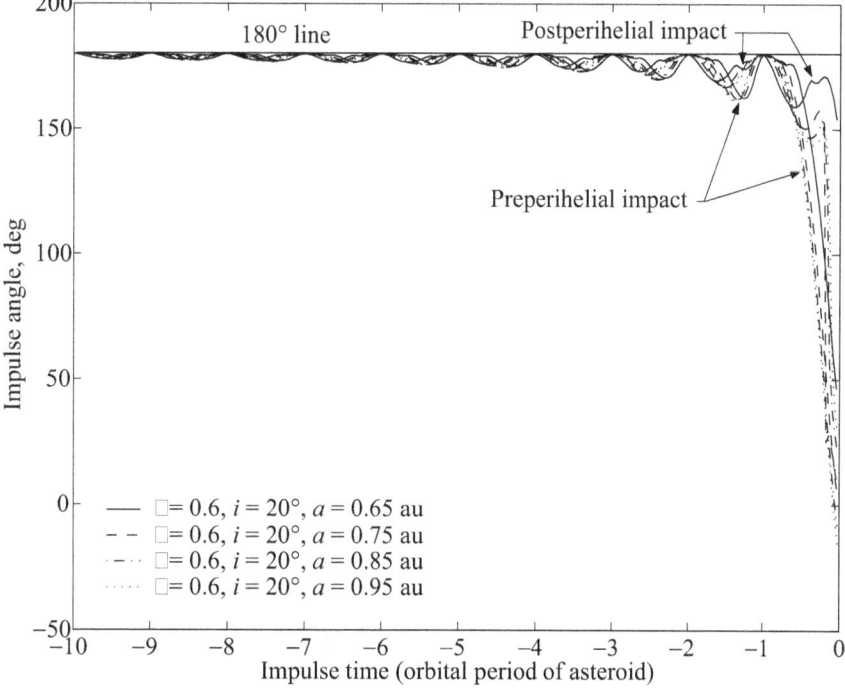

Figure 14. Impulse angle: Aten-type asteroids with $e = 0.6$, $i = 20°$, $a = 0.65, 0.75, 0.85, 0.95$ au.

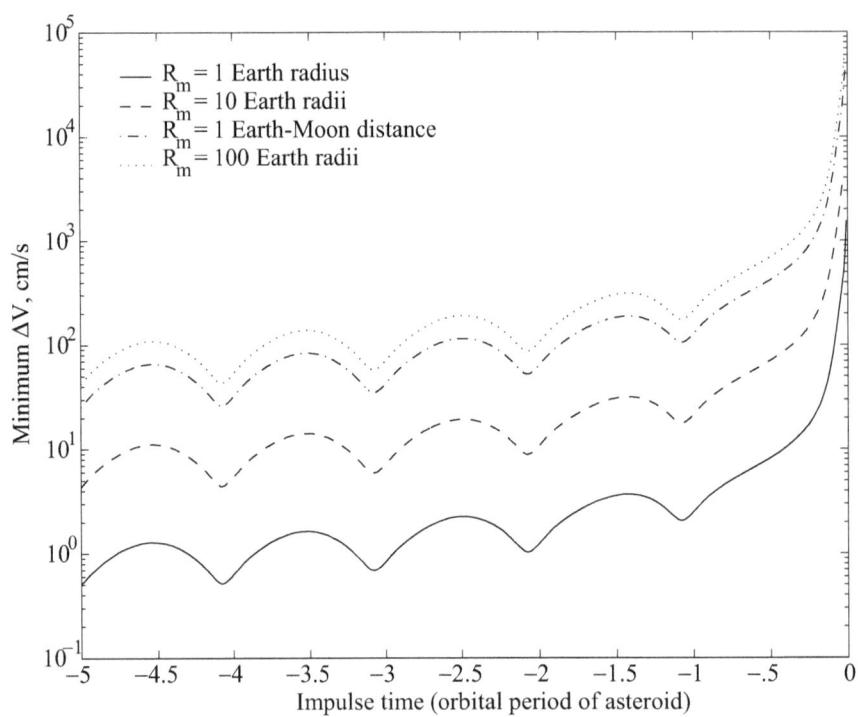

Figure 15. Minimum ΔV: with respect to miss distance (R_m).

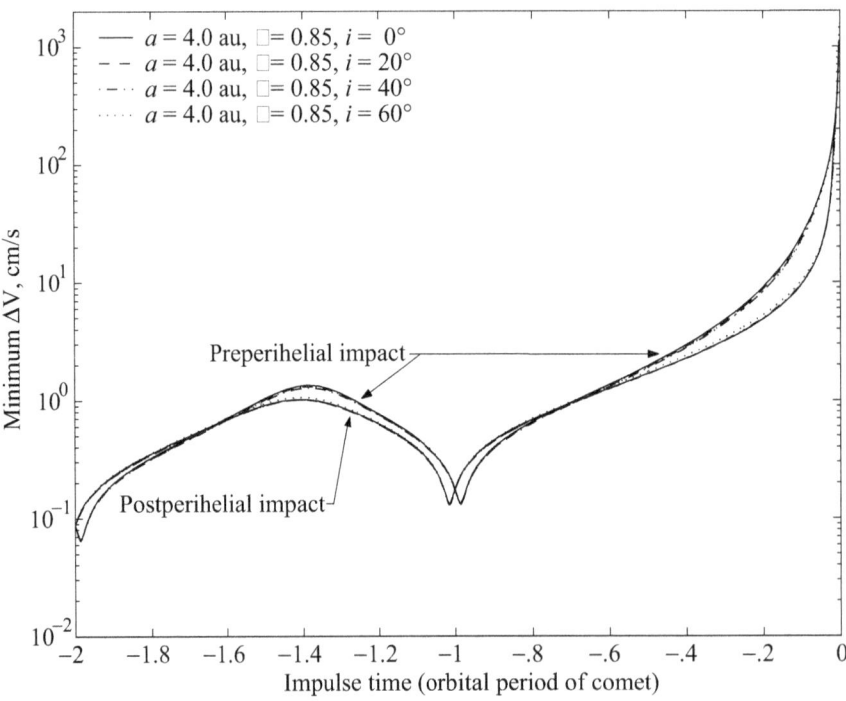

Figure 16. Minimum ΔV: short-period comets with $a = 4.0$ au, $e = 0.85$, $i = 0°, 20°, 40°, 60°$.

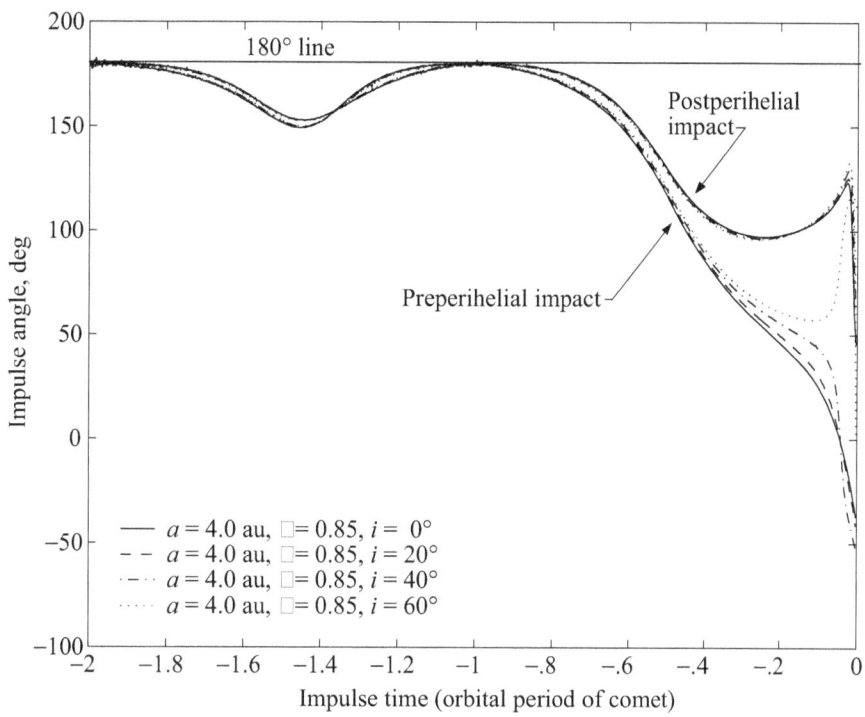

Figure 17. Impulse angle: short-period comets with $a = 4.0$ au, $e = 0.85$, $i = 0°, 20°, 40°, 60°$.

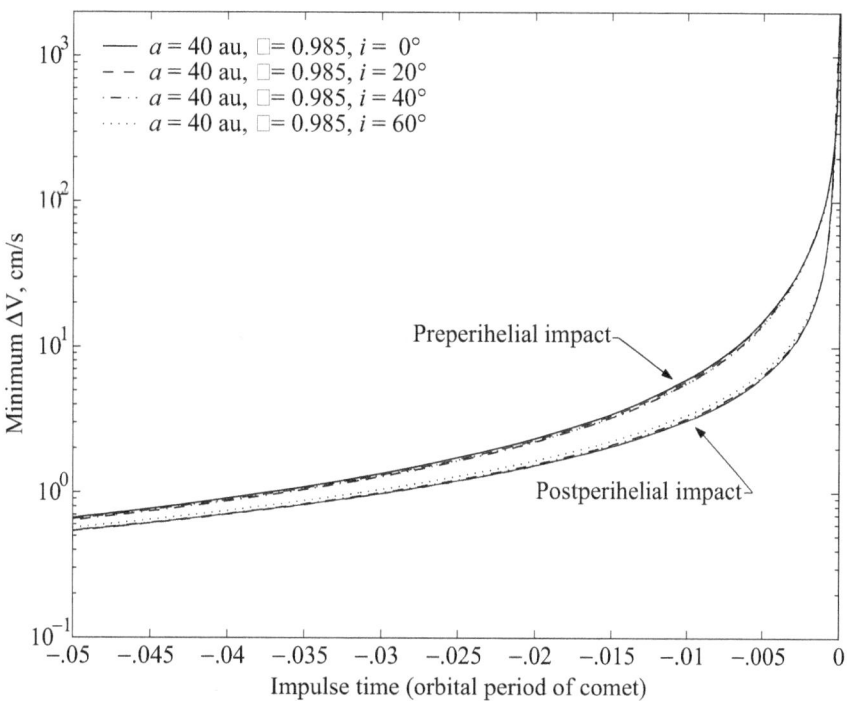

Figure 18. Minimum ΔV: long-period comets with $a = 40$ au, $e = 0.985$, $i = 0°, 20°, 40°, 60°$.

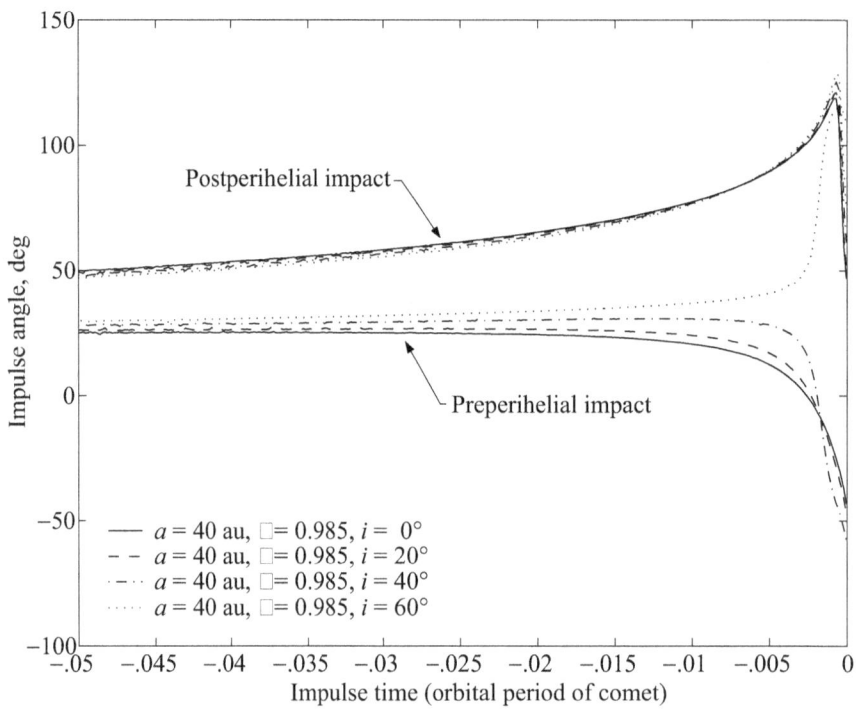

Figure 19. Impulse angle: long-period comets with $a = 40$ au, $e = 0.985$, $i = 0°, 20°, 40°, 60°$.

LPCs. In the figures, the semimajor axis and the eccentricity are fixed at 40 au (orbital period ≈253 years, perihelial distance ≈0.6 au, aphelial distance ≈79.4 au) and 0.985, respectively, and the inclination is varied from 0° to 60°.

For ECOs, the data in figures 20 and 21 show ΔV and impulse angle histories of less than a 10-year impulse time. The ΔVs for the fictitious LPC (0.05 orbital period ≈12.7 years) are monotonically decreasing functions of impulse time, while the ΔVs for the fictitious NEA have the cyclic component varying with respect to orbital period. For a given miss distance and impulse time, the slower ECOs with respect to the Earth usually require a larger ΔV. The order of magnitude of ΔV is not significantly different for either the asteroids or the comets considered here. For an impulse time less than 10 years, the impulse angle of the LPC is mostly monotonically decreasing with respect to impulse time while the angle of the NEA has fluctuations due to its short orbital period. At the final stage, where impulse time approaches impact time, asteroids and comets have no significant differences in optimal ΔV and impulse angles. Additionally, the ΔV required for deflecting Earth-approaching asteroids and comets is dramatically increased, and impulse angle is quickly decreased.

Interceptor Strategies

It is important to estimate the required interceptor mass or energy to deflect or disrupt asteroids/comets on a collision course with Earth. With the estimated mass or energy, we can investigate the appropriate strategies for the deflection missions. The interceptor parameter of M_{ht}, E_{laser}, M_k, M_{nso}, and M_{nss} required for a given ECO mass and ΔV are expressed for high-thrust engine, laser ablation,

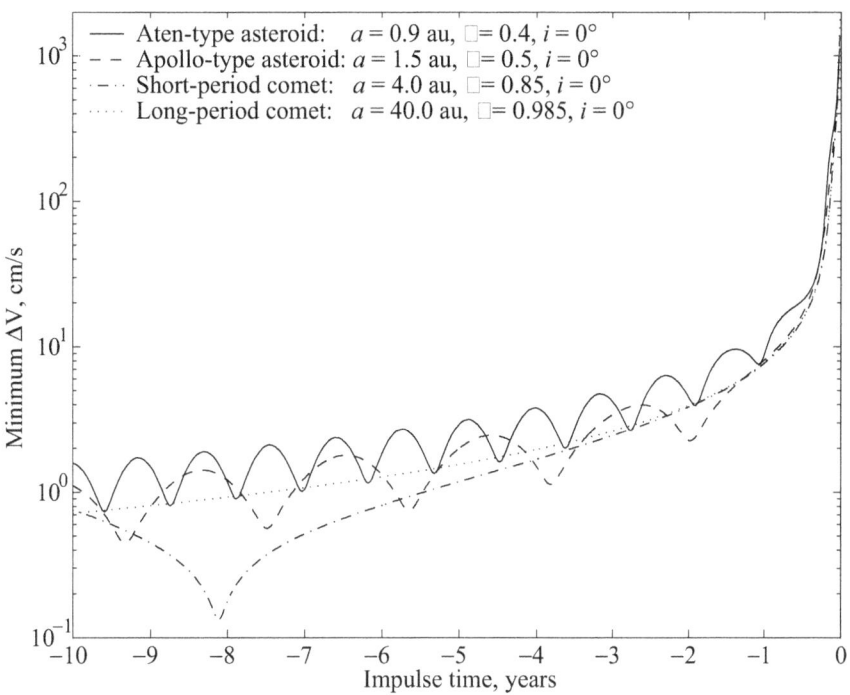

Figure 20. For comparison: minimum ΔV of postperihelial impact for various ECOs.

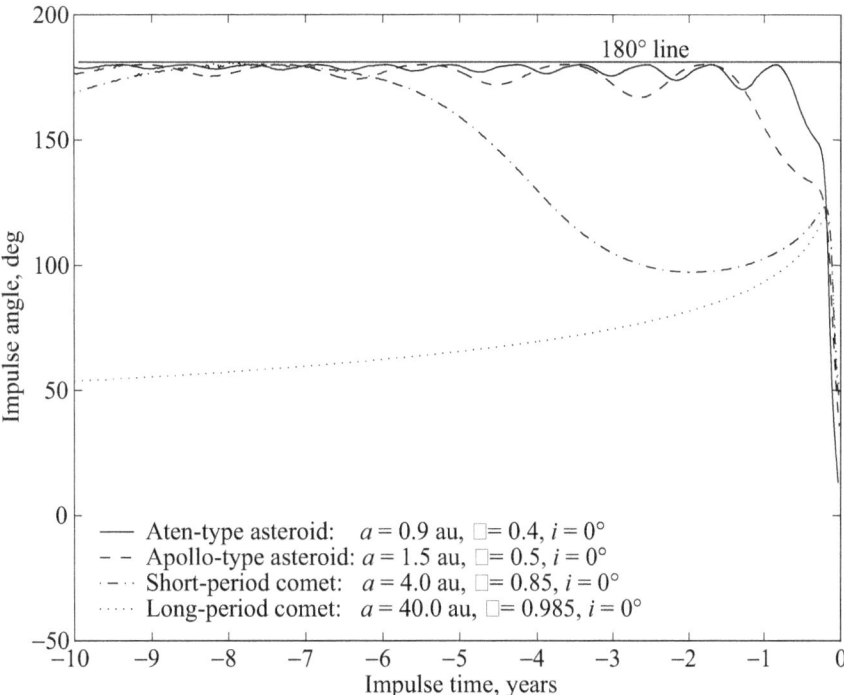

Figure 21. For comparison: impulse angle of postperihelial impact for various ECOs.

kinetic-energy deflection, standoff nuclear detonation, and slightly subsurface nuclear burst (refs. 3, 16, and 17) as follows:

$$M_{ht} \approx \frac{m\Delta V}{c_e} \quad \{kg\} \tag{15}$$

$$E_{laser} \approx \frac{m\Delta V}{c_m} \quad \{J\} \tag{16}$$

$$M_k \approx \frac{m}{\varepsilon}\left(\frac{v_e \Delta V}{v^2}\right) \quad \{kg\} \tag{17}$$

$$M_{nso} \approx \frac{c_p}{\varepsilon\phi} m\Delta V \quad \{kg\} \tag{18}$$

$$M_{nss} \approx \frac{m}{2\varepsilon}\frac{v_e \Delta V}{\phi} \quad \{kg\} \tag{19}$$

The interceptor parameter indicates mass (kg) of fuel for a high-thrust engine, energy (J) for a laser ablation technique, spacecraft total mass (kg) for a kinetic-energy deflection, mass (kg) of nuclear explosive for a standoff nuclear detonation, and a subsurface nuclear detonation. In these equations, other parameter values are defined or assumed as follows (ref. 17). The velocity of the ejecta is assumed as $v_e \approx 100$ m/s for subsurface bursts, m is the mass of the asteroid/comet, $m = \frac{4}{3}\pi\rho\left(\frac{D}{2}\right)^3$, density is assumed as 3×10^3 kg/m^3 for stony asteroids (200 kg/m^3 for comets), D is diameter of the asteroid/comet, about a fraction $\varepsilon \approx 0.5$ of the total energy could go into the kinetic energy of ejected material, nuclear explosions provide a specific energy $\phi \approx 8 \times 10^{12}$ J/kg, a compression wave velocity in the asteroid/comet material is $c_p \approx 2$ km/s, c_e is the exhaust velocity of a high-thrust engine, c_m is the laser momentum coupling coefficient (assumed in ref. 16 as $c_m = 5$ dyn-s/J), v is the orbital velocity of an asteroid/comet, and ΔV is the orbital velocity increment to deflect an asteroid/comet on a collision course with Earth by 1 Earth radius. It is assumed that the velocity of the interceptor is much less than that of celestial objects and can be neglected.

The optimization method mentioned previously yields the ΔV to be used in equations (15) through (19). Note that equations (15) through (19) implicitly have estimation errors to give the final interceptor parameter, and the impulse is assumed to be applied with the optimal impulse angle discussed previously. Two fictitious asteroids (Aten-type asteroid: $a = 0.9$ au, $e = 0.4$, $i = 0°$, orbital period = 0.85 years; Apollo-type asteroid: $a = 1.5$ au, $e = 0.5$, $i = 0°$, orbital period = 1.84 years) and two comets (SPC: $a = 4.0$ au, $e = 0.85$, $i = 0°$, orbital period = 8.0 years, LPC: $a = 40$ au, $e = 0.985$, $i = 0°$, orbital period = 253 years) are used. With ΔV calculated with respect to impulse time, equations (15) through (19) can be applied to estimate the final interceptor parameter to deflect the asteroid/comet by at least 1 Earth radius. Inclined asteroid/comet orbits ($i \neq 0°$) should give approximately the same interceptor parameter as that of a planar orbit because there are no significant differences in ΔV histories for both the inclined and planar orbits.

Figures 22 and 23 include the histories of the final interceptor parameter in order to deflect 1-km diameter stony asteroids (Aten-type or Apollo-type) by 1 Earth radius, while figures 24 and 25 include

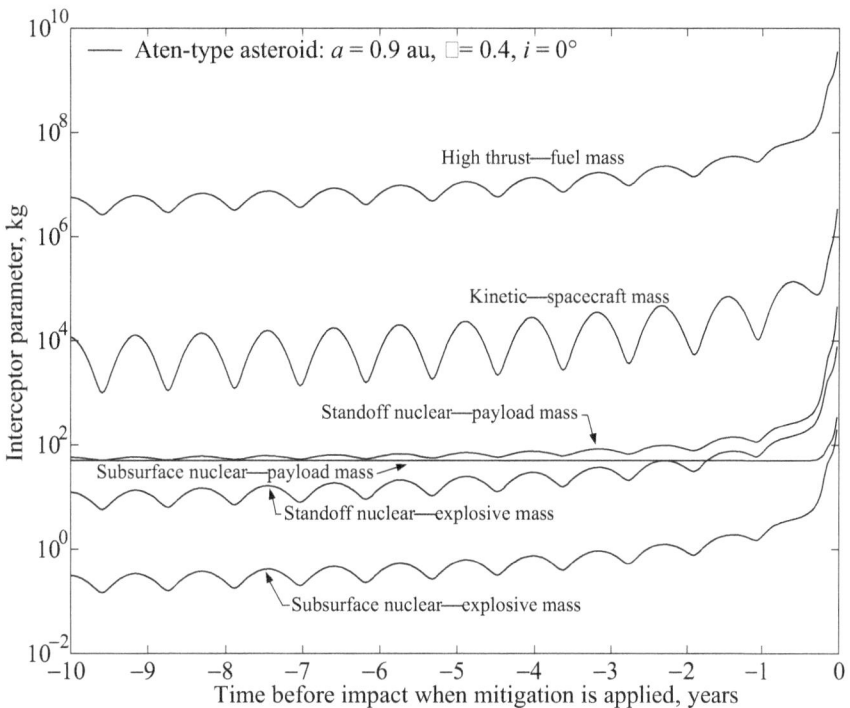

Figure 22. Various interceptor masses for 1-km stony Aten-type asteroid.

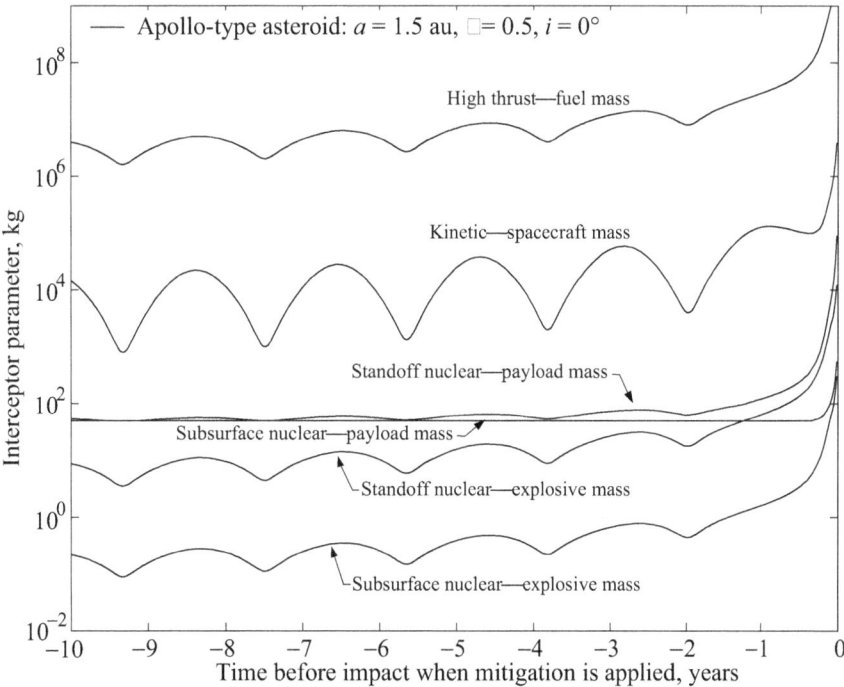

Figure 23. Various interceptor masses for 1-km stony Apollo-type asteroid.

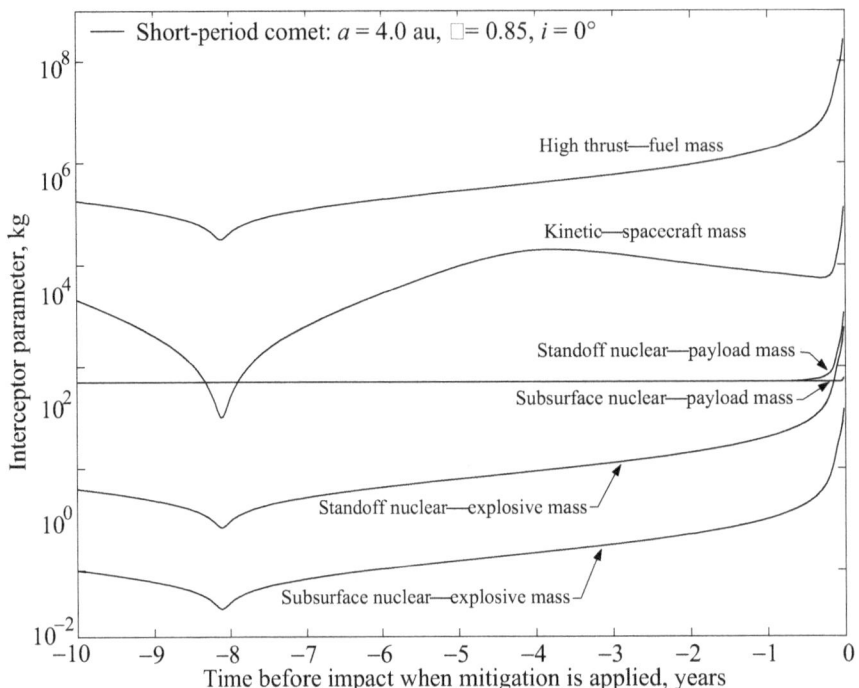

Figure 24. Various interceptor masses for 1-km short-period comet.

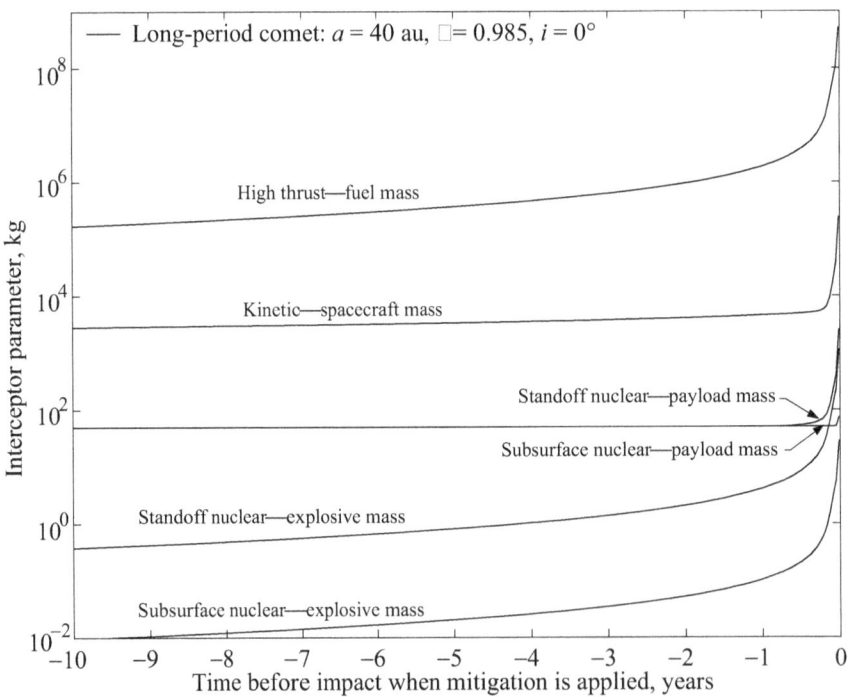

Figure 25. Various interceptor masses for 1-km long-period comet.

similar histories for 1-km diameter comets (an SPC and an LPC). The final intercepter parameter for Apollo-type asteroids shows the same pattern as those for Aten-type asteroids, with less fluctuations with respect to impulse time. For these figures, the payload masses of nuclear equipment are estimated in table 1, which aligns the data to show the relationship between nuclear yield and nuclear device weight (ref. 18). Note that a minimum payload mass of 50 kg is assumed when nuclear yield is less than 10 kt. The final interceptor parameters also vary according to the fluctuations of minimum ΔV required for the space deflection missions. For both the asteroid and comet, a high-thrust method requires much more mass than other methods identified (kinetic-energy deflection, standoff nuclear detonation, and subsurface nuclear burst) because of the fuel expenses for a soft landing and smaller exhaust velocity (ref. 3). Kinetic-energy deflection methods require hundreds of times less mass than high-thrust methods for asteroid deflection, and several orders of magnitude less mass for comet deflection. Nuclear detonation methods reduce the final payload mass by several orders of magnitude compared with the kinetic-energy deflection method. The explosive mass of a slightly subsurface nuclear burst is less than that for standoff nuclear detonation by several factors of 10 for both asteroid and comet deflection. However, the payload mass of both nuclear methods could be 50 kg when impulse time is greater than 1 year before impact. An interesting phenomenon is that, for the SPC case, spacecraft mass for kinetic-energy deflection could be less than nuclear payload mass (fig. 24) when deflection is provided at the ΔV's local minimum. The reason for this phenomenon is that the velocity of the comet is relatively high at perihelion, and final mass of kinetic deflection is inversely proportional to the square of orbital velocity. Generally, the interceptor masses for comet deflection are less than those for asteroid deflection, because densities of comets are assumed to be much less than those of asteroids. Because the mass of an asteroid and comet is linearly proportional to the cube of its diameter, it is easy to estimate the final interceptor masses for different sizes of asteroids and comets. For example, 0.1-km stony asteroids require one thousand times less final mass than 1-km stony asteroids. For asteroid missions, the "optimal time" for application of the ΔV is the earliest possible perihelion for impulse times greater than 1 orbital period, while for LPC missions, the optimal time is simply the earliest possible time. Surprisingly, for an LPC mission, the interceptor mass for the kinetic-energy deflection is approximately the same when the impulse time goes from 1 to 10 years before collision with Earth, although ΔV is increased during the period (see fig. 18). The reason for this phenomenon is that the orbital velocity of the comet is increased accordingly as it travels toward its perihelion, and the interceptor mass of kinetic-energy deflection is inversely proportional to the square of orbital velocity.

Table 1. Approximate Nuclear Yield Versus Nuclear Device Weight[a]

Nuclear yield, kt	Payload weight, kg
10	50
100	100
1 000	1 000
100 00	20 000

[a]From reference 18.

Use of a Hybrid Propellant Module

A hybrid propellant module (HPM) is a future reusable in-space transportation concept being studied under NASA's Revolutionary Aerospace Systems Concepts (RASC) Program. The HPM can store indefinitely both chemical and electrical propellants and provide propulsion with attached modular orbital transfer/engine stages. The HPM can utilize a chemical propellant (liquid hydrogen/liquid oxygen) for fast orbital transfers and can use an electrical propellant (low thrust) for prepositioning or to return the HPM for reuse and refueling. A chemical transfer module (CTM) serves as a high energy injection stage when attached to the HPM, while a solar electric propulsion (SEP) module serves as a low-thrust transfer stage when attached to the HPM. Although propulsive ΔV is much less efficient for deflecting asteroids and comets, the cost of utilizing an already available infrastructure might be favorable compared with developing a separate defensive capability. A preliminary investigation of how an HPM-based transportation system could be utilized is provided. It is assumed that the dry mass of an HPM is 4000 kg, the dry mass of a CTM is 4400 kg, the chemical propellant mass of a CTM is 30000 kg, and the effective exhaust velocity, c_e, is 4.4 km/s for the CTM. An HPM/CTM/SEP vehicle stack can travel to an ECO by using the low-thrust, high-specific impulse provided by the SEP and the high thrust provided by the CTM and HPM chemical propellant. The idea of having abundant propellant sources in deep space may offset the disadvantages of a chemical propellant when combined with electric propulsion.

After rendezvous with the ECO, the combined vehicle could use its CTM as a high-thrust engine with a full 30000 kg load of chemical propellant to change the orbit of the asteroid/comet. According to figures 22 through 25, by using a CTM as a high-thrust engine, it can be estimated that ≈0.13-km asteroid (≈0.3-km comet) defense mission could be performed 2 years before collision. The same CTM could be used for ≈0.06-km stony asteroid (≈0.15-km comet) deflection just 6 months before collision. An asteroid of about a 1-km (0.2-km) diameter can be deflected by direct spacecraft (HPM with CTM, total dry mass is about 8400 kg) impact on the target for longer than a 2-year (6-month) impulse time if the impulse is applied to an asteroid at its perihelion. An approximate 1-km comet can be deflected by kinetic energy using HPM with CTM for longer than 6 months of impulse time. These cases assume that all the fuel of a combined HPM with CTM can be consumed to travel longer or faster to dangerous objects. With an impulse time of longer than 2 years, HPM (dry mass of 4000 kg) itself could perturb orbits of asteroids/comets of 0.3 km in size using direct kinetic impact on the objects. By assuming that the HPM/CTM stack could deliver a 20000-kg nuclear warhead, this in-space transportation system could be used for a 1-km asteroid or a 2-km comet deflection mission for any short impulse time.

An HPM-based in-space transportation system could be available in 10 to 20 years. For comparative purposes, it is useful to determine what deflection strategies are currently possible. Payload mass is directly related to propellant mass, so the final mass of an interceptor can be converted into a rough estimate of interceptor cost. An interceptor payload would require 3 times more mass into deep space and 10 times more mass into low-Earth orbit (ref. 19). The Russian Energia launch vehicle was the most powerful launch vehicle and had a payload capacity of 80000 to 90000 kg for low-Earth orbits (ref. 12). With this launch capability, about 9000 kg of maximum final interceptor mass can be reasonably assumed. The data in figures 22 through 25 indicate that a 0.1-km asteroid or 0.3-km comet defense mission could be performed for any short impulse time by using kinetic-energy technology. A 1-km asteroid would require at least 2 years of impulse time if the ΔV is applied at the asteroid's perihelion, while a 1-km comet would require at least 1 year. These results coincide with previous studies stating that kinetic-energy impacts are suitable to deflect rocky asteroids of up to about 0.1 km in size (ref. 5). It was also shown that today's technology is sufficient to defend asteroids of less than 0.1 km in diameter for warning times on the order of several years (ref. 12). A 1-km asteroid or a 2-km comet defense mission could be prepared for a few months impulse time by using nuclear detonations. Nuclear explosives offer energies great enough for much larger asteroids/comets and shorter impulse times. The studies in this paper may be slightly different from previous studies due to different approximations and assumptions. Generally, for any warning time, we can roughly say that the kinetic-energy deflection would be preferable for smaller ECOs (less than a few hundred meters in size); the nuclear strategy would be required for large objects (greater than 1 km in size) or for very short impulse times.

Use of a Laser Ablation Technique

Although nuclear detonations and kinetic-energy impacts appear to be the most practical near-term impact protection options, both methods have uncertainties in the momentum change imparted and can potentially fragment the asteroid/comet. This fragmentation could create multiple impacts, which ultimately could be more hazardous than the original object. Additionally, it is envisioned that comets and asteroids could provide a virtually limitless supply of resources for the future exploration and colonization of space. An orbit modification technique that could also facilitate the utilization of nonthreatening comets and asteroids would likely function in a controlled manner. One approach to alter the trajectory of the object in a highly controlled manner is to use pulsed laser ablative propulsion. This technique has been studied to remove space debris from low-Earth orbit (ref. 20). A sufficiently intense laser pulse ablates the surface of the space debris by causing plasma blowoff. The momentum change from a single laser pulse is very small. However, the cumulative effect is very effective because the laser can yield 10 to 1000 pulses/s over several minutes (ref. 21). The dynamic reaction from multiple laser hits reduces the perigee altitude of the orbiting debris. Once perigee is reduced sufficiently ($\approx$200 km), atmospheric drag deorbits the debris. The highest pulse energy currently available is about 20 kJ/pulse operated at under 0.02 Hz (ref. 20). In the future, a much more advanced version of this same technique may be used for the deflection/orbit modification of asteroids/comets on a collision trajectory with Earth. The laser ablation technique could overcome the mass penalties associated with other nondisruptive approaches because no propellant is required to generate the ΔV (the material of the celestial object is the propellant source). Additionally, laser ablation is effective against a wide range of surface materials and does not require any landing or physical attachment to the object.

For deflection of distant asteroids and comets, the power and optics requirements of a laser ablation system on the ground or near Earth may be too extreme to contemplate in the next decades. One hybrid solution would be to permit a spacecraft to carry a laser as a payload to a particular celestial body. The spacecraft would require an advanced propulsion system capable of rapid rendezvous with the asteroid/comet and an extremely powerful energy generation system. The spacecraft would orbit or station-keep with the object at a "small" standoff distance. In this section, it is assumed that a spacecraft with a laser ablation tool has already rendezvoused with the asteroid/comet. Figure 26 includes the required laser energy for various ECOs to deflect them by 1 Earth radius; it is demonstrated that the required laser energy also varies according to the required ΔV. The size of the celestial bodies in figure 26 is assumed to be 1 km. Again, it is easy to estimate the final interceptor energy for different sizes of ECO because the energy of an ECO is linearly proportional to the cubic of its diameter. Figure 27 houses an explanatory detail of figure 26 for an Apollo-type asteroid and an LPC, using fixed impulse times, and describes the required energy for laser ablation to deflect the given size of an asteroid and comet by 1 Earth radius. Figure 28 includes the achieved energy for a given laser power and operation period. From figures 27 and 28, we can easily estimate the power of laser for a selected case. For instance, if we can use a 1-year impulse time, about 2.5×10^3 GJ of energy (fig. 27) is required for a 0.1-km stony asteroid to be deflected by 1 Earth radius. The data in figure 28 can be used to determine what laser power is required and how long the laser should be continuously operated in order to achieve 2.5×10^3 GJ of energy. If we choose a laser system with 1 MW of power (100 kJ/pulse and 10 Hz laser repetition frequency), it would take about 30 days of operation to provide 2.5×10^3 GJ of energy. If we choose a laser system with 10 MW of power, it would take about 3 days of operation to provide 2.5×10^3 GJ of energy. If a laser system with average power of 1 MW could be continuously used for asteroid deflection 2 years before collision with Earth, it would take about 8 days to provide a 0.1-km stony Apollo-type asteroid with enough energy to avoid the collision. If the same equipment is available 6 months before the collision, it would take 60 days for the same asteroid. For deflection of a 1-km LPC, a 25-MW laser system needs to be operated

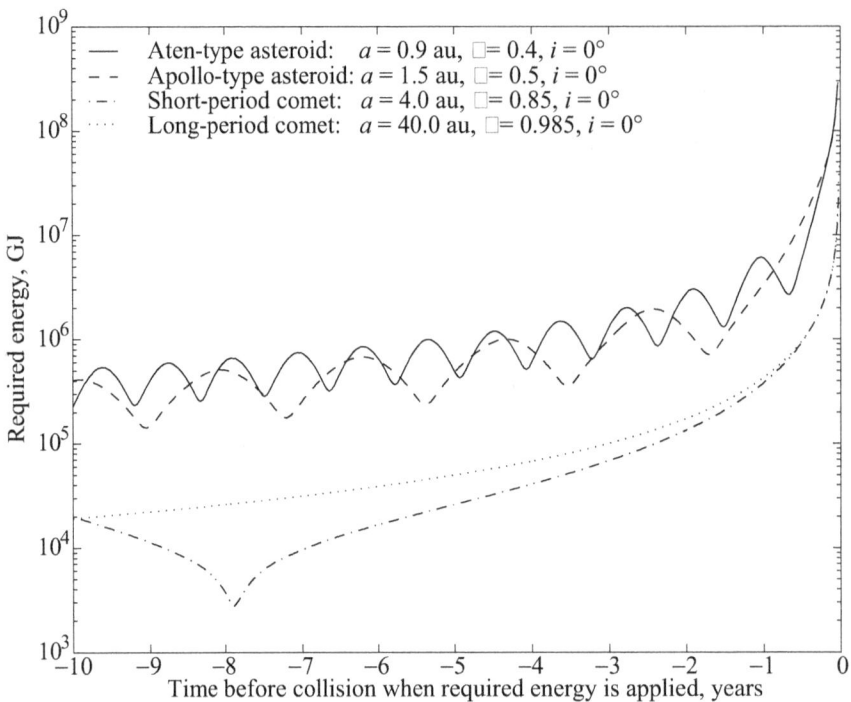

Figure 26. Required laser energy of postperihelial impact for 1-km ECOs.

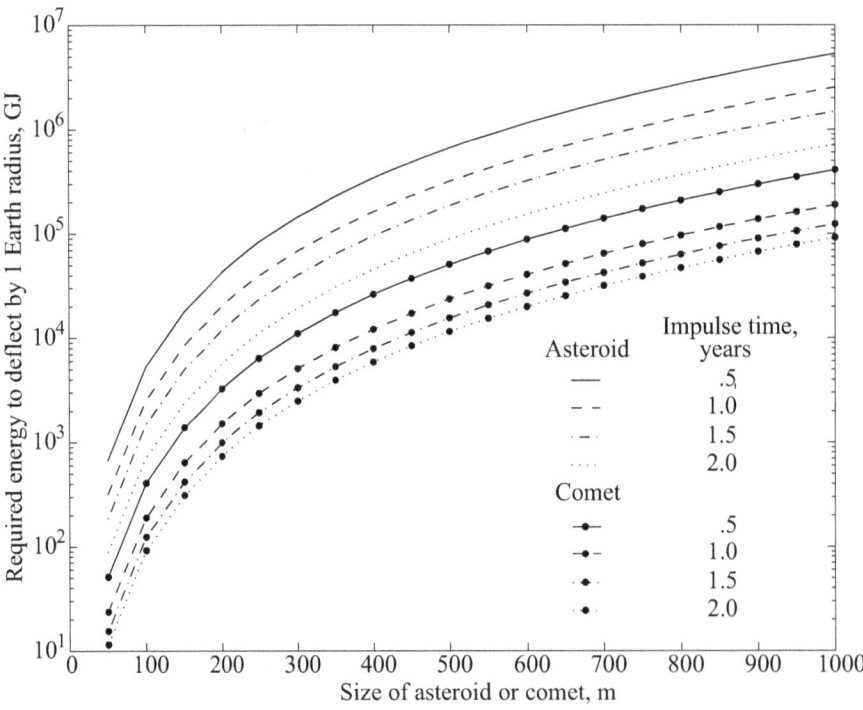

Figure 27. Required laser ablation energy versus size for Apollo-type asteroid and long-period comet.

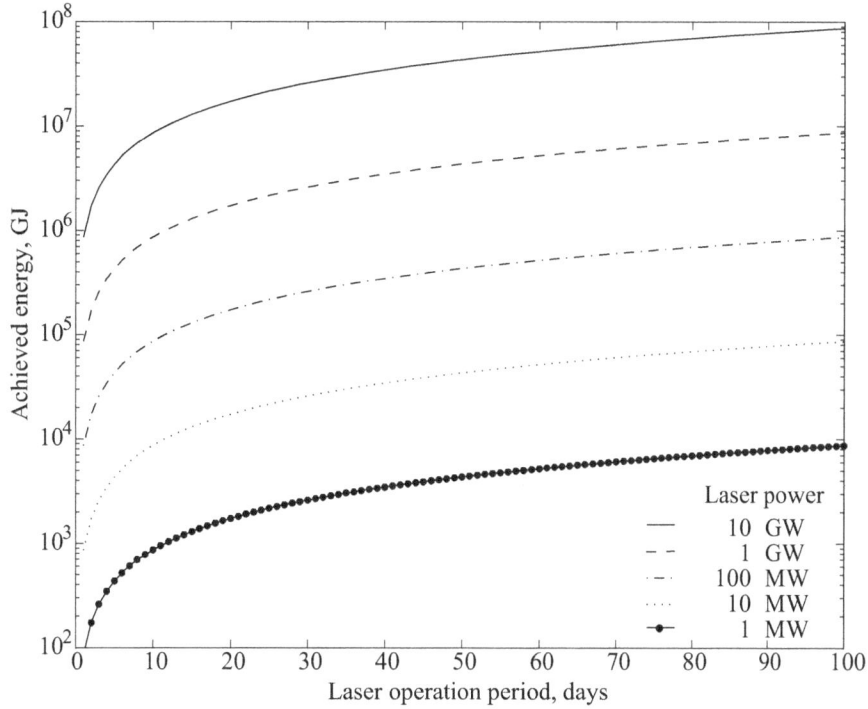

Figure 28. Achieved energy for given laser power and operation period.

continuously for about 3 months beginning 1 year before collision. Although these power levels may not be readily achievable today, they are consistent with the power levels needed for high-thrust, high-specific impulse propulsion systems envisioned for rapid in-space transportation in the future. This power needed for distant rendezvous with an asteroid/comet would then be available for the orbit modification phase of the defense mission.

Overview of Results

The minimum ΔV vector has finer structures associated with the impulse time to deflect an ECO. Both asteroids and comets have similar fluctuation structures of ΔV and impulse angle histories with respect to their orbital periods. For a given miss distance and impulse time, the slower Earth-crossing objects relative to the Earth usually require a larger ΔV. When the gravitational effects of Earth are considered, the minimum ΔV is linearly proportional not to the miss distance but the impact parameter, b_i. The effects of orbital inclination on ΔV and impulse angles are very small for the deflection problem. The ΔV and impulse angles are more sensitive to eccentricity and orbital period rather than the inclination. The velocity increment, ΔV_W, normal to orbital plane of Earth-crossing objects is negligible for most deflection missions. The case of preperihelial impact has slightly larger local minima in the magnitude of impulse than the case of postperihelial impact, and preperihelial impact cases have more fluctuations in optimal impulse angle than postperihelial impact cases. At the final stage, where impulse time approaches impact time, the optimal impulse has been changed dramatically and the ΔV is logarithmically increased. Earth-crossing asteroids and comets may have no significant differences in optimal ΔV and impulse angles when impulse time is within a few months.

The final interceptor mass (or energy) also changes according to the fluctuations in the minimum ΔV required for the space deflection missions. The optimal deflection strategy is dependent on the size and composition of an asteroid/comet as well as the amount of warning time. The potential of several deflection approaches, including high-thrust engine, kinetic deflection, nuclear detonation, and laser ablation, is discussed. A high-thrust method requires much more mass than other methods. Kinetic-energy deflection needs about a few tens or a few hundreds times less mass than a high-thrust method. Nuclear detonations reduce final payload mass by several orders of magnitude compared with the kinetic-energy deflection method. For a given diameter target, the interceptor mass for comet deflection is usually less than the mass needed for asteroid deflection. Although not as mass or energy efficient as a nuclear detonation or kinetic-energy impact, a laser ablation system has significant advantages and the potential to provide controlled orbit modification of Earth-crossing celestial bodies.

Concluding Remarks

Deflection or orbit modification of potentially Earth-impacting asteroids/comets has been considered. Several techniques were assessed.

These analyses demonstrate that the optimal deflection strategy is highly dependent on the size and the orbital elements of the asteroid/comet, as well as the amount of warning time. It is concluded that the best deflection strategy and optimal impulse change in velocity (ΔV) should be carefully investigated if an impacting object is discovered, provided that sufficient time is available. Depending on when the ΔV can be applied to an Earth-crossing object (ECO), the optimal impulse angle is not necessarily along the velocity or antivelocity vector, particularly when the impulse time is less than 1 orbit before impact. Although not as mass or energy efficient as a nuclear detonation or kinetic-energy impact, a laser ablation system has significant advantages including the potential to provide controlled orbit modification of ECOs. Regardless of the method employed, early application of the impulse dominates the magnitude of the deflection effort. Impacting objects confirmed years or decades in advance are much easier to deflect and allow a detailed assessment and development of a planetary defense strategy. However, the use of operational space-based elements in the event of a confirmed threat could provide a reliable, rapid response to ECOs, even with relatively short warning times.

References

1. Ahrens, T. J.; and Harris, A. W.: Deflection and Fragmentation of Near-Earth Asteroids. In *Hazards Due to Comets & Asteroids*, T. Gehrels, ed., The University of Arizona Press, 1994, pp. 897–928.

2. Solem, J. C.: Interception of Comets and Asteroids on Collision Course With Earth. *J. Spacecr. & Rockets*, vol. 30, no. 2, 1993, pp. 222–228.

3. Ivashkin, V. V.; and Smirnov, V. V.: An Analysis of Some Methods of Asteroid Hazard Mitigation for the Earth. *J. Planet. Space Sci.*, vol. 43, no. 6, 1995, pp. 821–825.

4. Simonenko, V. A.; Nogin, V. N.; Petrov, D. V.; Shubin, O. N.; and Solem, J. C.: Defending the Earth Against Impacts From Large Comets and Asteroids. In *Hazards Due to Comets & Asteroids*, T. Gehrels, ed., The University of Arizona Press, 1994, pp. 929–953.

5. Tedeschi, W. J.: Mitigation of the NEO Impact Hazard Using Kinetic Energy. *Proceedings of the Planetary Defense Workshop*, Lawrence Livermore National Laboratory, Livermore, CA, May 1995, URL: <http://www.llnl.gov/tid/lof/documents/toc/232015.html> [cited 5 Oct. 2001].

6. Park, S.-Y.; and Ross, I. M.: Two-Body Optimization for Deflecting Earth-Crossing Asteroids. *J. Guid., Control, & Dyn.*, vol. 22, no. 3, 1999, pp. 415–420.

7. Ross, I. M.; Park, S.-Y.; and Porter, S. D.: Gravitational Effects of Earth in Optimizing Delta-V for Deflecting Earth-Crossing Asteroids. *J. Spacecr. & Rockets*, vol. 38, no. 5, Sept.–Oct. 2001, pp. 759–764.

8. Conway, B. A.: Near-Optimal Deflection of Earth-Approaching Asteroids. *J. Guid., Control, & Dyn.*, vol. 24, no. 5, 2001, pp.1035–1037.

9. Guelman, M.; and Harel, D.: Power Limited Soft Landing on an Asteroid. *J. Guid., Control, & Dyn.*, vol. 17, no. 1, 1994, pp. 15–20.

10. Remo, J. L.; and Sforza, P. M.: Subsurface Momentum Coupling Analysis for Near-Earth-Object Orbital Management. *Acta Astronautica*, vol. 35, no. 1, 1995, pp. 27–33.

11. Melosh, H. J.; Nemchinov, I. V.; and Zetzer, Yu. I.: Non-Nuclear Strategies for Deflecting Comets and Asteroids. In *Hazards Due to Comets & Asteroids*, T. Gehrels, ed., The University of Arizona Press, 1994, pp. 1111–1132.

12. Meissinger, H. F.: Technology Assessment for Defense Against Asteroids or Comet. *Proceedings of the Planetary Defense Workshop*, Lawrence Livermore National Laboratory, Livermore, CA, May 1995.

13. Hale, F. J.: *Introduction to Space Flight*. Prentice Hall, Englewood Cliffs, NJ, 1994, p. 89.

14. Battin, R. H.: *An Introduction to the Mathematics and Methods of Astrodynamics*. American Institute of Aerodynamics and Astronautics, Inc., New York, 1987, pp. 153–178.

15. Grace, A.: *Optimization TOOLBOX User's Guide*, The MathWorks, Inc., 1992, pp. 1.13–3.12.

16. Phipps, C. R.: Lasers Can Play an Important Role in the Planetary Defense. *Proceedings of the Planetary Defense Workshop*, Lawrence Livermore National Laboratory, Livermore, CA, May 1995.

17. Canavan, G. H.; and Solem, J. C.: Near-Earth Object Interception Workshop. In *Hazards Due to Comets & Asteroids*, T. Gehrels, ed., The University of Arizona Press, 1994, pp. 93–124.

18. Canavan, G. H.; Solem, J. C.; and Rather, J. D. G., eds.: *Proceedings of the Near-Earth-Object Interception Workshop*, Los Alamos National Laboratory, Feb. 1993, pp. 202–205.

19. Canavan, G. H.: Cost and Benefits of Near-Earth Object Defenses. *Proceedings of the Planetary Defense Workshop*, Lawrence Livermore National Laboratory, Livermore, CA, May 1995.

20. Campbell, J. W.: *Project ORION: Orbital Debris Removal Using Ground-Based Sensors and Lasers*. NASA TM 108522, MSFC, AL, 1996.

21. Campbell, J. W.; and Mazanek, D. D.: Laser Solutions for Reducing the Environment Risks Associated With Orbital Debris and Near Earth Objects. *Proceedings of 52nd International Astronautical Congress*, IAF-01-C.2.01, 1–5 Oct. 2001, Toulouse, France.

Orbit Modification of Earth-Crossing Asteroids/Comets Using Rendezvous Spacecraft and Laser Ablation[7]

SANG-YOUNG PARK
Swales Aerospace

DANIEL D. MAZANEK
NASA Langley Research Center

Introduction

Fast space trips are important to intercept and rendezvous with an impacting asteroid or comet, particularly those not detected many years in advance. Fast trajectories can shorten space flight times and allow orbit modification efforts to begin earlier. The earlier the effort begins, the less change in velocity (ΔV) required to alter the object's trajectory. However, shorter trip times require more propellant to provide enough thrust if the travel distance is fixed. This additional propellant mass can be a burden to the structural architecture of spacecraft. Thus, it is necessary to trade off between flight time and propellant mass.

Background

This report describes the approach and results of an end-to-end simulation to deflect a long-period comet (LPC) by using a rapid rendezvous spacecraft and laser ablation system. The laser energy required for providing sufficient deflection ΔV and an analysis of possible intercept/rendezvous spacecraft trajectories are studied in this analysis. These problems minimize a weighted sum of the flight time and required propellant by using an advanced propulsion system. The optimal thrust-vector history and propellant mass to use are found in order to transfer a spacecraft from the Earth to a targeted celestial object. One goal of this analysis is to formulate an optimization problem for intercept/rendezvous spacecraft trajectories.

One approach to alter the trajectory of the object in a highly controlled manner is to use pulsed laser ablative propulsion (ref. 1). A sufficiently intense laser pulse ablates the surface of a near-Earth object (NEO) by causing plasma blowoff. The momentum change from a single laser pulse is very small. However, the cumulative effect is very effective because the laser can interact with the object over long periods of time. The laser ablation technique can overcome the mass penalties associated with other nondisruptive approaches because no propellant is required to generate the ΔV (the material of the celestial object is the propellant source). Additionally, laser ablation is effective against a wide range of surface materials and does not require any landing or physical attachment to the object. For diverting distant asteroids and comets, the power and optical requirements of a laser ablation system on or near the Earth may be too extreme to contemplate in the next few decades. A hybrid solution would be for a spacecraft to carry a laser as a payload to a particular celestial body. The spacecraft would require an advanced propulsion system capable of rapid rendezvous with the object and an extremely powerful electrical generator, which is likely needed for the propulsion system as well. The spacecraft would station-keep with the object at a "small" standoff distance while the laser ablation is performed.

---

[7]Chapter nomenclature available in chapter notes, p. 217.

Trajectory Optimization Algorithm

For intercept and rendezvous trajectories, optimization problems (ref. 2) in three dimensions are formulated to minimize flight time with moderate propellant mass. Many problems in the design of modern guidance and control systems require optimization of the trajectory, which minimizes (or maximizes) some performance criterion. Using the theory of the calculus of variations, the formulation of these problems yields a two-point boundary-value problem (TPBVP). The resulting optimal trajectory also satisfies the physical constraints and the given differential equations. The open-loop optimal trajectory can be used as a reference trajectory for intercept or rendezvous with Earth-crossing asteroids or comets. To simplify the presentation and focus more on the inequality constraint, we first present the necessary conditions for an optimal control problem without the inequality constraint and then discuss the inequality constraint separately. A general optimal control problem can be stated as follows:

Given the performance index, J, and radial velocity, u,

$$J(u) = \phi[x(t_f), t_f] + \int_{t_0}^{t_f} L(x, u, t) \, dt \tag{1}$$

subject to the dynamic equations, $\dot{x}$, and boundary conditions

$$\dot{x} = f(x, u, t), \, x(t_0) \equiv x_0, \, t_0 \text{ given} \tag{2}$$

and with free final time t_f, find the control history $u(t)$ to minimize J(u) with the prescribed terminal constraints

$$\Phi[x(t_f), t_f] = 0 \tag{3}$$

Here $x(t) \in R^n$ are the state variables, $u(t) \in R^l$ are the control components, and $\Phi \in R^k$. The Hamiltonian function is defined with Lagrange multipliers $\Lambda(t) \in R^n$ as

$$H \equiv L + \Lambda^T f \tag{4}$$

The performance index in equation (1) is augmented and rewritten as

$$J' = \phi[x(t_f), t_f] + \nu^T \Phi[x(t_f), t_f] + \int_{t_0}^{t_f} (H - \Lambda^T \dot{x}) \, dt \tag{5}$$

where ν is a constant multiplier vector of the dimension of the constraint Φ. The Minimum Principle requires that the optimal controls minimize the Hamiltonian function, H:

$$u^*(t) = \min_{u \in \Omega} \arg H(x^*, \Lambda^*, u, t) \tag{6}$$

where Ω is the set of admissible controls and x^*, Λ^*, and u^* are the extremal states, costates, and controls, respectively. The initial time and the initial states are known. The conditions to be satisfied to minimize J' are found by taking the first variation of J' and setting relations equal to zero. From this, the states, costates, and the Hamiltonian function satisfy the following conditions:

$$\dot{x}^T = H_\Lambda \tag{7a}$$

$$\dot{\Lambda}^T = -H_x \tag{7b}$$

$$\Lambda^T(t_f) = \left(\phi_x + \nu^T \Phi_x\right)_{t=t_f} \tag{7c}$$

$$H(t_f) = -\left(\phi_t + \nu^T \Phi_t\right)_{t=t_f} \tag{7d}$$

$$H_u = 0 \tag{7e}$$

Equation (7e) can be solved for the control, so that the control is removed from equations (7a) and (7b). There are n + k + 1 unknown values; n Λ's; k ν's; and the final time. These values can be solved by using the k terminal constraints, equation (3), the n equations, equation (7c), and equaton (7d).

The control inequality constraint is represented as

$$u_{min} \leq u(t) \leq u_{max} \quad 0 \leq t \leq t_f \tag{8}$$

Control variable inequality constraint is augmented to the cost function and additional necessary conditions are obtained as a result (refs. 3 and 4). The optimal trajectory is composed of two types of control—nominal control [$u_{min} < u(t) < u_{max}$] and boundary control [$u(t) = u_{min}$ or u_{max}]. The nominal control satisfies the same necessary conditions as the unconstrained problem. For boundary control, the inequality constraint becomes an equality constraint. Many classical problems in the calculus of variations treat constraints of this form very well. A new Hamiltonian function with control variable inequality constraint is redefined as

$$\tilde{H} = H + \mu_1(u - u_{max}) + \mu_2(u_{min} - u) \tag{9}$$

where

$$\left.\begin{array}{l} \mu_1(u - u_{max}) = 0 \quad \mu_1 \geq 0 \\ \mu_2(u_{min} - u) = 0 \quad \mu_2 \geq 0 \end{array}\right\} \tag{10}$$

The necessary conditions and controls for active constraints are

$$\dot{x}^T = \tilde{H}_\Lambda \tag{11}$$

$$\dot{\Lambda} = -\tilde{H}_x \tag{12}$$

$$\tilde{H}_u = 0 \tag{13}$$

We use the control $u(t)$ from the condition $H_u = 0$ when the control constraints are not active. For the problem with active inequality constraints, equations (11) and (12) are the equations of state and costate variables. For the Lagrange multiplier μ_i, i = 1,2 must necessarily satisfy that

$\mu_i = 0$ if the associated constraint is not active

$\mu_i \geq 0$ if the associated constraint is active

and μ_i can be obtained from solving $\tilde{H}_u = 0$ for μ_i. Hence, when $u(t) < u_{min}$, the control and the Lagrange multipliers are $u(t) = u_{min}$, $\mu_1 = 0$, $\mu_2 = H_u$. When $u(t) > u_{max}$, the control and the Lagrange multipliers are $u(t) = u_{max}$, $\mu_2 = 0$, $\mu_1 = -H_u$.

Many numerical algorithms to solve optimal control problems have been developed. Indirect methods are theoretically based on the Minimum Principle, which characterizes the set of optimal states and controls in terms of the solution of a boundary value problem. One indirect method is the shooting method, which yields solutions of high precision. The shooting method is a second order method and hence is very sensitive to small changes of costate initial conditions. Shooting methods have the associated difficulties caused by instability of the initial value problem for the system of differential equations and by the requirement for good initial guesses for the iterative solutions of nonlinear problems. In this analysis, a shooting method is used to solve the comets or asteroids intercept or rendezvous trajectory problems.

Problem Statement

Propulsion

Many future propulsion systems have been proposed and analyzed. One potential propulsion approach that has been examined for a Comet/Asteroid Protection System (CAPS) deflection capability is the Variable Specific Impulse Magnetoplasma Rocket (VASIMR). VASIMR is a high power magnetoplasma rocket that gives continuous and variable thrust at constant power (ref. 5). Hydrogen plasma is heated by radio frequency (RF) power to increase exhaust velocity up to 300 km/s. The power output of the engine is kept constant, thus thrust and specific impulse, I_{sp}, are inversely related. Thrust is increased proportional to the power level. The engine can optimize propellant usage and deliver a maximum payload in minimum time by varying thrust and I_{sp} (ref. 6). Therefore, VASIMR can yield the fastest possible trip time with a given amount of propellant by using constant power throttling (CPT). A 10-kW space demonstrator experiment has been completed, and a VASIMR engine with 200-MW power could be available around the year 2050. The specific impulse range of the engine would be 3000 s to 30000 s, and the corresponding thrust range would be approximately 13600 N to 1360 N (assuming 100 percent power efficiency of 200 MW). To calculate acceleration, a, and spacecraft mass flow rate, the following relationships are used. The thrust, $T = |\dot{m}|v_e$, and exhaust velocity, $v_e = I_{sp}g_o$, are described by specific impulse, I_{sp}, and the acceleration due to gravity at the Earth's sea level, g_o. Mass flow rate, $\dot{m}$, is itself negative value.

The power, p, required to expel mass at the mass flow rate, $\dot{m}$, is $\varepsilon p = \frac{1}{2}|\dot{m}|v_e^2$. ε is the efficiency of the propulsion system. Thus, we know

$$\varepsilon p = \frac{1}{2}Tv_e \Rightarrow T = \frac{2\varepsilon p}{v_e} \Rightarrow T = \frac{2\varepsilon p}{I_{sp}g_o} \tag{14}$$

Using equation (14), acceleration due to VASIMR can be derived:

$$a = \frac{T}{m} = \frac{2\varepsilon p}{mg_o}\frac{1}{I_{sp}} \tag{15}$$

where m is spacecraft mass at any time. Finally, the mass flow rate is calculated as

$$\dot{m} = \frac{-T}{v_e} = -\frac{2\varepsilon p}{g_o^2} \frac{1}{I_{sp}^2} \qquad (16)$$

Equations of Motion

The spacecraft is considered to fly in three-dimensional interplanetary space. The three-degree-of-freedom equations of motion are the following:

$$\dot{r} = u \qquad (17a)$$

$$\dot{u} = \frac{v^2}{r} + \frac{w^2}{r} + a\sin\alpha\cos\beta - \frac{1}{r^2} \qquad (17b)$$

$$\dot{v} = -\frac{uv}{r} + \frac{vw\sin\phi}{r\cos\phi} + a\cos\alpha\cos\beta \qquad (17c)$$

$$\dot{w} = -\frac{uw}{r} - \frac{v^2\sin\phi}{r\cos\phi} + a\sin\beta \qquad (17d)$$

$$\dot{\theta} = \frac{v}{r\cos\phi} \qquad (17e)$$

$$\dot{\phi} = \frac{w}{r} \qquad (17f)$$

$$\dot{m} = -\frac{2\varepsilon p}{g_o^2 I_{sp}^2} \qquad (17g)$$

where r is the radial distance from the Sun to spacecraft, u is the radial velocity, v is the tangential velocity, w is the normal velocity, θ is the angle measured from the x-axis (defined as vernal equinox) in the x-y plane, ϕ is the angle measured from x-y plane, m is the mass of spacecraft, *a* is acceleration of spacecraft, p is the power of spacecraft, g is the gravitational parameter at Earth's sea level, and I_{sp} is the specific impulse of spacecraft engine. The control variables are thrust direction angle in plane (α), thrust direction angle of out-of plane (β), and the specific impulse (I_{sp}).

The Hamiltonian function is

$$H = w_t + \lambda_r u + \lambda_u \left(\frac{v^2}{r} + \frac{w^2}{r} + a\sin\alpha\cos\beta - \frac{1}{r^2} \right) + \lambda_v \left(-\frac{uv}{r} + \frac{vw\sin\phi}{r\cos\phi} + a\cos\alpha\cos\beta \right)$$
$$+ \lambda_w \left(-\frac{uw}{r} - \frac{v^2\sin\phi}{r\cos\phi} + a\sin\beta \right) + \lambda_\theta \frac{v}{r\cos\phi} + \lambda_\phi \frac{w}{r} - \lambda_m \frac{2\varepsilon p}{g_o^2 I_{sp}^2} \qquad (18)$$

The costate equations are

$$\dot{\lambda}_r = \lambda_u\left(\frac{v^2+w^2}{r^2}-\frac{2}{r^3}\right)+\lambda_v\left(\frac{vw\sin\phi}{r^2\cos\phi}-\frac{uv}{r^2}\right)+\lambda_w\left(-\frac{v^2\sin\phi}{r^2\cos\phi}-\frac{uv}{r^2}\right)+\lambda_\theta\frac{v}{r^2\cos\phi}+\lambda_\phi\frac{w}{r^2} \quad (19a)$$

$$\dot{\lambda}_u = -\lambda_r+\lambda_v\frac{v}{r}+\lambda_w\frac{w}{r} \quad (19b)$$

$$\dot{\lambda}_v = -\lambda_u\frac{2v}{r}+\lambda_v\left(\frac{u}{r}-\frac{w\sin\phi}{r\cos\phi}\right)+\lambda_w\frac{2v\sin\phi}{r\cos\phi}-\lambda_\theta\frac{1}{r\cos\phi} \quad (19c)$$

$$\dot{\lambda}_w = -\lambda_u\frac{2w}{r}-\lambda_v\frac{v\sin\phi}{r\cos\phi}+\lambda_w\frac{u}{r}-\lambda_\phi\frac{1}{r} \quad (19d)$$

$$\dot{\lambda}_\theta = 0 \quad (19e)$$

$$\dot{\lambda}_\phi = -\lambda_v\frac{vw}{r\cos^2\phi}+\lambda_w\frac{v^2}{r\cos^2\phi}-\lambda_\theta\frac{v\sin\phi}{r\cos^2\phi} \quad (19f)$$

$$\dot{\lambda}_m = \lambda_u\frac{2\varepsilon p}{m^2 gI_{sp}}\sin\alpha\cos\beta+\lambda_v\frac{2\varepsilon p}{m^2 gI_{sp}}\cos\alpha\cos\beta+\lambda_w\frac{2\varepsilon p}{m^2 gI_{sp}}\sin\beta \quad (19g)$$

Performance Index

Optimal control theory is concerned with finding the control history to optimize a measure of the performance index of the following general form:

$$J(u) = -m_f + \int_{t_0}^{t_f} w_t \, dt \quad (20)$$

where m_f represents the final mass of spacecraft, and w_t is weight for flight time (w_t is set as 10 in this analysis). For the problem at hand, it is required to find the optimal trajectory that maximizes the final mass of spacecraft and minimizes the flight time.

Initial Conditions and Terminal Conditions

The spacecraft departs from the Earth with the following initial conditions at $t = 0$ s:

$r(t_0) = 1$ au, $\quad u(t_0) = 0, \quad v(t_0) =$ Earth's velocity, $\quad w(t_0) = 0,$

$\theta(t_0) =$ obtained from spacecraft departure time before collision,

$\theta(t_0) = 0°, \quad m(t_0) =$ free (unknown), $\quad t_f =$ free (unknown)

The initial mass of spacecraft is a free parameter to include propellant mass. Thus, initially the costate of mass is set as $\lambda_m(t_0) = -1$. The other costate values $[\lambda_r(t_0), \lambda_u(t_0), \lambda_v(t_0), \lambda_w(t_0), \lambda_\theta(t_0), \lambda_\phi(t_0)]$ at t_0 are unknown.

Final boundary conditions are specified to satisfy the position and velocity of asteroids/comets. The spacecraft must intercept or rendezvous with the targeted asteroid/comet with specified orbit. To achieve the desired trajectory, final conditions should be satisfied. These are the positions of spacecraft for intercept trajectory and the positions and velocities of spacecraft for rendezvous trajectory. Hence, for intercept trajectory, the terminal state conditions are

$$\Phi[x(t_f),t_f] \equiv \begin{Bmatrix} r(t_f) - r_{target}(t_f) \\ \theta(t_f) - \theta_{target}(t_f) \\ \phi(t_f) - \phi_{target}(t_f) \\ m(t_f) - m_{dry} \end{Bmatrix} = 0 \qquad (21)$$

Rendezvous trajectory has the following terminal state conditions:

$$\Phi[x(t_f),t_f] \equiv \begin{Bmatrix} r(t_f) - r_{target}(t_f) \\ u(t_f) - u_{target}(t_f) \\ v(t_f) - v_{target}(t_f) \\ w(t_f) - w_{target}(t_f) \\ \theta(t_f) - \theta_{target}(t_f) \\ \phi(t_f) - \phi_{target}(t_f) \\ m(t_f) - m_{dry} \end{Bmatrix} = 0 \qquad (22)$$

where subscript "target" denotes the state of targeted celestial object, and m_{dry} is the spacecraft dry mass. From transversality conditions, we obtain the following costate terminal conditions for intercept trajectory:

$$\lambda_r(t_f) = v_1 \qquad (23a)$$

$$\lambda_u(t_f) = \lambda_v(t_f) = \lambda_w(t_f) = 0 \qquad (23b)$$

$$\lambda_\theta(t_f) = v_2 \qquad (23c)$$

$$\lambda_\phi(t_f) = v_3 \qquad (23d)$$

$$\lambda_m(t_f) = v_4 \qquad (23e)$$

where v_i is the Lagrange multiplier. From transversality conditions, we obtain the following costate terminal conditions for rendezvous trajectory:

$$\lambda_r(t_f) = v_1 \qquad (24a)$$

$$\lambda_u(t_f) = v_2 \qquad (24b)$$

$$\lambda_v(t_f) = v_3 \qquad (24c)$$

$$\lambda_w(t_f) = \nu_4 \tag{24d}$$

$$\lambda_\theta(t_f) = \nu_5 \tag{24e}$$

$$\lambda_\phi(t_f) = \nu_6 \tag{24f}$$

$$\lambda_m(t_f) = \nu_7 \tag{24g}$$

Furthermore, from equation (7d) the following condition is also satisfied at t_f:

$$H(t_f) = 0 \tag{25}$$

Equation (25) becomes another boundary equation at the final time. For the intercept problem, there are 14 differential equations describing the states and costates, with 15 unknowns [$m(t_0)$, $\lambda_r(t_0)$, $\lambda_u(t_0)$, $\lambda_v(t_0)$, $\lambda_w(t_0)$, $\lambda_\theta(t_0)$, $\lambda_\phi(t_0)$, t_f, ν_1, ν_2, ν_3, ν_4, $u(t_f)$, $v(t_f)$, and $w(t_f)$] and 15 boundary conditions [$r(t_0)$, $u(t_0)$, $v(t_0)$, $w(t_0)$, $\theta(t_0)$, $\phi(t_0)$, $\lambda_m(t_0)$, eqs. (21), $\lambda_u(t_f)$, $\lambda_v(t_f)$, $\lambda_w(t_f)$, and eq. (25)]. For the rendezvous problem, there are 14 differential equations describing the states and costates, with 15 unknowns [$m(t_0)$, $\lambda_r(t_0)$, $\lambda_u(t_0)$, $\lambda_v(t_0)$, $\lambda_w(t_0)$, $\lambda_\theta(t_0)$, $\lambda_\phi(t_0)$, t_f, ν_1, ν_2, ν_3, ν_4, ν_5, ν_6, and ν_7] and 15 boundary conditions [$r(t_0)$, $u(t_0)$, $v(t_0)$, $w(t_0)$, $\theta(t_0)$, $\phi(t_0)$, $\lambda_m(t_0)$, eqs. (22), and eq. (25)]. Thus, the two-point boundary problem can be completely solved with these boundary conditions.

Controls

The control variables are thrust direction angle in-plane (α), thrust direction angle of out-of-plane (β), and specific impulse (I_{sp}). A second-order necessary condition, the Legendre condition, states that the second derivative of the Hamiltonian, with respect to the controls, must be greater than or equal to zero for the performance index to be at a minimum. Thus, $H_{\alpha\alpha}$, $H_{\beta\beta}$ must be greater than or equal to zero for the performance index to be at a minimum. The first derivative of H with respect to α and convexity condition yields a control variable of α as follows:

$$\sin \alpha = \frac{-\lambda_u}{\sqrt{\lambda_u^2 + \lambda_v^2}} \tag{26a}$$

$$\cos \alpha = \frac{-\lambda_v}{\sqrt{\lambda_u^2 + \lambda_v^2}} \tag{26b}$$

The first derivative of H with respect to β and convexity condition yields a control variable of β as follows:

$$\sin \beta = \frac{-\lambda_w}{\sqrt{\lambda_u^2 + \lambda_v^2 + \lambda_w^2}} \tag{27a}$$

$$\cos \beta = \frac{\sqrt{\lambda_u^2 + \lambda_v^2}}{\sqrt{\lambda_u^2 + \lambda_v^2 + \lambda_w^2}} \tag{27b}$$

The first derivative of H with respect to I_{sp} and convexity condition yields a control variable of I_{sp} as follows:

$$I_{sp} = \frac{-2m\lambda_m}{g_o\sqrt{\lambda_u^2 + \lambda_v^2 + \lambda_w^2}} \qquad (28)$$

$$I_{sp\,min} \leq I_{sp} \leq I_{sp\,max} \qquad (29)$$

If the I_{sp} constraint is not active and $\lambda_m < 0$, the solution would be optimal. If the unconstraint control $I_{sp} < I_{sp\,min}$, then $I_{sp} = I_{sp\,min}$, whereas $I_{sp} = I_{sp\,max}$ if the unconstraint control $I_{sp} > I_{sp\,max}$. For $\lambda_m > 0$, the optimal solution is going to be $I_{sp} = I_{sp\,min}$.

Interplanetary optimal trajectories have been computed to maximize the final mass of the vehicle and to minimize the flight time. The optimal problem is a free final time problem to find the three controls satisfying the state and costate equations.

End-to-End Simulation

Here we consider a fictitious impacting LPC whose orbital parameters are given by aphelial distance $r_a = 100$ au and perihelion distance $r_p = 0.7$ au with inclination $i = 50°$. These orbital parameters yield semimajor axis $a = 50.35$ au, eccentricity $e = 0.98609732$, and orbital period of 357.27 years. It is assumed that the LPC has its density as $\rho = 1000$ kg/m^3. The minimum required impulses for deflecting an impactor by 3 Earth radii are solved in this analysis, and the calculation is always performed to move the LPC's trajectory from crossing the Earth's orbit at the Earth's center. The solutions represent impulse vectors that can be described by the magnitude of the minimum impulse and the optimal impulse angle. The gravitation effects of Earth are considered by using a three-dimensional optimization problem to calculate the impulse vectors. Figure 1 includes the minimum ΔV with respect to the impulse time that is defined as time before collision when ΔV is applied. Figure 2 includes the optimal impulse angle with respect to the impulse time. The impulse angle is described as being in an asteroid's/comet's orbital plane and is defined as the angle from the asteroid's/comet's original velocity vector to the impulse vector toward the Sun-asteroid/comet line. The dotted line in these figures explains the preperihelial collision case (a collision occurs before an asteroid/comet passes its perihelion), whereas the solid line in these figures explains the postperihelial collision case (a collision occurs after an asteroid/comet passes its perihelion). Figure 3 shows an estimate of the typical energy required for laser ablation to deflect the 1-km comet by 3 Earth radii, when preperihelial collision with Earth is considered. The required laser energy is also a function of the object's density and the required ΔV, which varies depending on the object's orbit and when the deflection occurs. It is easy to estimate the required laser energy for any size comet (or asteroid) because the minimum ΔV is linearly proportional to the cube of its diameter. Figure 4 explains a detail of figure 3 for an LPC, using fixed impulse times, and describes the required energy for laser ablation to deflect the given size of a comet by 3 Earth radii. The estimated energy is calculated assuming that the cumulative energy generated by the laser is applied as an equivalent impulsive ΔV at some time before collision. Because the laser ablation occurs over a significant period of time, the laser interaction must be complete prior to the time specified for each curve in order to assure that the deflection could be accomplished. Figure 5 shows a preliminary estimate of the achieved energy for a given laser power and operation period. Figures 4 and 5 can be used to estimate the nominal laser power required for a deflection mission. To illustrate a preperihelial case, figure 4 shows that approximately 5×10^4 GJ of energy is required to deflect a 0.2-km comet by 3 Earth radii if applied 1 year before

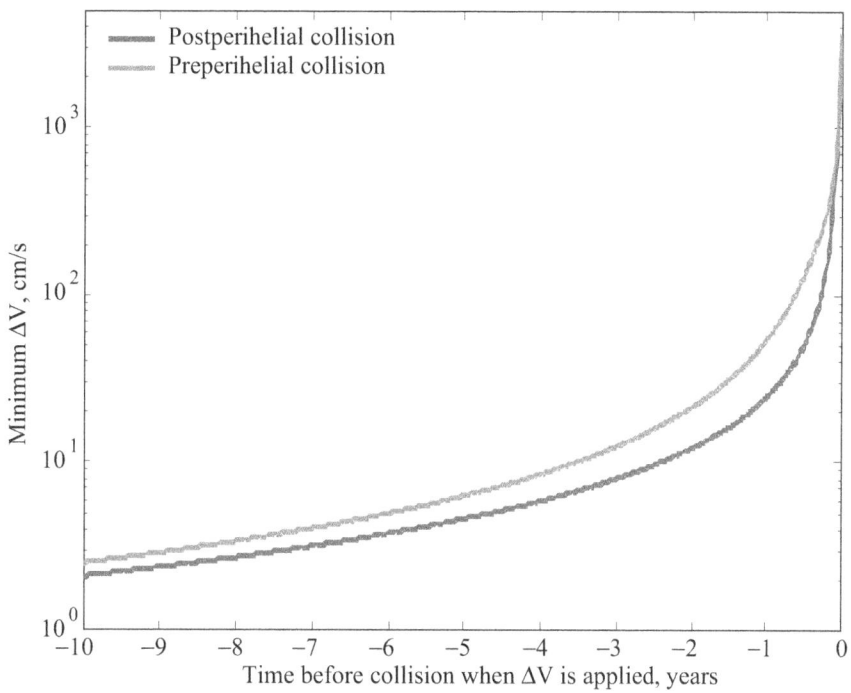

Figure 1. Minimum ΔV to deflect 1-km LPC by 3 Earth radii.

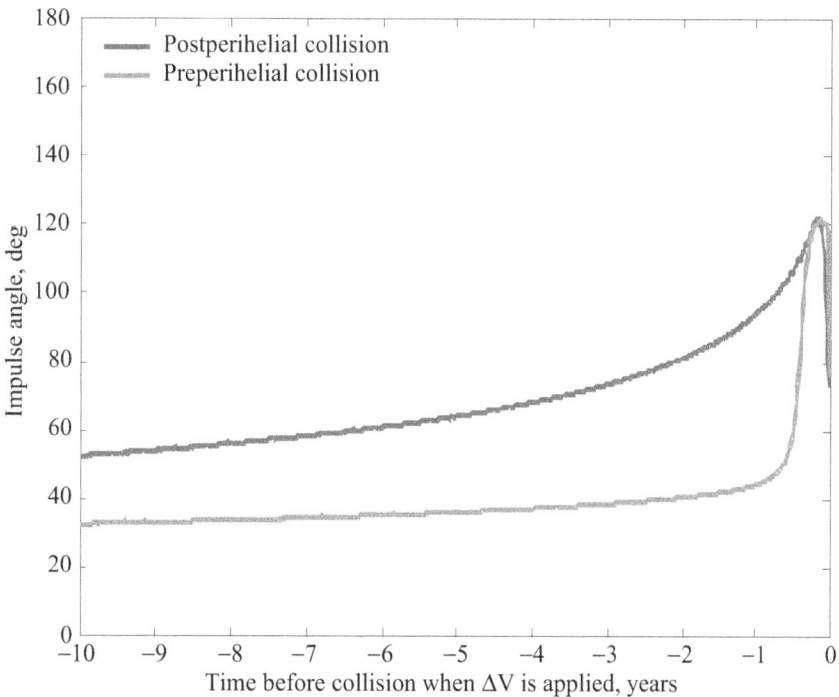

Figure 2. Optimal impulse angle with respect to impulse time.

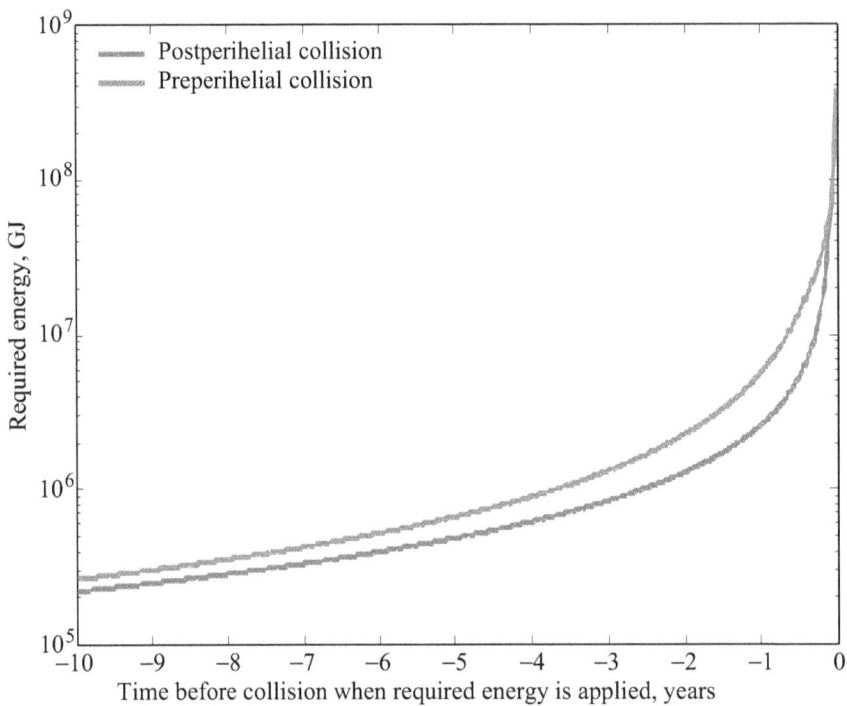

Figure 3. Required laser energy to deflect 1-km LPC by 3 Earth radii.

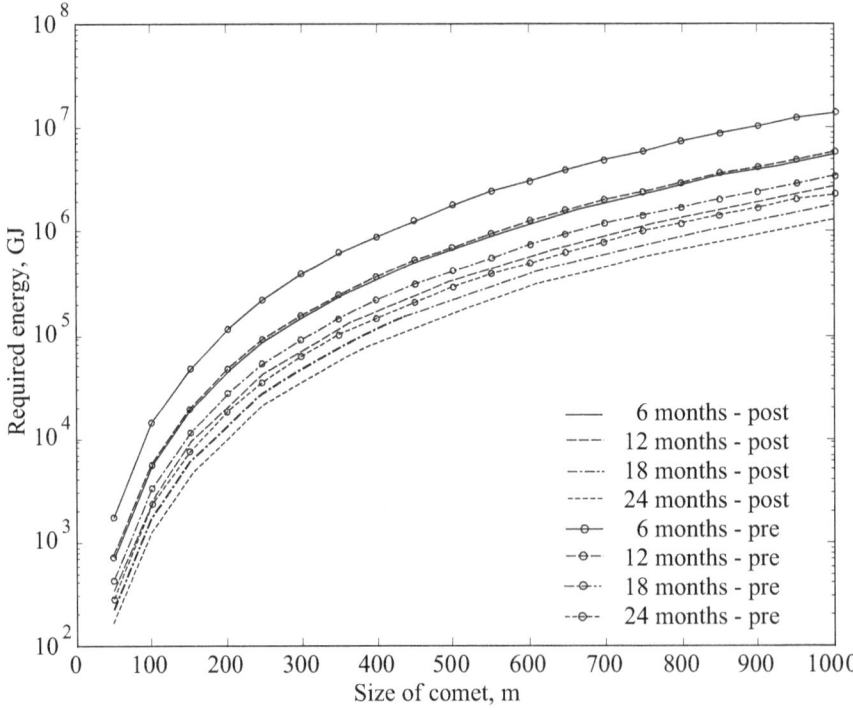

Figure 4. Estimated energy required for laser ablation versus diameter for LPCs ($\rho = 1000$ kg/m^3).

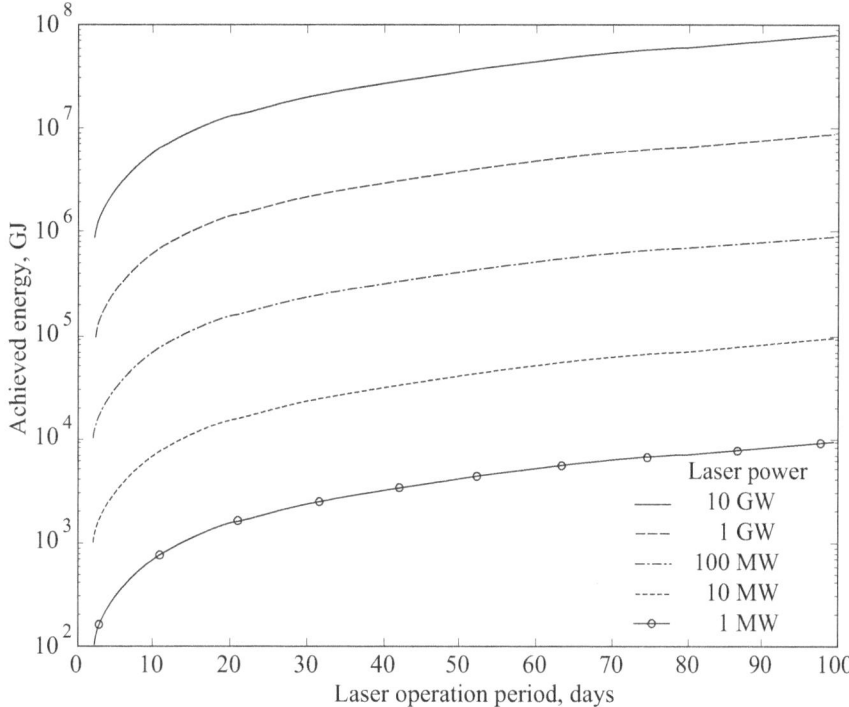

Figure 5. Estimate of achieved energy for given laser power and operation period.

collision. Figure 5 shows that a 100-MW laser (10000-kJ pulse and 10-Hz laser repetition frequency) would be required to operate continuously for approximately 5 days in order to achieve this cumulative energy. Figure 6 illustrates the trajectories and thrust vector for a 100-metric ton (t) spacecraft (including 10-t payload) to intercept or rendezvous with this particular LPC for a departure time of 7 months before a preperihelial collision with Earth.

The propulsion system is assumed to operate at 90-percent power efficiency. Figures 7 and 8 show flight time and required propellant for intercept and rendezvous trajectories for various departure times when a 100-t spacecraft with 1 GW of power is assumed. For this specific LPC, there is a peak at 11 months departure time because the spacecraft must fly in the reverse direction with respect to Earth's orbital velocity. Because there are local minima and maxima in the propellant required and flight time, as shown in the figures, it can be concluded that the values are dependent upon the orbital geometry relationship as well as distance between the Earth and the comet. For a given departure time, the rendezvous trajectory requires more propellant and longer flight time than the intercept trajectory. This is because the terminal velocity of rendezvous spacecraft must be matched with the target's velocity, which is not required for the intercept trajectory. Asteroids and comets with different orbital elements will have different flight times and propellant requirements. Even for the same celestial object, a postperihelial impact would have different results from those of a preperihelial impact.

Once the spacecraft has rendezvoused with the LPC, a laser ablation system makes use of the same electrical power system that the propulsion system uses for the orbital transfer. For example, if the laser ablation operation can be completed 12 months before collision, approximately 3×10^6 GJ of energy would be required for a 0.8-km LPC to be deflected by 3 Earth radii. If we choose a 500-MW laser

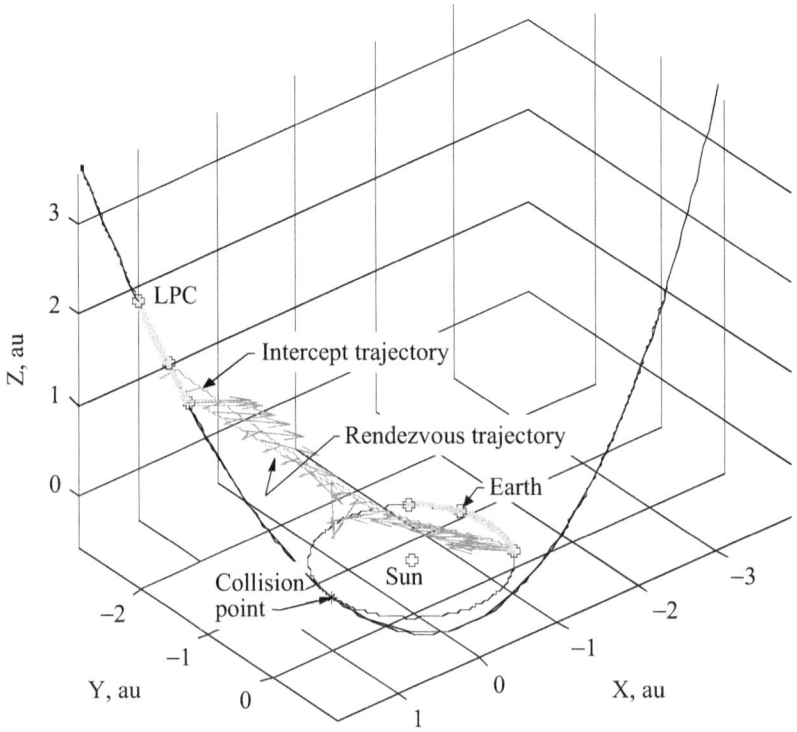

Figure 6. Example long-period comet intercept and rendezvous trajectories.

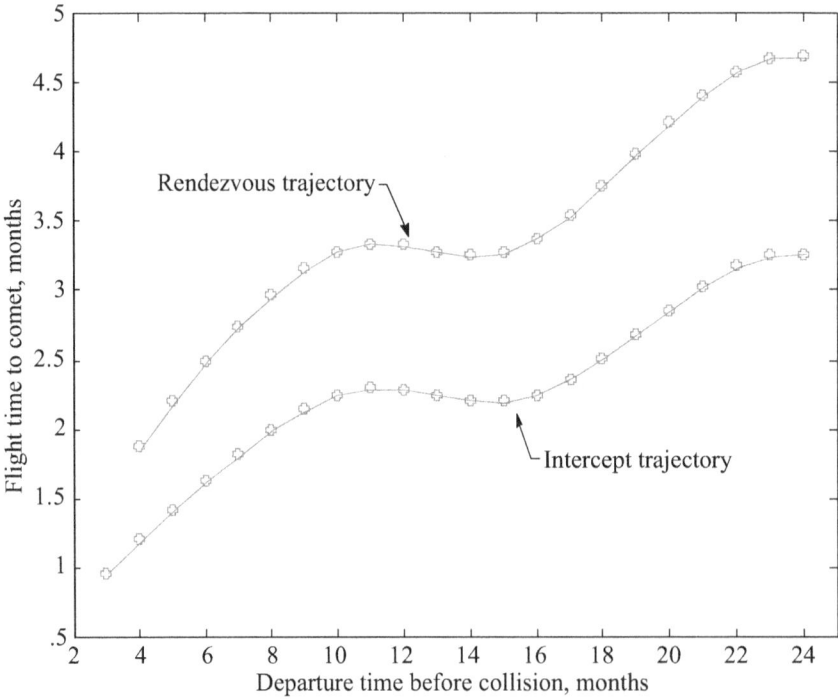

Figure 7. Flight time for each departure time.

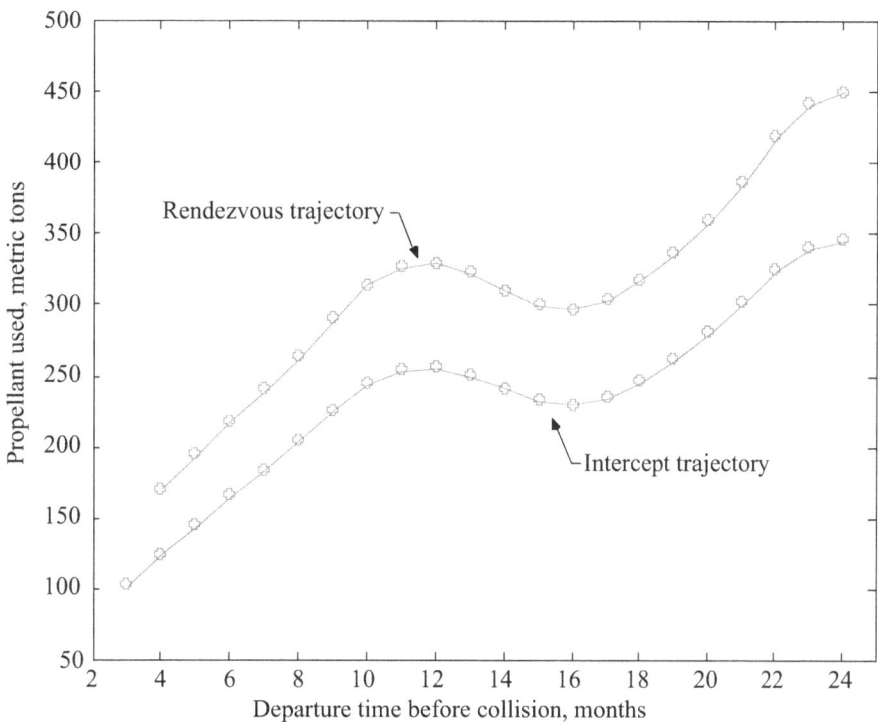

Figure 8. Propellant used for each departure time.

system (assuming 1 GW of supplied power and 50-percent laser system efficiency), it would take about 90 days of continuous operation to provide this amount of energy. From figure 7, the rendezvous spacecraft with a laser ablation system should depart from Earth approximately 19 months before the collision, with a trip time of about 4 months. For the mission, approximately 340 t of propellant would be required, assuming 100-t total dry mass of the spacecraft and payload. Depending on the payload mass capability of the propulsion system, multiple deflection devices could be delivered to the impactor, including a fallback option to the laser ablation system. In case the rendezvous deflection mission was unsuccessful, a similar spacecraft with a different payload (e.g., nuclear explosive device) could be sent to deflect the target using an intercept trajectory. If we assume that the intercept spacecraft departs from Earth 9 months before the collision with a 10-t payload, the spacecraft can arrive at the target approximately 7 months before impact and may require less than 230 t of propellant (as shown in figs. 7 and 8).

A multimegawatt nuclear electric propulsion (NEP) system using a VASIMR engine is currently estimated to have a maximum overall specific mass of 1.0 kg/kW (ref. 7). For a 1-GW system, this would result in a total spacecraft dry mass of 1000 mt (neglecting payload mass). This is ten times the total spacecraft mass assumed for this analysis, and the mass of a future laser ablation payload is presently not well understood. The assumed power efficiency of 90 percent is also optimistic. More capable power generators (gigawatt class) with lower specific masses could provide the power needed to reduce trip times and provide more powerful lasers. Deflecting an impactor by only 3 Earth radii would be sufficient for a deflection effort. More powerful lasers would be capable of providing a greater miss distance, and thus more margin for uncertainty in the object's orbit. NEOs with greater densities would also require a more capable laser ablation system. Longer warning times would reduce the requirements on the orbit modification system but would make the CAPS detection system more challenging to implement. For the 0.8-km LPC assumed in this analysis, the detection system would need to determine the comet's trajectory at least 19 months prior to its preperihelial collision with the Earth. For impactors with extremely short warning times, an intercept trajectory may be the only feasible scenario for diverting the object.

Technological advances that can significantly reduce the specific mass of the rendezvous spacecraft and laser payload may permit this type of deflection approach to become a reality. A tiered planetary defense approach using rapid rendezvous and intercept spacecraft could provide a feasible scenario to protect the Earth from an impacting LPC, as well as other classes of impacting NEOs. A rendezvous spacecraft with a laser ablation payload could also provide a capable and robust orbit modification approach for altering an NEO's orbit for resource utilization.

Concluding Remarks

This report presents intercept/rendezvous trajectories for an advanced spacecraft that is designed to deliver laser ablation energy to an Earth-crossing long-period comet. The trajectory optimization problem is solved using the shooting method, which yields highly accurate solutions. The open-loop optimal solutions can be used as reference spacecraft trajectory for the deflection problem. The end-to-end simulation in this report demonstrates a conceptual approach to altering the orbit of an Earth impacting long-period comet, particularly one which represents an immediate threat.

References

1. Phipps, C. R.: Lasers Can Play an Important Role in the Planetary Defense. *Proceedings of the Planetary Defense Workshop*, Lawrence Livermore National Laboratory, Livermore, CA, May 1995.
2. Bryson, A. E.: Dynamic Optimization, Addison-Wesley, Menlo Park, CA, 1999.
3. Maurer, H.; and Pesch, H. J.: Solution Differentiability for Parametric Nonlinear Control Problems With Control-State Constraints. *J. Optim. Theory & Appli.*, vol. 86, no. 2, 1995, pp. 285–309.
4. Kreim, H.; Kugelmann, B.; Pesch, H. J.; and Breitner, M. H.: Minimizing the Maximum Heating of a Re-Entering Space Shuttle: An Optimal Control Problem With Multiple Control Constraints. *Opt. Control Appli. & Methods*, vol. 17, 1996, pp.45–69.
5. Chang-Diaz, F. R., et al.: The Physics and Engineering of the VASIMR Engine. 36th AIAA/ASME/SAE/ASEE Joint Propulsion Conference, 17–19 July 2000, Huntsville, AL, AIAA 2000-3756.
6. Chang-Diaz, F. R.: The VASIMR Rocket. *Sci. American*, Nov. 2000, pp. 90–97.
7. Smith, B.; Knight, T.; and Anghaie, S.: Multimegawatt NEP With Vapor Core Reactor MHD. Space Technology and Applications International Forum (STAIF 2002), Albuquerque, NM, Feb. 3–6, 2002.

CAPS Simulation Environment Development[8]

DOUGLAS G. MURPHY
Analytical Mechanics Associates, Inc.

JAMES A. HOFFMAN
Analytical Mechanics Associates, Inc.

Introduction

The final design for an effective Comet/Asteroid Protection System (CAPS) will likely come after a number of competing designs have been simulated and evaluated. Because of the large number of design parameters involved in a system capable of detecting an object, accurately determining its orbit, and diverting the impact threat, a comprehensive simulation environment will be an extremely valuable tool for the CAPS designers. A successful simulation/design tool will aid the user in identifying the critical parameters in the system and eventually allow for automatic optimization of the design once the relationships of the key parameters are understood.

A CAPS configuration will consist of space-based detectors whose purpose is to scan the celestial sphere in search of objects likely to make a close approach to Earth and to determine with the greatest possible accuracy the orbits of those objects. Other components of a CAPS configuration may include systems for modifying the orbits of approaching objects, either for the purpose of preventing a collision or for positioning the object into an orbit where it can be studied or used as a mineral resource. The Synergistic Engineering Environment (SEE) is a space-systems design, evaluation, and visualization software tool being leveraged to simulate these aspects of the CAPS study. The long-term goal of the SEE is to provide capabilities to allow the user to build and compare various CAPS designs by running end-to-end simulations that encompass the scanning phase, the orbit determination phase, and the orbit modification phase of a given scenario.

Herein, a brief description of the expected simulation phases is provided, the current status and available features of the SEE software system is reported, and examples are shown of how the system is used to build and evaluate a CAPS detection design. Conclusions and the roadmap for future development of the SEE are also presented.

Comet/Asteroid Protection System Simulations

The scanning phase of the Comet/Asteroid Protection System (CAPS) activity is referred to in the Synergistic Engineering Environment (SEE) as the "Survey Mode" simulation. In this mode a configuration of space-based detectors is scanning the celestial sphere, identifying and cataloguing potentially hazardous objects, and performing preliminary orbit determination. The object of the survey simulation is to evaluate the effectiveness of a given telescope configuration and scanning pattern in identifying potentially hazardous near-Earth objects (NEOs).

The process of evaluating the ability of the CAPS system to properly identify the orbit of an object on a collision course with Earth is referred to as the "precision orbit determination" (POD) simulation. In this mode, one or more simulated NEOs will be placed on a collision course with Earth. The results of these simulations will show how well these orbits were able to be determined at a selected time prior to impact,

---

[8]Chapter nomenclature available in chapter notes, p. 217.

or the results will return a warning time that indicates how many days prior to impact the system was able to determine that the object was on a collision course.

The process of using the CAPS configuration to evaluate the composition of a given NEO is referred to as the "physical characteristics" simulation. In this mode the simulation will determine the ability of the system to report what the object is made of in order to determine its value as a resource.

Modeling the procedures for averting an impact is referred to as the "orbit modification" simulation. The orbit modification procedures can be divided into two types: deflection mode and resource utilization mode. In both cases the simulation will determine the ability of a given CAPS configuration to make desired orbital adjustments to an approaching object. A deflection case will model the modification of an NEO orbit from an Earth-impacting trajectory to a nonimpacting trajectory. The resource utilization case is not restricted to impacting NEOs and will have additional constraints on the target orbit (e.g., that the new orbit must leave the object more easily accessible for rendezvous missions).

The near-term goals of the SEE CAPS module were to support survey and impact simulations and schedule the physical characteristics and orbit modification simulations for longer term development.

Software Architecture and Current Feature Set

The SEE is a cross-platform application that may be run on Windows, IRIX, and Linux operating systems. The main feature of the SEE is the ability to create and view objects in orbit or on the surface of planets and moons in the solar system. The user interface is designed to allow ease of navigation to points and times of interest in the three-dimensional (3-D) scene. All models including both planets and spacecraft are rendered by default at their actual scales. An interactive scaling command allows the user to create views in which all objects in the scene are to be visible regardless of size.

Running a simulation generally involves creating SEE objects to represent all of the real objects being simulated (e.g., planets, moons, telescopes) and then populating the SEE objects with appropriate parameters (e.g., orbital parameters, detector performance models). An SEE object refers to an entity in the simulation that has properties that can be updated as the simulation progresses. The simulation time is controlled by an event loop, which will cause each of the SEE objects to update its time dependent parameters whenever the current simulation time is changed. The output of the simulation can be in the form of an interactive graphics window that shows the geometry of the scene for a given simulation time, or in data plots, data files, and reports describing the outcome of the run. Figure 1 shows a control flow diagram of an impact analysis loop connected to the main application.

In the current revision of the SEE application, the user must launch the program in an interactive mode; that is, launching the application brings up an application window with graphical user interface (GUI) components. The user must create the simulation objects by adding various craft and minor planets or other natural objects using the GUI or by loading previously saved versions of these objects from files. Once created, the interactive mode will allow the user to vary the current simulation time using a graphical time controller. Once launched, however, the SEE will also support noninteractive analysis by allowing the user to select from a list of data-generating routines such as a survey analysis. An analysis wizard will guide the user through the process of inputting parameters required to run a given analysis, always including a start time, stop time, and time step at which to record data. When the user accepts the analysis parameters and closes the wizard, the event loop is controlled automatically for the duration of the analysis. The event loop controller is responsible for changing the simulation time to the

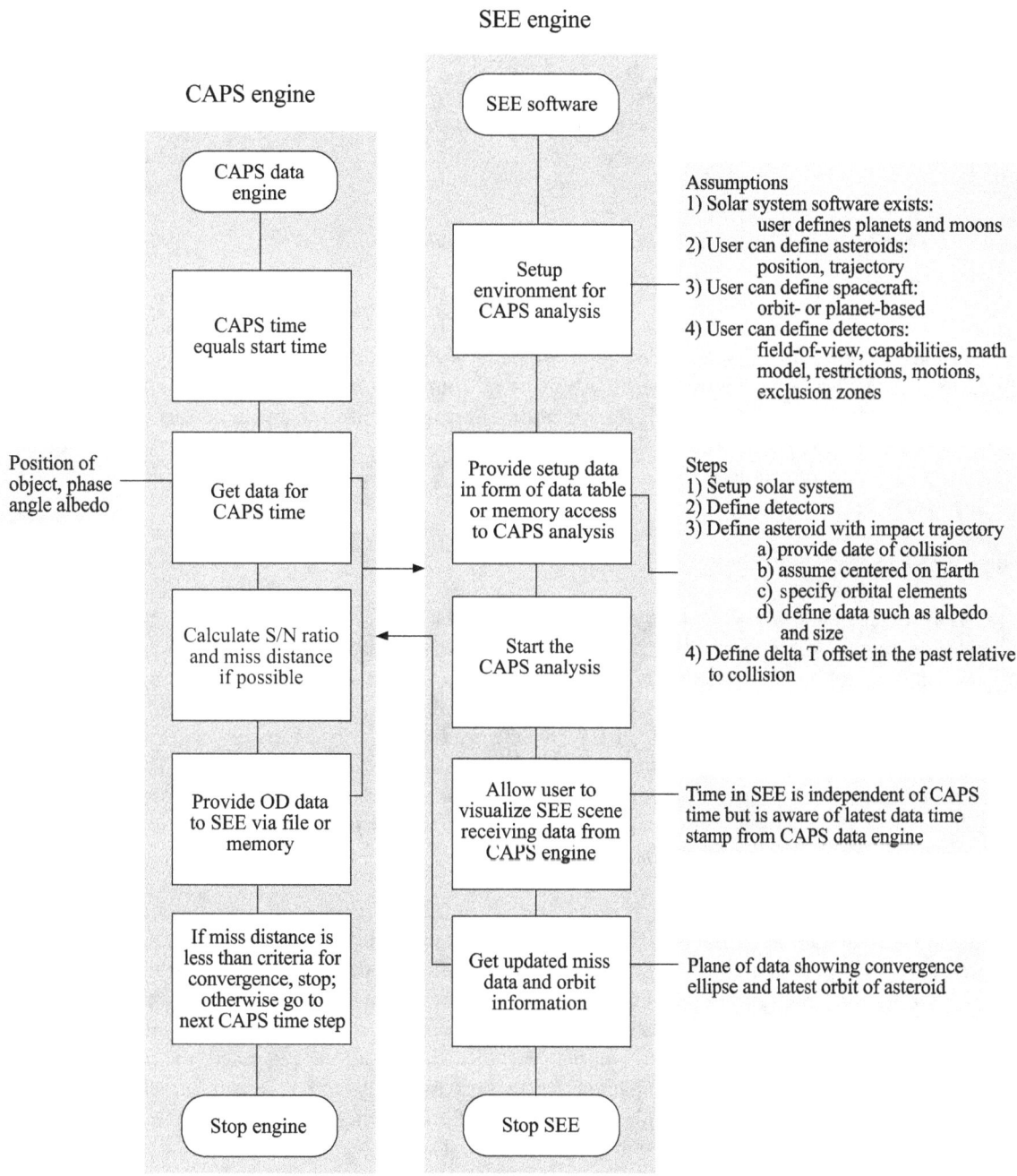

Figure 1. SEE CAPS module control flow.

analysis start time. The controller then polls the SEE objects for data needed for the analysis, processes and saves the data, and advances the simulation time by one analysis time step. The process completes when the analysis stop time is reached. Data files and summary reports containing the results of the simulation can be viewed once the analysis is completed. All interactive navigation controls (except for the time controls) and scaling controls are active and available to the user while the analysis is underway.

The currently implemented features of the SEE support the survey mode simulation and can also be used in the preparation of POD simulations. The survey mode simulation utilizes a helper application called the CAPS Coverage/Sensitivity Tool. A brief review of those capabilities is described subsequently.

SEE and The Coverage/Sensitivity Tool Survey Mode Capabilities

The SEE survey tools comprise features within the main SEE application and a separate coverage visualization application called the CAPS Coverage/Sensitivity Tool. The Coverage/Sensitivity Tool is used to gain insight into the success of a survey pattern with respect to total coverage and measured signal-to-noise (S/N) ratios for approaching objects at a uniform target distance without modeling the flight paths of those objects. The SEE application can then be used to evaluate the survey program against an asteroid/comet data set in which the motion of both the NEOs and the telescopes is modeled. A future goal of this project is to merge the features of the CAPS Coverage/Sensitivity Tool directly into the SEE main execution environment.

The CAPS Coverage/Sensitivity Tool

Overview. Placement of the CAPS survey telescopes will be influenced by the need to efficiently survey the celestial sphere at the desired target radius and the need to maximize the S/N ratio for the observed objects. In this section, a description is given of the CAPS Coverage/Sensitivity Tool and the simplified coverage/sensitivity analyses that have been performed to test the software and begin to understand the system tradeoffs regarding placement of CAPS detection assets. The input to the CAPS Coverage/Sensitivity Tool is a schedule of telescope positions and bore sight direction vectors. The output of the tool is an interactive 3-D visualization of the coverage and sensitivity results. This tool facilitates a trade-off study among architectures and scanning strategies. The time dependent 3-D animations in the CAPS analysis module are useful for refining scanning strategies.

For this analysis the position and bore sight data for the scanning telescopes were generated using the commercially available Satellite Tool Kit (STK) space system modeling software. The format of the position and bore sight data are straightforward ASCII file listing times, positions, and direction vectors for each scope. This file can also be generated using standard spreadsheet software or other tools capable of saving the data in ASCII format. STK was used here instead of the SEE because this analysis took place prior to completion of the SEE software. Once the schedule file is generated, the user may import these data, specify the telescope design parameters, and run the analysis (figs. 2 and 3). After the analysis is run, the user may adjust various rendering properties (fig. 4). The user can display the data as a time dependent animation or control the time manually. The mouse can be used to control the vantage point of the data visualization.

Coverage analysis approach. Coverage analysis is accomplished by choosing a radius, such as 7 astronomical units (au), and examining how well a scanning strategy covers the heliocentric sphere of that radius. In the analysis, one measures where and how often the viewing frusta of the scanning telescopes intersect a given heliocentric sphere. To accomplish this, the sphere is "pixelized" using the algorithm in reference 1. Each pixel represents an approximately equal area region of the sphere (fig. 5).

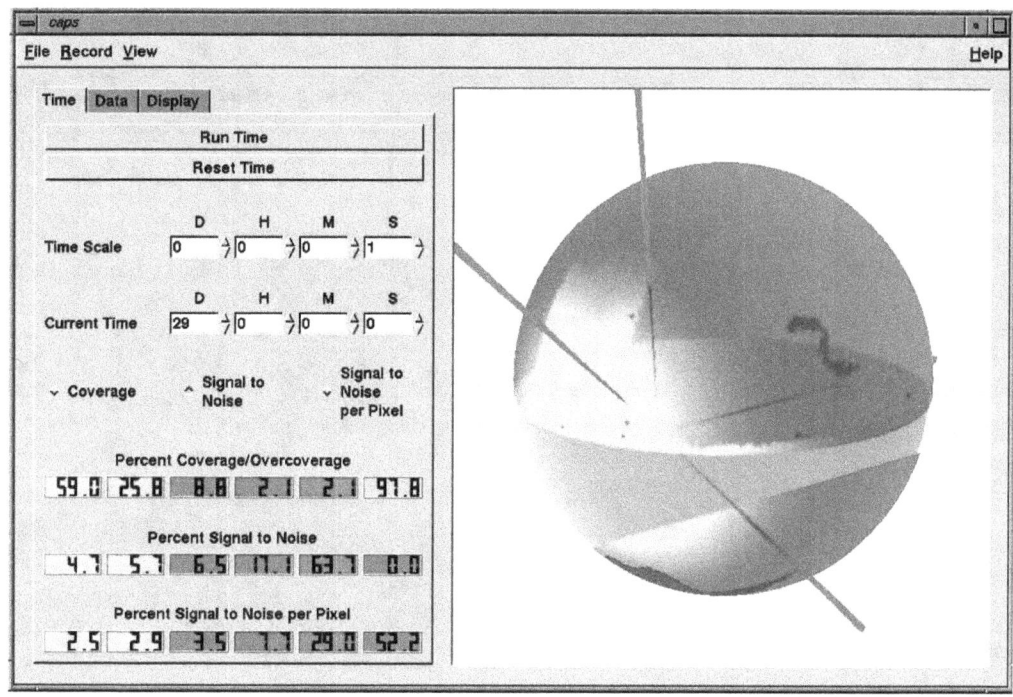

Figure 2. CAPS coverage/sensitivity tool, time inputs.

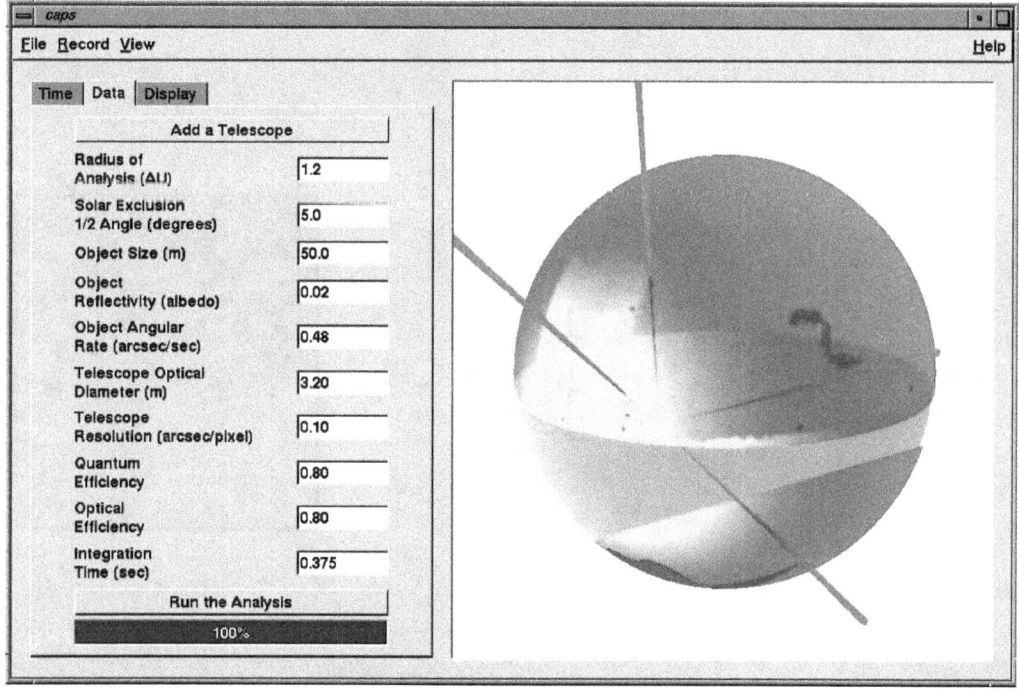

Figure 3. Telescope and NEO data input.

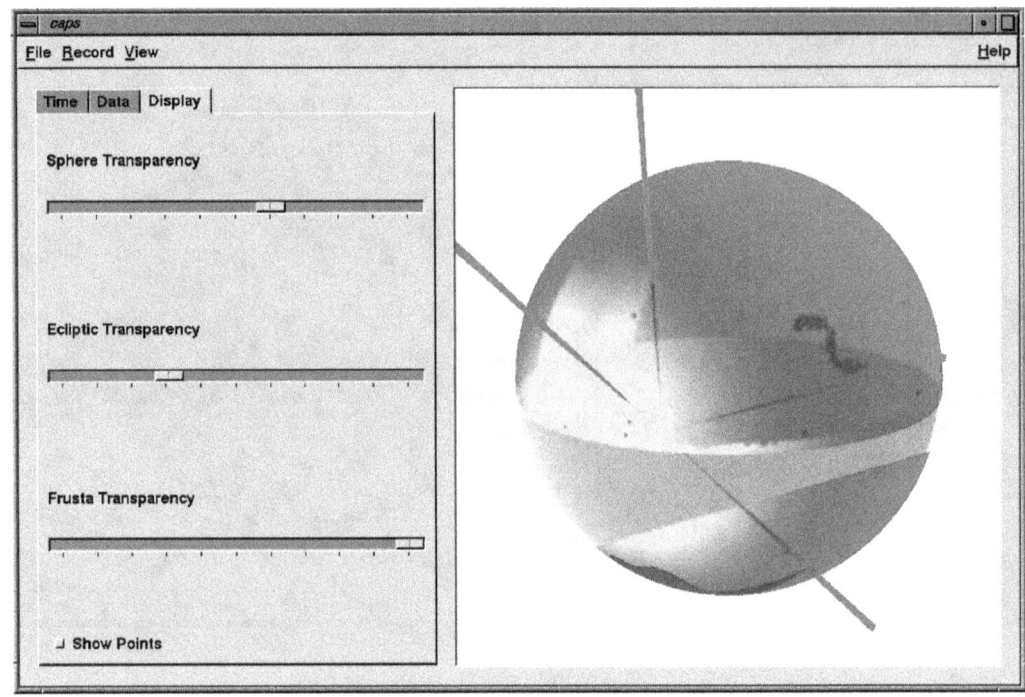

Figure 4. Rendering options.

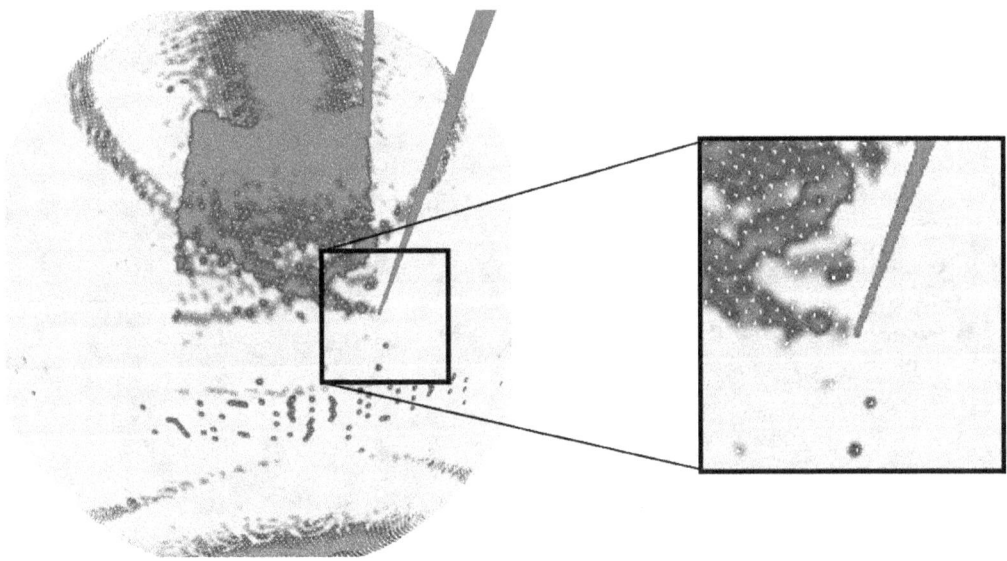

Figure 5. "Pixelizing" the celestial sphere.

The area surrounding a pixel is colored according to how often that pixel is viewed by the scanning telescopes. Initially, each pixel is colored blue to represent an unscanned pixel. A scanned pixel is colored from green to red, a green pixel having been scanned precisely once and a red repeatedly scanned five or more times. In this way, one can differentiate which portions of the sphere are scanned or overscanned. For example, in figure 5 one sees the reddish-green bands where the polar and equatorial scopes overlap in their scans. One might also note that there is a scattering of unscanned (blue) pixels. This is due to the fact that the displayed scanning strategy allowed for absolutely no overlap in frames. Precision errors likely caused these pixels to lie just outside the scanning pattern of the frustum.

The STK toolkit was used to produce 29 days of position and bore sight data for four lunar-based telescopes. Figure 6 shows a 1.2-au heliocentric sphere (semitransparent) and the viewing frusta of the four telescopes as seen from slightly above the ecliptic plane (in light blue). Figure 7 gives a zenith view and figure 8 gives a view from near Earth (slightly above the ecliptic).

Coverage analysis results. For long-period comets (LPCs), coverage for the four-telescope, lunar-based architecture was tested on a 7-au heliocentric sphere with a 90° solar exclusion half-angle. The analysis was repeated with a 5° solar exclusion half-angle so that one may consider the scanning strategy under the assumption that some idea or device may eventually allow for such an improvement.

For near-Earth asteroids (NEAs), coverage was tested on heliocentric spheres of radii 2, 1.2, and 0.8 au with a solar exclusion half-angle of 45°. Once again, the analysis was repeated with a 5° solar exclusion half-angle. Of particular interest in these cases are the near-Earth views as they describe coverage for objects that are 1 or 0.2 au from Earth.

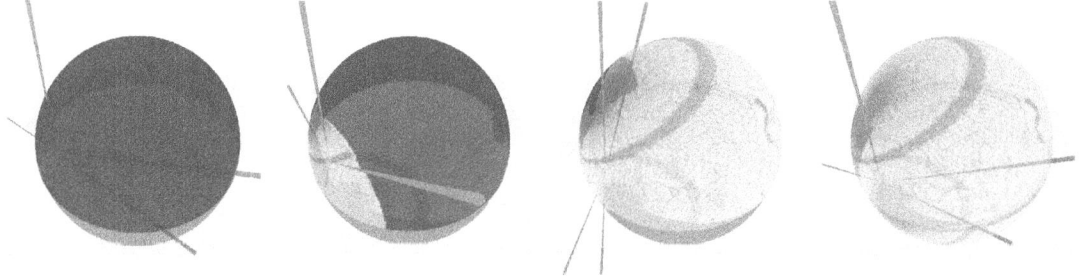

Figure 6. Telescope placement relative to a 1.2-au heliocentric sphere.

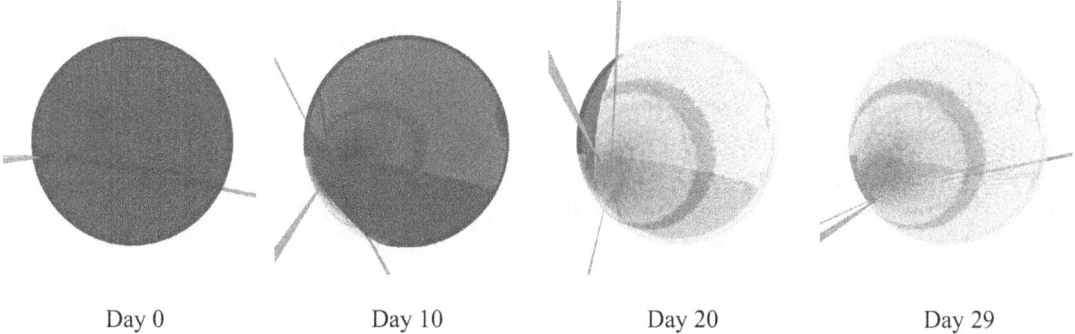

Day 0   Day 10   Day 20   Day 29

Figure 7. Zenith view of a 1.2-au heliocentric sphere.

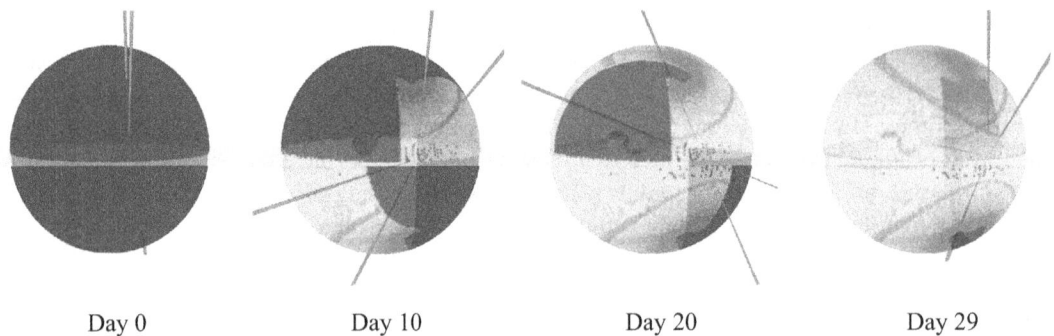

Figure 8. Near-Earth view of a 1.2-au heliocentric sphere.

Tables 1 and 2 give the percentage of the sphere covered by the scanning strategy. It also indicates what percentage of the sphere received single or multiple coverage. It is clear from the percent coverage shown in the tables that solar exclusion is the primary factor for a lunar-based architecture. If one is to achieve a nearly complete scan in 30 days, then either the solar exclusion angle must be reduced or more telescopes are needed in some heliocentric orbit. The multiple coverage in this simple scanning strategy is surprisingly low. One would expect some double coverage if frames are to overlap by a few pixels. Employing a more sophisticated scanning algorithm can reduce the triple (and greater) coverage. One also expects to see some overcoverage on a lunar-based architecture as lunar months overlap, as they do in a 29-day scan.

Table 1. Coverage Analysis With a 90° (7 au only) or 45° Solar Exclusion Half-Angle

Radius of sphere, au	Coverage, percent	Single coverage, percent	Double coverage, percent	Triple coverage, percent	Quadruple coverage, percent	Over-coverage (5 or more), percent
7	43.2	24.2	11.8	4.2	1.4	1.6
2	72.4	37.8	21.1	8.2	2.5	2.8
1.2	58.8	29.8	17.5	7.4	2.0	2.1
0.8	33.6	18.3	10.6	3.6	1.0	0.1

Table 2. Coverage Analysis With a 5° Solar Exclusion Half-Angle

Radius of sphere, au	Coverage, percent	Single coverage, percent	Double coverage, percent	Triple coverage, percent	Quadruple coverage, percent	Over-coverage (5 or more), percent
7	99.8	56.2	27.7	9.6	3.0	3.3
2	99.2	57.5	26.9	9.4	2.6	2.9
1.2	97.8	59.0	25.8	8.8	2.1	2.1
0.8	95.0	64.4	23.8	5.4	1.2	0.1

Figures 9 through 12 give the visual representation of this analysis. Screen shots were taken from zenith, nadir, near-Earth, and the opposing far side (see figs. 7 and 8). The reddish-green bands in each of these images identify the overlap between two scanning telescopes. The reddish-green sector in the near-Earth (upper hemisphere) and far side (lower hemisphere) identify the overlap is scanning due to lunar orbit. Also, note the gap in coverage seen in the nadir views of the 1.2 and 0.8 au spheres that is also attributable to the lunar orbit. Either a modification in the scanning strategy is needed or one must accept a scanning period of greater than 29 days.

Sensitivity analysis approach. A sensitivity analysis is performed in a fashion similar to the coverage analysis described previously. First, a heliocentric sphere is chosen and pixelized. For each pixel scanned by a telescope, the signal-to-noise (S/N) ratio is computed based on a user input set of telescope parameters. Analysis of the survey telescope design with blur centroiding incorporated demonstrated that the S/N ratio for each scan can be reduced to 3.6 and still provide positive indication of a target. The area representing that pixel is then colored according to the value of the S/N ratio. Values of 1 or less are colored gray with lighter shades close to 1 and darker shades close to 0. Ratios greater than 1 and less than 3.6 are colored on a continuous scale between red and green. Values near 1 are colored red and values near 3.6 are green. Ratios greater than 3.6 are also colored green. If a pixel is scanned multiple times, then the largest S/N ratio found for that pixel is used.

Sensitivity analysis results. For all cases, the telescope was assumed to have an optical diameter of 3.2 m with quantum and optical efficiencies of 0.8. The integration time for the LPC is assumed to be 15 s. An integration time of 0.375 s is assumed for NEAs.

Tables 3 and 4 give the percentage of the sphere scanned within a given range of sensitivity, or S/N ratio. The sensitivity for LPCs is well within the tolerance limit of 3.6.

Table 3. Sensitivity Analysis With a 90° (7 au only) or 45° Solar Exclusion Half-Angle

Radius of sphere, au	S/N > 3.6, percent	1.0 < S/N < 3.6, percent	S/N ≤ 1.0, percent
7	43.2	0	0
2	0	7.5	64.9
1.2	4.7	54	0
0.8	3.5	30.3	0

Table 4. Sensitivity Analysis With a 5° Solar Exclusion Half-Angle

Radius of sphere, au	S/N > 3.6, percent	1.0 < S/N < 3.6, percent	S/N ≤ 1.0, percent
7	99.8	0	0
2	0	7.5	91.7
1.2	4.7	93.1	0
0.8	3.8	91.1	0.1

For NEAs, coverage was tested on heliocentric spheres of radii 2, 1.2, and 0.8 au with a solar exclusion half-angle of 45°. The object was assumed to be 50 m in diameter with an albedo of 0.02. As before, the analysis was repeated with a 5° solar exclusion half-angle. A schematic of the orbits assumed for NEA sensitivity analysis is shown in figure 13.

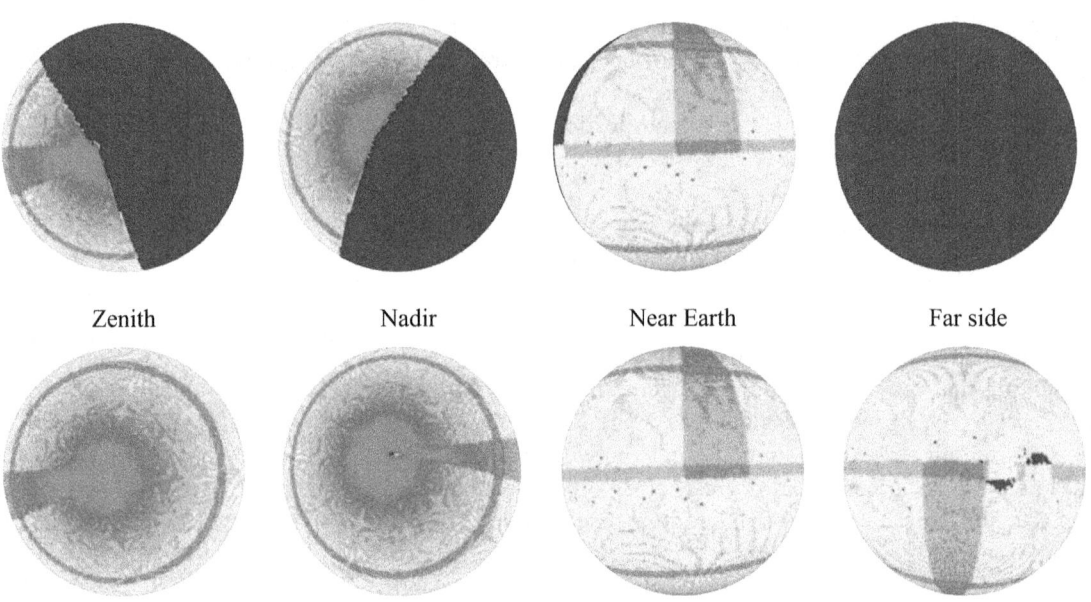

Figure 9. Coverage of a 7-au heliocentric sphere with 90° half-angle (top) and 5° half-angle (bottom).

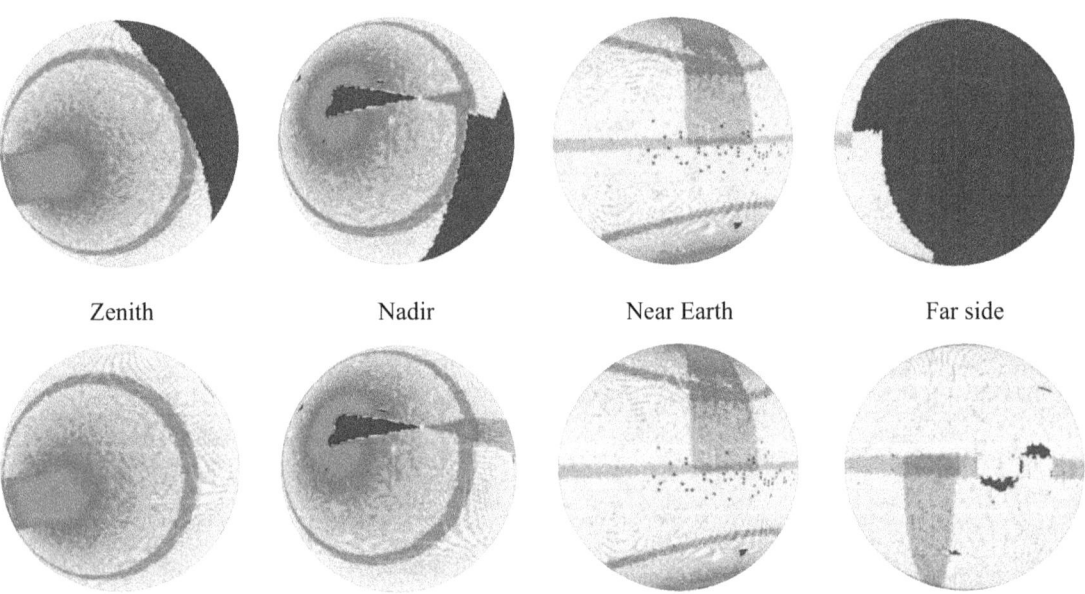

Figure 10. Coverage of a 2-au heliocentric sphere with 45° half-angle (top) and 5° half-angle (bottom).

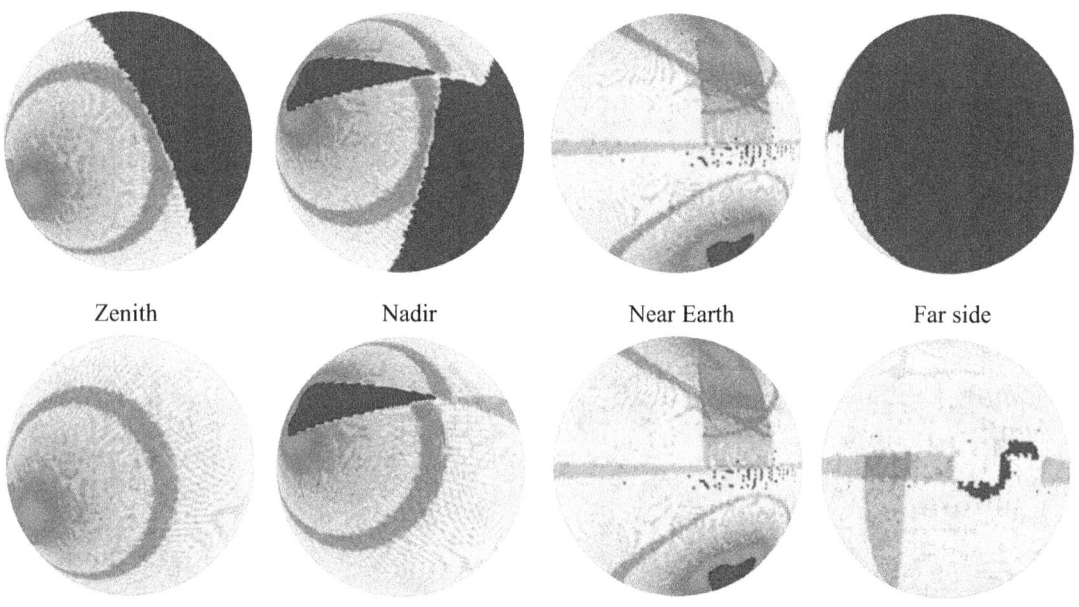

Figure 11. Coverage of a 1.2-au heliocentric sphere with 45° half-angle (top) and 5° half-angle (bottom).

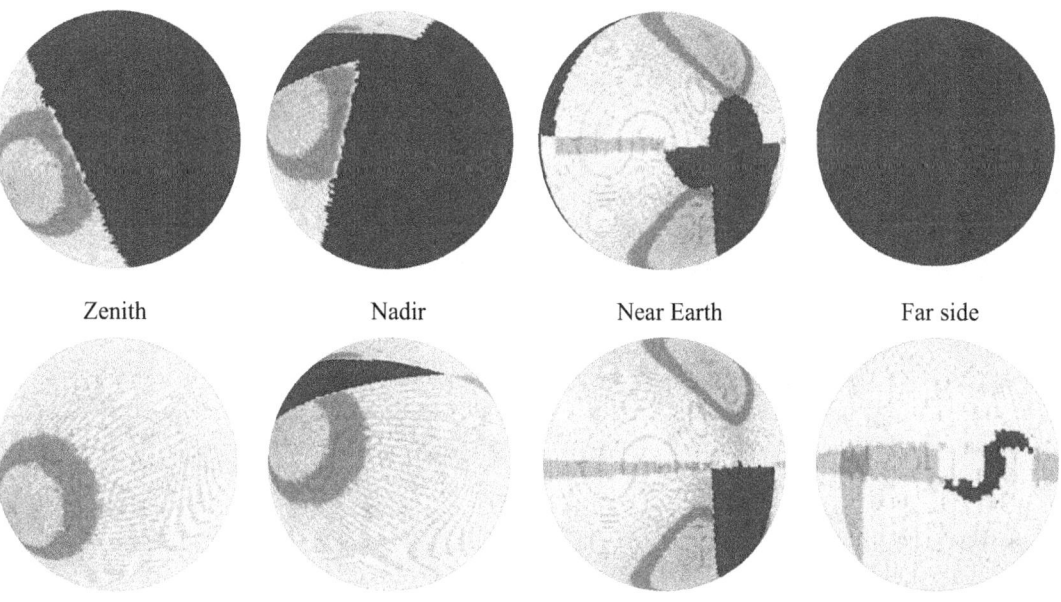

Figure 12. Coverage of a 0.8-au heliocentric sphere with 45° half-angle (top) and 5° half-angle (bottom).

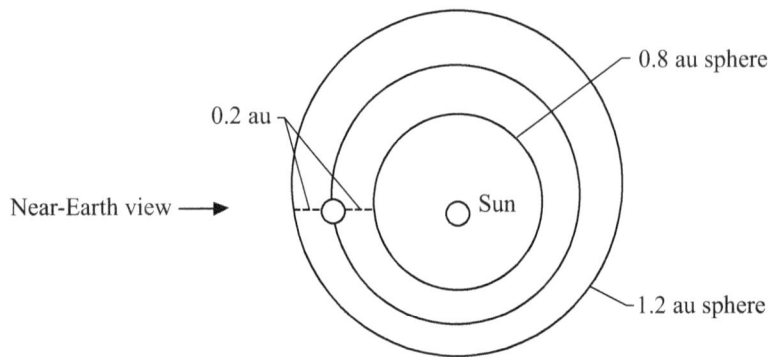

Figure 13. Orbit diagram for near-Earth asteroid sensitivity analysis.

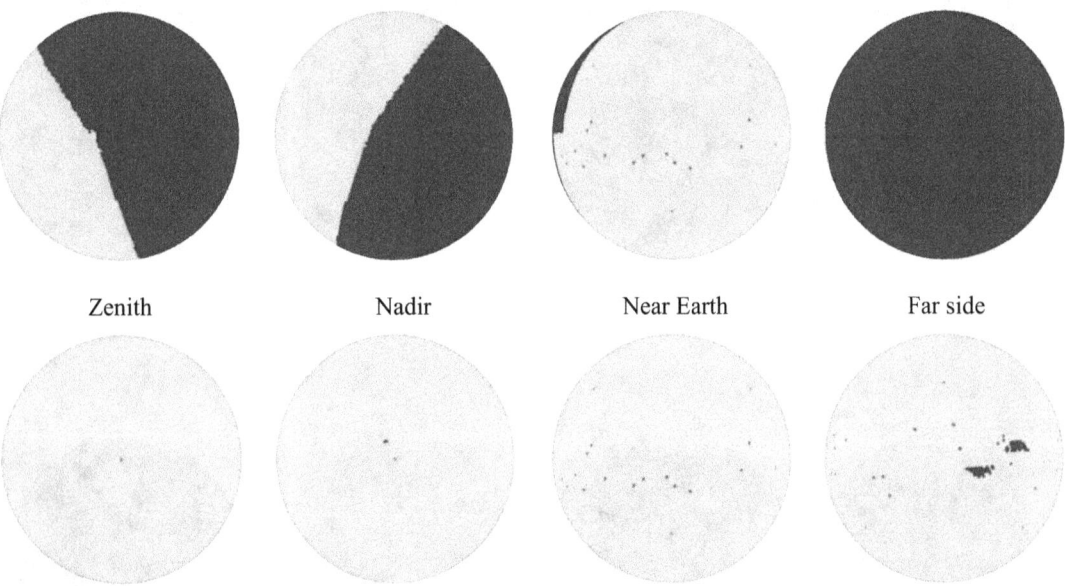

Figure 14. Sensitivity for a 7-au heliocentric sphere with 90° half-angle (top) and 5° half-angle (bottom).

For LPCs, sensitivity for the four-telescope, lunar-based architecture was tested on a 7-au heliocentric sphere with a 90° solar exclusion half-angle. The object was assumed to be 1000 m in diameter with an albedo of 0.02. The analysis was repeated with a 5° solar exclusion half-angle so that one may consider the scanning strategy under the assumption that some idea or device may eventually allow for such an improvement. The results of these simulations are shown in figure 14.

The NEAs within 0.2 au of the Earth are also within this tolerance. This can be noted from the near-Earth views of the 1.2- and 0.8-au heliocentric spheres in figures 15 and 16. The NEAs at 1 au from the Earth are generating S/N ratios close to 1, as can be seen by the bright red splotch on the near-Earth view in figure 17.

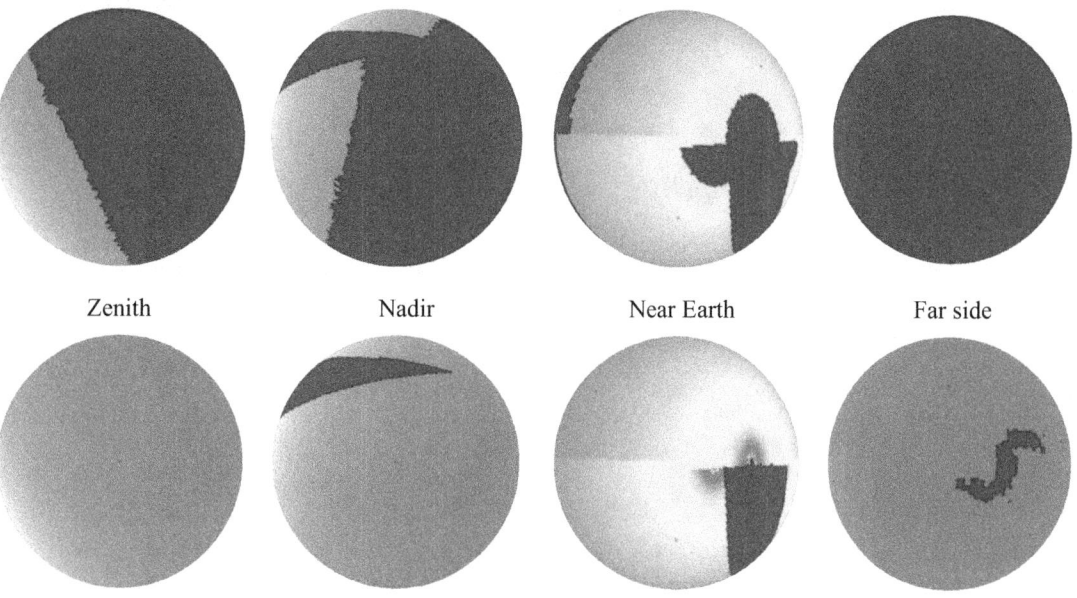

Figure 15. Sensitivity for a 0.8-au heliocentric sphere with 45° half-angle (top) and 5° half-angle (bottom).

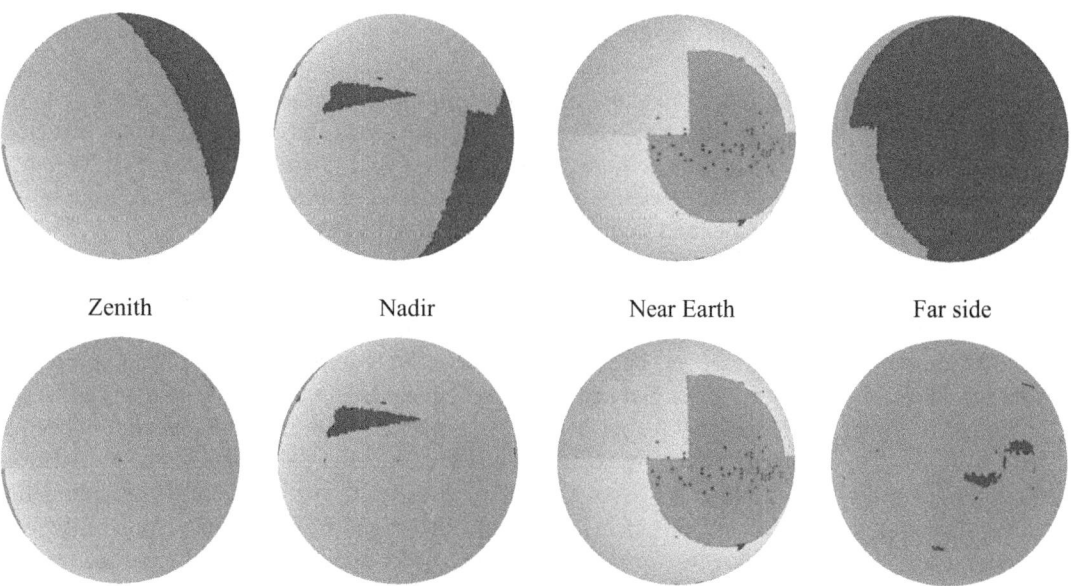

Figure 16. Sensitivity for a 2-au heliocentric sphere with 45° half-angle (top) and 5° half-angle (bottom).

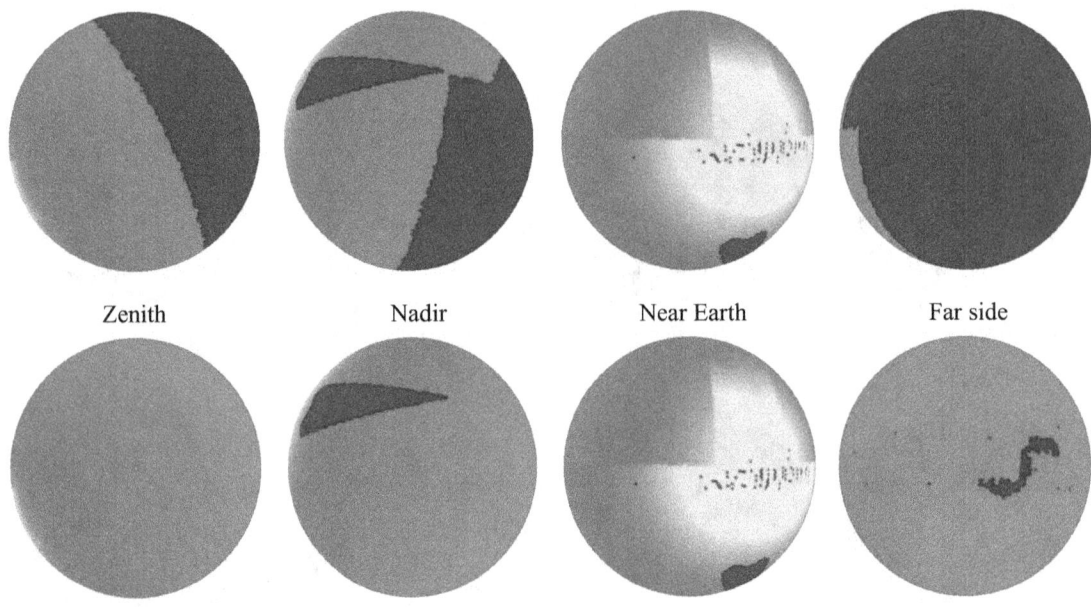

Figure 17. Sensitivity for a 1.2-au heliocentric sphere with 45° half-angle (top) and 5° half-angle (bottom).

The Synergistic Engineering Environment. The limitation of the Coverage/Sensitivity Tool is that the NEO motion cannot be modeled. Also, the position and orientation of the detectors is fixed once the detector schedule is read in. The SEE addresses these issues by allowing the user to import data sets of classical orbital elements to define two-body keplerian orbits for many thousands of objects. The SEE user also has interactive control of the detector placement in orbit or on the surface of a planet or moon.

In figure 18, a screen capture of a survey simulation in progress in the SEE is shown. The current release supports the ability to import up to 25 000 minor planets using orbit data read from a text file (the upper limit of the object population may be higher depending on the system hardware). All these objects can be scaled and tethered to for visualization purposes. Figure 19 is a view from the bore sight of one of the detectors, with several NEOs (pyramid shaped objects) in view.

At present, the list of objects to be loaded into the scene is specified at application launch time by selecting an SEE mission from a file browser dialog box. The mission file also specifies the number and type of crafts to be loaded into the scene. The ability to interactively import a single object or a collection of objects into the scene at any time is a planned near-term enhancement. The craft objects can be placed in orbit around or on the surface of any planet in the solar system. Code for interactively modifying the orbits of various craft is in place. The properties of the detectors, physical properties of the NEOs, and the parameters of the scanning algorithm have been hard coded into this version of the application in order to conduct testing. Further enhancements are planned to support the automatic generation of efficient scanning algorithms for a given configuration of detectors and to modify the parameters of the detectors.

A short report that lists by name the NEOs currently in the field-of-view (FOV) of each of the telescopes is available at any time step (fig. 20). Additionally, a complete survey analysis can be launched via the survey simulation wizard. The wizard prompts the user for the survey start time, survey end time, frame integration time, and a radius at which to record the coverage map. Once the parameters are

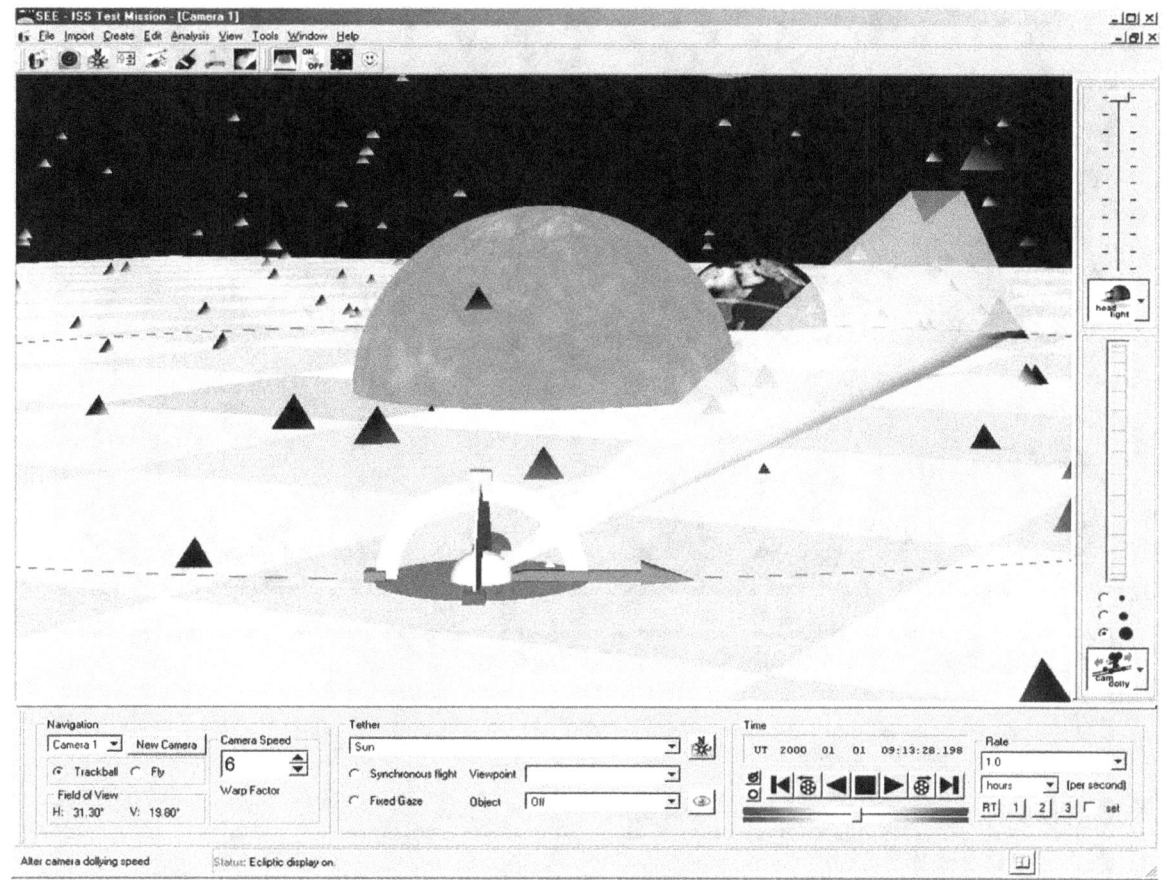

Figure 18. SEE application screen: survey simulation in progress.

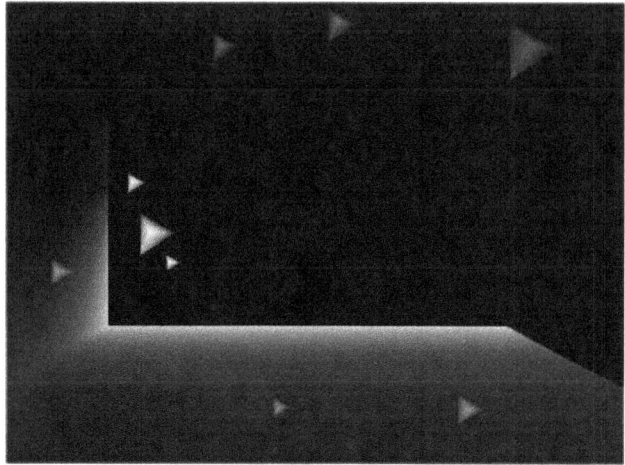

Figure 19. View from detector.

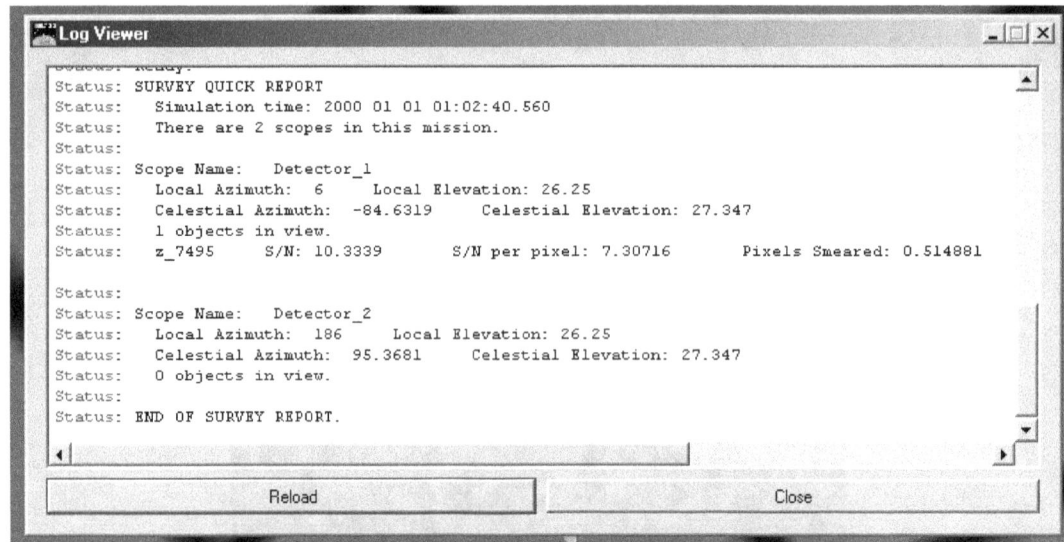

Figure 20. Quick survey report with one NEO in FOV.

accepted, a survey analysis thread is launched, and a progress bar displays the completion percentage. During the full survey analysis an interactive coverage map is displayed (fig. 21). The map depicts areas of the target sphere that have been covered by the detectors in the survey. The dotted grey lines represent lines of latitude and longitude of the target sphere in a sinusoidal projection. The central horizontal line represents the target sphere equator, and the top and bottom points of the projection represent the north and south poles of the target sphere. Blue rectangles with black outlines are drawn in areas of the map corresponding to areas of the target sphere that have been "covered" by the detectors. An area is considered covered if it has been inside the detector view frustum while the detector was collecting data. The map view may be browsed while the analysis is running by using pan and zoom functions in the map window. The native resolution of the bitmap is 3600×1800 pixels.

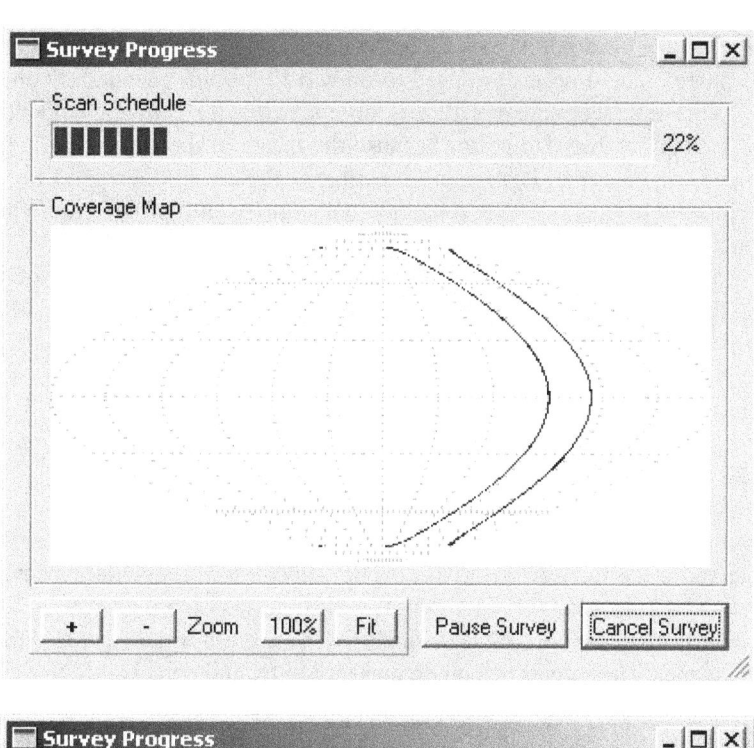
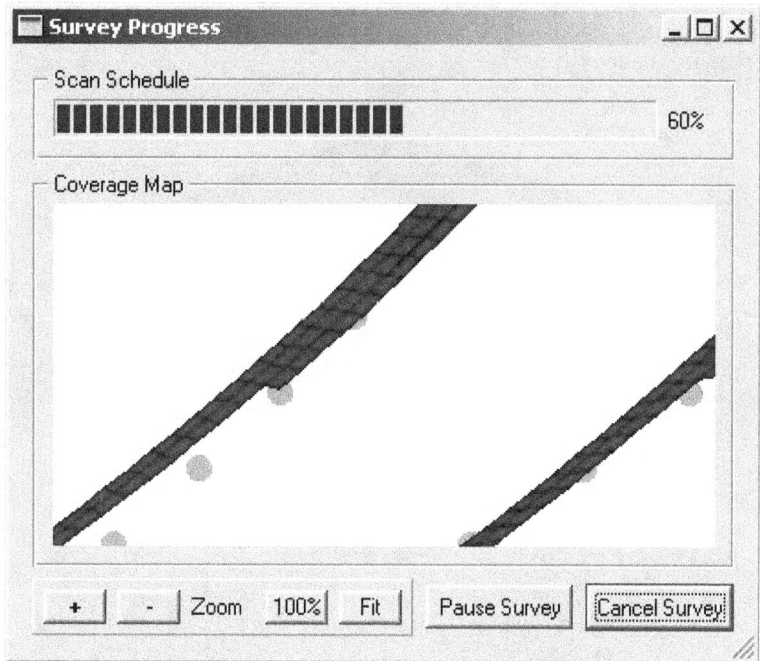

Figure 21. Interactive coverage map.

The purpose of the interactive coverage map is to give the user immediate qualitative feedback on the effectiveness of the proposed scanning strategy. The algorithm for drawing the coverage map employs an intersection technique to find the points at which the corners of the rectangular detector view frustum intersect the target sphere. These points are used to draw a filled blue rectangle with a black outline on the coverage map. This intersection algorithm is computationally efficient and does not noticeably impact the speed of the simulation. However, because the edges of the filled rectangle are straight lines rather than sphere segments, the map is an approximation of the actual coverage. For view-frusta larger than those being considered here, a sphere pixelization technique would provide more accurate results at a much greater computational expense. For efficiency reasons, the overcoverage of the target sphere is not directly measured in this algorithm. However, a qualitative sense of the amount of overcoverage may be gained by examining the density of the black outlines drawn around the blue-filled rectangles (see fig. 22). Areas of substantial overcoverage will have more closely spaced black lines.

The coverage map does not account for data-taking exclusion zones caused by light from the Sun or by physical occlusions of the detector view by planets and moons. The Coverage/Sensitivity Tool can be used to examine these effects if desired.

When the survey is complete a series of reports are written. These include one report file for each detector in the scene and one summary report file. The detector reports contain a review of the detector and survey parameters and a list of observed NEOs at each time step in which at least one NEO was in the telescope FOV. The summary report lists how many NEOs were never detected during the survey and lists those objects by name. These reports include measured S/N and per-pixel S/N values for each observation, as well as survey statistics. Future enhancements include the ability to consider exclusion zones for a detector FOV caused by bright objects such as the Sun or Jupiter, or by physical blockages by objects obstructing the view cone or frustum.

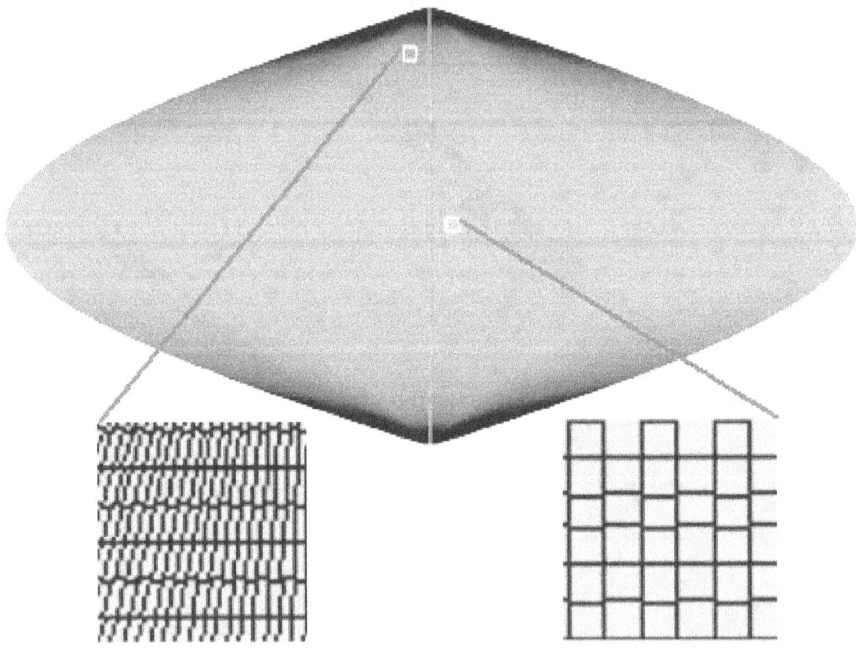

Figure 22. Case 1 target sphere coverage map.

Impact Simulations

To follow is a description of the current and planned capabilities of the SEE that relate to conducting a CAPS impact simulation. A simulated Earth impactor being viewed from two vantage points as part of an impact analysis is shown in figure 23 (note that the impactor scale is greatly exaggerated in the right-hand image). At present, the SEE allows a user to create an impacting asteroid by loading a trajectory file that contains position versus time, or position and velocity versus time data. It is assumed that if the object being imported is identified as an impactor, the trajectory data describe a path that intersects the ecliptic at 1 au. If the Earth is defined as having a zero eccentricity, zero inclination orbit with the semi-major axis at exactly 1 au, algorithms in the SEE will be able to assign an epoch to the NEO trajectory data that will cause an Earth impact in an arbitrary simulation year. If the simulated Earth orbit does not exactly meet the above parameters, the epoch will still be resolved for a closest Earth approach but may not generate an impact.

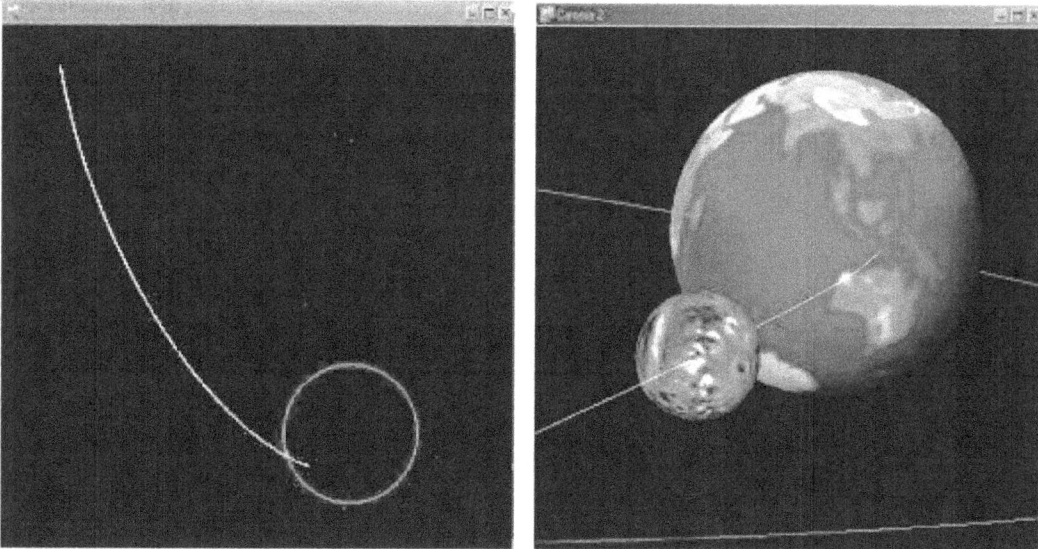

Figure 23. Earth impactor.

Future enhancements are planned that will allow the user to create a minor planet that impacts the Earth or another planet at any selected date, without requiring the user to supply trajectory data.

A POD analysis is conducted in the current version of the SEE by first creating one or more CAPS observatories and then one impacting NEO. Selecting "precision orbit determination" from the analysis pull-down menu will then launch the POD wizard, which guides the user through the process of inputting the required parameters. Among the required parameters is the date of the first observation of an NEO by observatory. If the user prefers to gauge the first observation opportunity in terms of distance to the object from the observatories rather than by selecting a date, a range GUI is provided under the tools pull-down menu. The range GUI displays the current distance between any two objects in the scene. The user may initially adjust the simulation time controls until the range GUI reports the desired distance at which to begin the observations; then the user includes the corresponding date to populate the date field of the POD wizard.

When the POD wizard is completed, a POD analysis thread is launched to collect the data required for the POD report. The POD report contains time stamped data describing the location of the observatories, the direction of the NEO with respect to these observatories, and the position and velocity of the NEO. The POD report file can be read directly into the Matlab®/Simulink® POD system used to generate the reconstructed trajectories and the erroneous predicted miss distance (EPMD). Future enhancements are planned that will allow the reconstructed trajectories and an error ellipsoid (a metric designed to indicate the margins of error associated with the collections of reconstructed trajectories) to be imported into the SEE and visualized.

Preliminary Results

To follow are results given for some preliminary survey simulations. The basic scanning algorithms used in these examples are intended primarily to show how the SEE environment can be used to compare the performance of various CAPS configurations using different NEO data sets. The first configuration examined will be a nonrealistic configuration in which the CAPS detectors are fixed inertially in space at the center of the target sphere (this location is the center of the Sun in the SEE space.) This case is intended to serve as a baseline to isolate the effects of the craft location and orbit motion on the effectiveness of the scan. The second configuration consists of two detectors in orbit about the Sun at a radius of 1 au. The third configuration consists of two detectors located on the Moon.

Case 1: Nonorbiting Detectors

In this case, two CAPS telescopes are fixed inertially at the center of the Sun. Each telescope contains a detector with a 1° square FOV. The stare-slew scan proceeds by slewing the detector to the target location and holding the detector fixed for the stare duration. The scan pattern moves the detector in an elevation sweep from the north celestial pole to the south celestial pole. In between each elevation sweep, an azimuth clock is performed to move the detector to a new celestial longitude. Figure 24 illustrates the scanning pattern. The view at left is looking down toward the south celestial pole and shows the detector FOV cones as black lines. The view at right illustrates the motion of the detector boresight as it performs an elevation sweep. The black circles are positions at which a stare occurs. The pattern also provides for a 5-percent overlap of the adjacent areas of the scan at the celestial equator. This scan requires ≈45 days to cover the entire celestial sphere. This simple scanning algorithm produces a large amount of undesirable overcoverage as the detector moves toward the poles (an alternative scanning algorithm with less overcoverage is also be examined).

Scan geometry

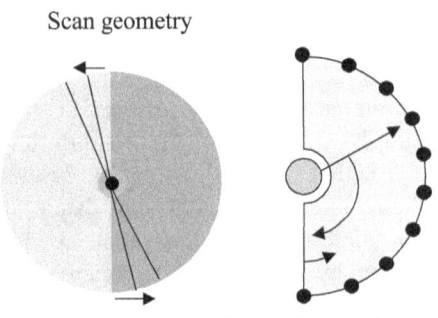

Elevation sweeps from pole to pole
with azimuth clock after each sweep

Overcoverage of 5 percent at equator

Figure 24. Case 1 initial scanning pattern.

The detector instrument is a telescope equipped with a charged-coupled device (CCD) array of pixel dimensions of 32 000 × 32 000. Additional detector properties are as follows:

Telescope optical diameter, m	3.20
Focal length/optical diameter, f/#	5.20
Resolution, arcsec/pixel	0.10
Quantum efficiency	0.80
Optical efficiency	0.80
Integration time, s, or dwell time	90.00

The first NEO data set used in this case consisted of 8988 external returning comets (ERCs) with semimajor axes ranging from 33 to 10 000 au and absolute magnitude ranging from 9.2 to 19. This data set was provided by Dr. William Bottke of Southwest Research Institute based on reference 2. These are simulated objects that were created using methods designed to produce realistic orbits based on known comets (ref. 3). None of the simulated objects is designed to necessarily impact the Earth. Each object was assigned an identical physical property as follows:

Diameter	1 km
Albedo	0.02
Photometric slope	0.15

The true anomaly of each object was set such that its heliocentric distance on the date that the scan begins was approximately 9 au (see fig. 25).

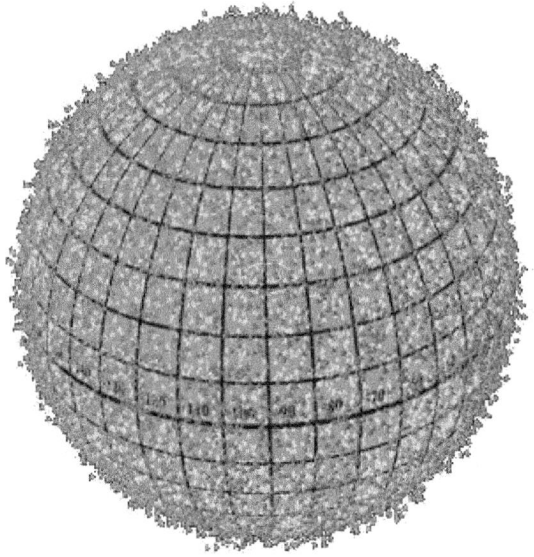

Figure 25. Visualization of ERC set at start of scan. The gridded sphere marks a heliocentric distance of 9 au. The density of the objects appears isotropic.

The summary report information generated by the SEE at the completion of this simulation is shown below:

> Number of targets: 8988
>   Targets not scanned: 1
>   Targets with low per-pixel S/N (<6): 0
>
> Total undetected targets: 1(0.0111259 %)
>
> Signal-to-Noise Ratios
>   Highest: 11.1409   Lowest: 8.63558   Average: 10.3474
>
> Per-Pixel Signal-to-Noise Ratios
>   Highest: 7.87783   Lowest: 6.10628   Average: 7.31668
>
> Number of scopes: 2
> Survey start date: 2000 01 01 00:00:00.000
> Survey duration, days: 44.3345
> Frame integration time, s: 90
> Run time: 4 hr, 9 min, 35 s
>
> Targets not scanned:
> z_6059

Figure 22 includes the final coverage map. All areas of the target sphere have been covered by the scan. The high density of black outlines near the poles of map relative to a similar area at the equator indicates the severe overcoverage at high and low latitudes (see insets).

The "targets not scanned" category in the survey summary report refers to the number of objects never found in the FOV of any of the CAPS detectors. In this case, because our scan covered the entire celestial sphere, this could happen only if the NEO moved into a previously visited area of sky before the NEO's original location had been visited. Review of the simulation graphics showed that object "z_6059" had indeed been located just west of the eastward-moving detector at the start of the simulation. By the end of the scanning period, the object had moved eastward into a previously covered area of sky. This object would have been observed within the first several hours if the scan had been allowed to repeat.

The "targets with low per-pixel S/N" category refers to objects that were scanned but would have gone undetected due to low per-pixel S/N ratio. In this example the cutoff value has been set to 6.0. The total S/N ratio for an object is a measure of the mean number of photons per unit time received by the detector. The per-pixel S/N ratio takes into account the number of pixels on the CCD over which this signal will be spread (a minimum of 1 pixel). An object with a higher angular rate relative to the detector will reduce the per-pixel S/N reading.

These preliminary results indicate that the detector parameters appear to be sufficient for observing 1-km ERCs with the given orbits and physical parameters. To examine the effect of faster moving NEOs, the case was repeated with a new object data set consisting of 8268 ERCs with true anomaly adjusted so that the heliocentric distance was approximately 5 au. An excerpt from the summary report for this run is below:

> Number of targets: 8268
> Targets not scanned: 21
> Targets with low S/N (<6): 0
>
> Total undetected targets: 21(0.253991 %)
>
> Signal-to-Noise Ratios
> Highest: 40.8559   Lowest: 28.6865   Average: 35.3438
>
> Per-Pixel Signal-to-Noise Ratios
> Highest: 28.8895   Lowest: 14.5187   Average: 22.7721

In this case, 21 objects move in such a way as to avoid the scan pattern. Figure 26 shows a view of the objects looking toward the south celestial pole. These objects move in a clockwise direction, effectively following behind the detectors as they also move clockwise. The circle marks a radius of 1 au.

The next experiment utilizes a collection of 9001 simulated NEAs as the target set. These objects were also provided by Dr. William Bottke. The semimajor axis of these objects ranges from 0.4 to 7.4 au, and visual magnitudes range from 13.2 to 22. Figure 27 is a visualization of the object distribution, which shows a majority of low-inclination orbits.

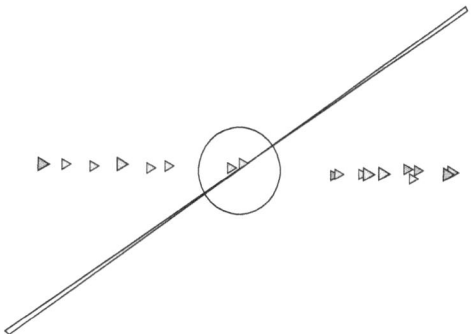

Figure 26. View of missed objects.

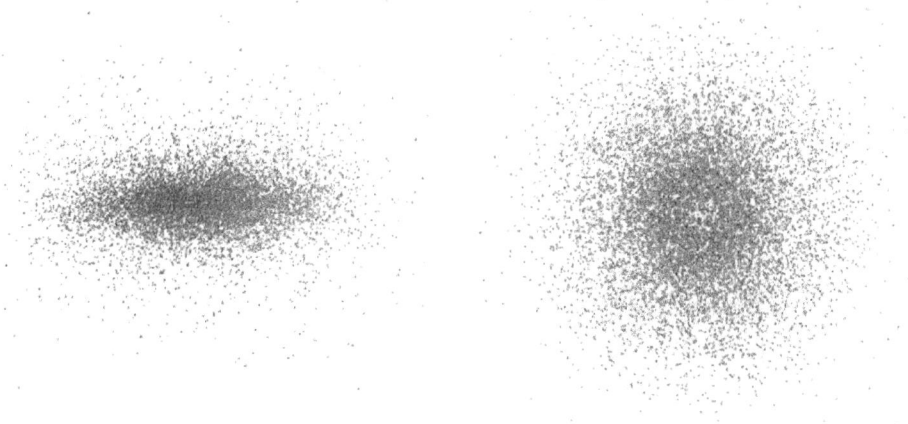

Figure 27. NEA distribution: View along ecliptic (left) and from the north celestial pole (right).

A repeat of the scanning simulation produced the following report:

>   Number of targets: 9001
>   Targets not scanned: 889
>   Targets with low per-pixel S/N (<6): 9
>
>   Total undetected targets: 898(9.97667 %)
>
>   Signal-to-Noise Ratios
>   Highest: 172158   Lowest: 5.28778   Average: 435.808
>
>   Per-Pixel Signal-to-Noise Ratios
>   Highest: 3402.03   Lowest: 3.73903   Average: 76.3751

A review of the visualization showed the majority of the nearly 900 unscanned objects were missed because their high velocities allowed them to pass between subsequent elevation sweeps in the scanning pattern. The nine objects with less than the minimum required per-pixel S/N ratio had absolute magnitudes greater than 19.

The last experiment in this case incorporated a new scanning strategy designed to reduce the overcoverage at areas near the poles. This strategy sweeps through arcs in azimuth with an elevation clock after every azimuth sweep. The scan starts at the north celestial pole and ends at the south celestial pole. The length of the azimuth arc is adjusted for the current elevation of the scan. The arc length is zero at the poles and largest at the celestial equator. The scan also incorporates a 5-percent coverage overlap at the equatorial azimuth sweep. A schematic of the scan strategy is shown in figure 28. The coverage map for this scan is shown in figure 29. Comparison of this map with figure 22 shows the reduced overcoverage of the poles for the azimuth-sweep strategy.

The improved scanning efficiency is reflected in the survey duration, which has been reduced from 44.3 days for full coverage for the elevation-sweep scan to 28.3 days for the azimuth-sweep scan. The results of a simulation using the improved scan strategy against the same NEA object data set are excerpted here from the summary report:

>   CAPS Survey Summary
>
>   Number of targets: 9001
>   Targets not scanned: 81
>   Targets with low per-pixel S/N (<6): 9
>
>   Total undetected targets: 90(0.999889 %)
>
>   Signal-to-Noise Ratios
>   Highest: 98052.7   Lowest: 5.29113   Average: 440.672
>
>   Per-Pixel Signal-to-Noise Ratios
>   Highest: 3093.17   Lowest: 3.74139   Average: 81.9013
>
>   Number of scopes: 2
>   Survey start date: 2000 01 01 00:00:00.000
>   Survey duration, days: 28.3514
>   Frame integration time, s: 90
>   Run time: 2 hr, 34 min, 30 s

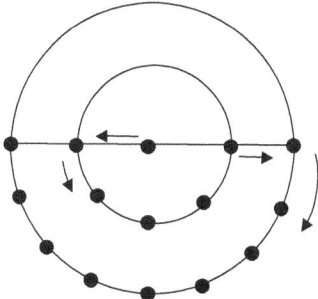

Figure 28. Scanning pattern using azimuth sweep. This view is looking down toward south pole of target sphere. The number of stare points in an azimuth sweep increases as elevation goes from pole to equator.

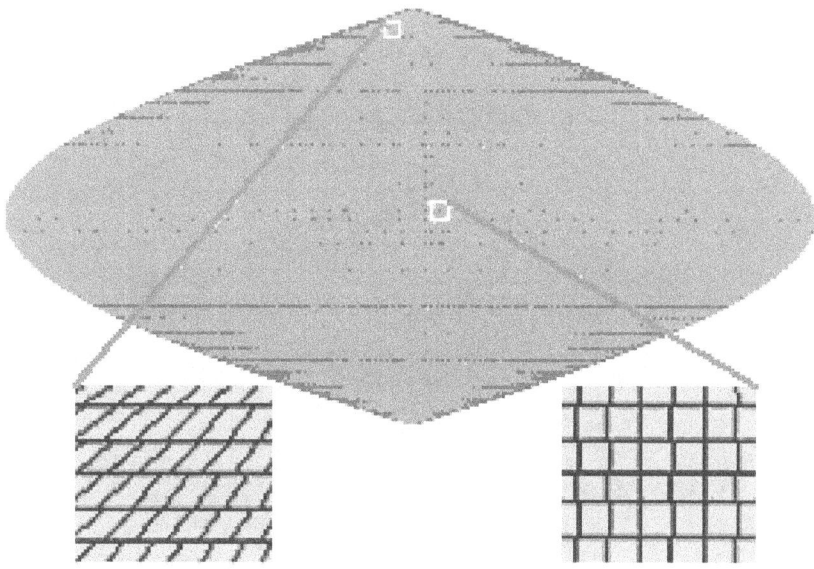

Figure 29. Survey map for azimuth-sweep scan pattern.

Also interesting is the substantially reduced number of unscanned objects (81 here versus 889 in the elevation-sweep strategy). Likely causes could be that the reduced time between subsequent passes of the azimuth sweep reduce the possibility of the NEO traveling far enough to evade detection, or that the azimuth sweeps are better aligned with the object trajectories. The nine low-signal objects are the objects with magnitudes greater than 19.

Case 2: Two-Location System

In this case, a detector is located on each of two crafts orbiting the Sun at 1 au. The scanning pattern is similar to the elevation-sweep method used at the start of case 1, but is adjusted for the fact that each detector in the two-location system needs to sweep over its local north and south poles in order to view the poles of the sun-centered target sphere. In figure 30, the left view shows the posigrade 1-au orbits of the two crafts, with the craft starting and ending positions as unfilled circles. The filled black circles indicate the positions of the crafts when the scan is at opposition. The right view shows that the elevation sweep spans more than 180° in order to cover the poles of the sun-centered target sphere.

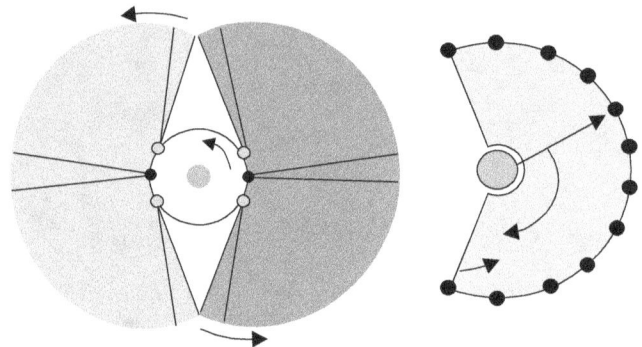

Elevation sweeps over local poles with azimuth clock after each sweep

Overcoverage of 5 percent at equator

Figure 30. Two-location system scanning pattern.

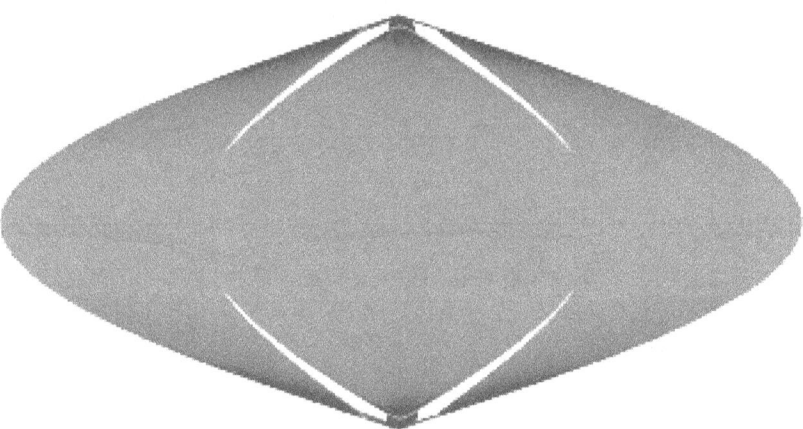

Figure 31. Coverage map for two-location system, elevation-sweep scan, 9-au target sphere.

The coverage map for this pattern is shown in figure 31. Note that because the detector locations are offset from the target sphere origin, the radius of the target sphere will affect the amount of coverage registered by the map. While this effect can be mitigated by selecting a "celestial" sphere with infinite radius, the coverage of such a sphere is not necessarily a desirable metric for the CAPS system. Because the NEOs of interest are within approximately 9 AU, the desirable metric will be one that indicates the volume of space covered by the detectors in this region. To examine the coverage at other heliocentric distances, the simulation can be programmed to produce multiple maps, although with increasing expense in simulation time and memory usage. Because of these difficulties, selecting a method for measuring the combined coverage of a two-location system in the space near the detectors will be an important design consideration for further CAPS survey simulations.

The NEOs used in this run were the ERCs in case 1 whose heliocentric distances were approximately 9 au at the beginning of the scan; therefore, a single 9-au sun-centered target sphere was used. The scan

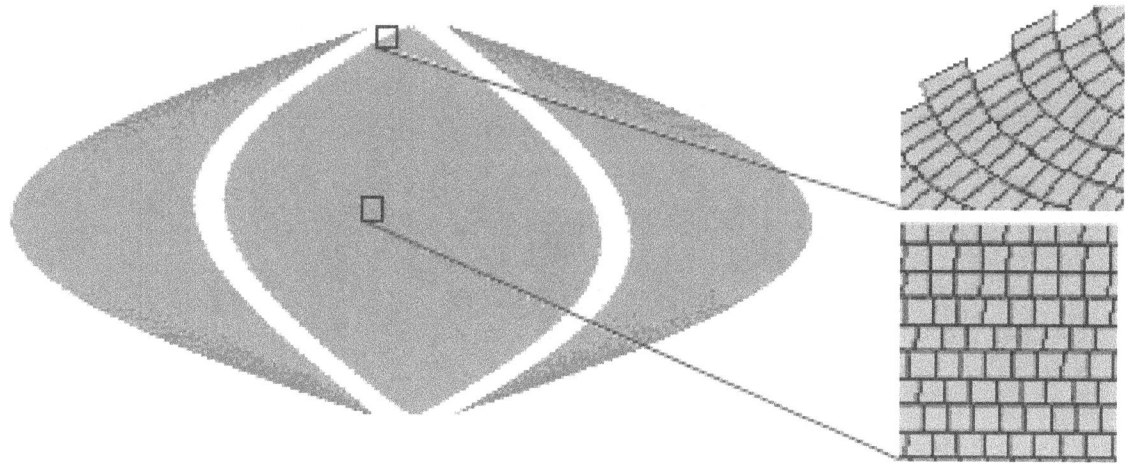

Figure 32. Coverage map for two-location system, 180° elevation span.

was terminated when the elevation sweeps had covered all longitudes along the equator. The wedge-shaped gaps in the map are an effect of the off-opposition scanning geometry, where the north-to-south motion of the detector was not corrected to maintain alignment with the lines of longitude on the target sphere. Subsequent repeats of the scan will tend to fill in these gaps as the motion of the craft causes the scan opposition point to move.

In this case, 204 objects were not scanned due to the gaps in coverage. No objects had less than the minimum S/N ratio of 6.0. The scan took 51 days to complete the coverage at the equator.

This simulation was also run using the 5-au ERCs and a subset of NEA objects used in case 1 (the reduced NEA set contains only objects with absolute magnitude greater than 19.1). Again it was seen that the faster moving objects caused an increasing number of missed sightings. Of the 8268 ERCs at 5 au, 461 were unscanned. Of the 2383 NEAS, 892 were not scanned and two objects had low per-pixel S/N measurements.

The azimuth-sweep scanning strategy used to reduce overcoverage at the celestial poles was also attempted in this case. Figures 17 and 15 show the coverage results for two versions of the azimuth-sweep strategy. In figure 32, the azimuth-sweep span is 180° of the elevation on a sphere centered at the detector. The 1-au distance from the detector location to the 9-au sun-centered target sphere origin causes the gaps in the target sphere coverage. Figure 33 includes the results of adjusting the scan to cover 200° of elevation at the local sphere in order to get better coverage of the 9-au sun-centered target sphere. The duration of the scans shown in figures 32 and 33 were 28 and 31 days, respectively.

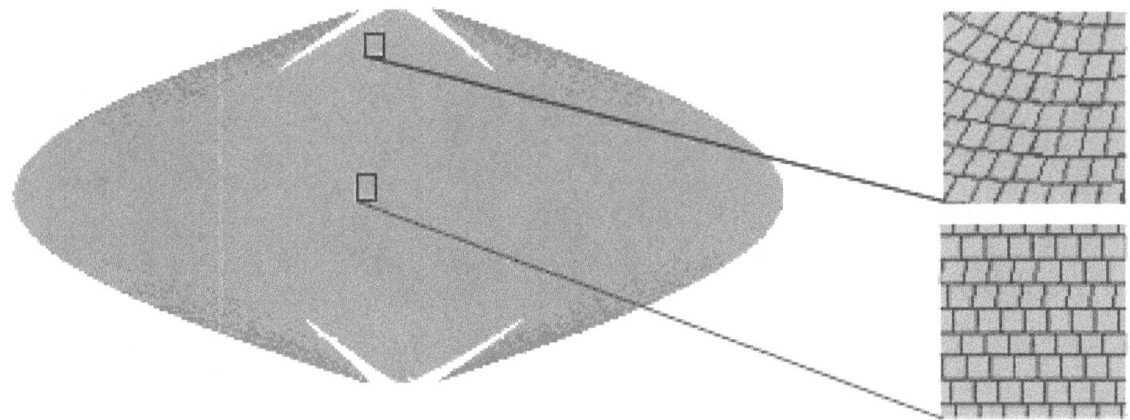

Figure 33. Coverage map for two-location system, 200° elevation span.

Case 3: Two Detectors on the Surface of the Moon

The geometry of this CAPS survey configuration is shown in figure 34. The view at left shows the view from the Moon's equator of the elevation sweeps. The view at right shows the visualization graphics of the scaled-up Moon bases and the detector view frusta. The two detectors are located on the Moon surface at 75° north and south latitude. The north detector is at 90° east longitude, and the south detector is at 90° west longitude. Each moon-based detector is performing an elevation sweep designed to cover the northern or southern hemisphere of a 9-au target sphere centered at the Sun. The rotation of the Moon itself is used to change the azimuth of the detector, so no azimuth clock is performed. The scan was terminated after 28.5 days. The coverage map (fig. 35) illustrates that the azimuth rate introduced by the Moon's rotation is too large to allow full coverage at the equator. To improve coverage at the equator, the elevation-sweep scan pattern must reduce coverage at the poles or use a smaller dwell time or a larger view frustum.

The coverage results for an alternative configuration with the detectors located directly on the lunar poles, each using an azimuth-sweep scan to cover half of the sun-centered 9-au target sphere, are given in figure 36. A full 360° azimuth sweep is performed at each elevation. After a complete azimuth sweep, an elevation clock of 0.95° is performed and a new azimuth sweep is started. The duration of the survey was 28.5 days. Overall coverage appears improved over the elevation-sweep scan. Small horizontal strips of missed coverage are caused by the rotation of the Moon. Revision of the scanning program to account for this effect could provide 100-percent coverage with little additional scanning time. The effect of occlusion of the detectors by the Earth will, however, substantially affect the real coverage results near the lower scan elevations.

The scan shown in figure 36 was performed on the collection of 2383 NEAs used in case 2. Sixty of the objects were unscanned and two were missed due to low signal. This result is consistent with the trend that patterns sweeping generally along lines of latitude (azimuth sweep) rather than lines of longitude (elevation sweep) have been more effective in reducing the number of missed NEAs.

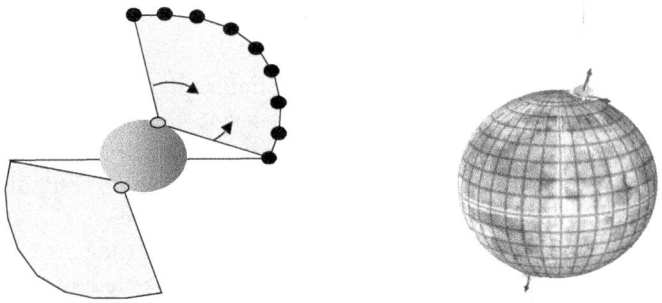

Elevation sweeps less than 180°

Overcoverage of 5 percent at equator

Figure 34. Moon-based system scanning pattern.

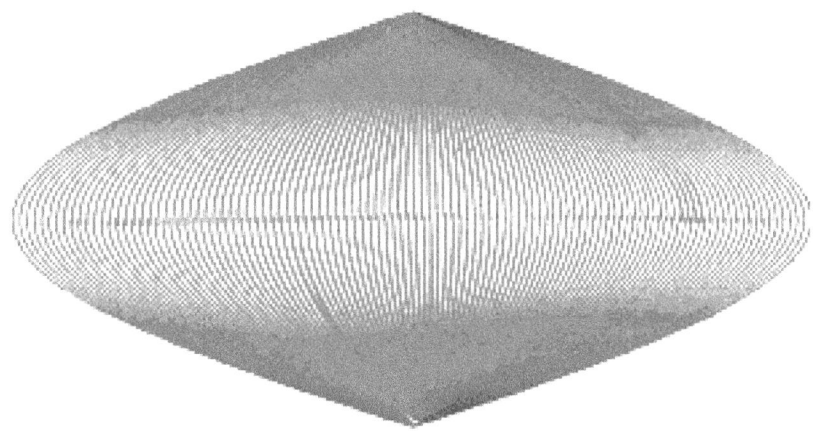

Figure 35. Coverage map for moon-based detectors with elevation-sweep scan.

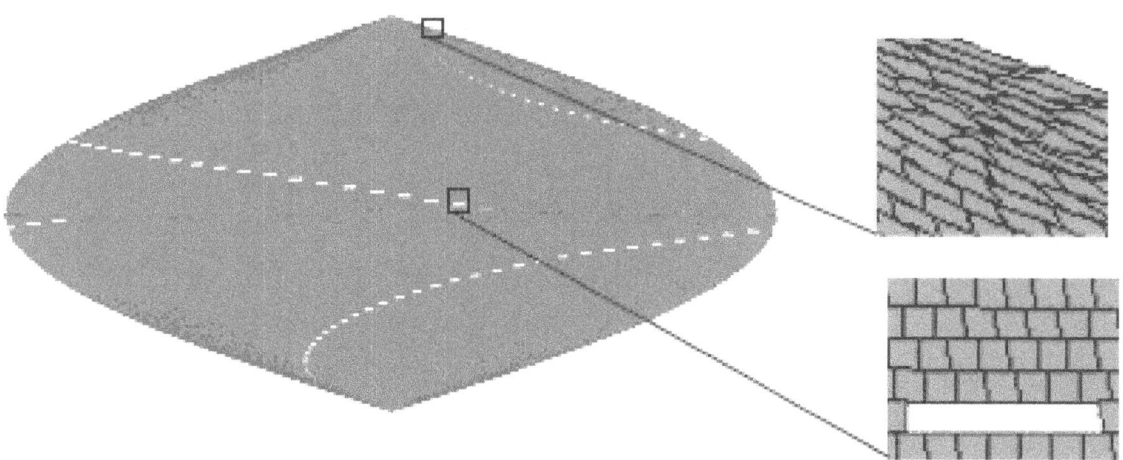

Figure 36. Coverage map for moon-based detectors with azimuth-sweep scan.

Concluding Remarks

The beginnings of an integrated Comet/Asteroid Protection System (CAPS) simulation and visualization tool have been constructed in the Synergistic Engineering Environment (SEE) application. The availability of interactive three-dimensional views of the scene geometry assists in evaluating the effectiveness of proposed detection systems and in improving their designs. The preliminary examination of several proposed survey configurations against simulated near-Earth object (NEO) data sets showed the ability of the SEE to compare the efficiency of different scanning algorithms. The problem of designing an effective scanning algorithm was shown to be closely tied to the problem of establishing a valid survey coverage metric when the system contains more than one detector location. It was demonstrated that a scan pattern achieving full coverage of a given area on a target sphere can miss substantial numbers of fast moving objects due to the objects' motion between subsequent passes of a sweeping-style scan. To counter this effect, a successful scanning pattern should consider the geometry of the object paths.

The future development of the SEE for CAPS would benefit from addition of the following features:

- Sophisticated scanning algorithms that allow multiple craft to work together to provide full coverage of the target volume with minimum overcoverage.

- Modeling of survey detector frustum occlusion by solid objects or light sources.

- Volume rendering of the coverage map showing overcoverage at all distances between a minimum and maximum radii of the target coverage volume.

- Near-Earth object-creation routine to automatically create object trajectories that will impact the Earth at a user specified date.

- Further integration with the Coverage Analysis Tool separate from the SEE used in the orbit determination and orbit modification sections of this report.

References

1. Tegmark, M.: An Icosahedron-Based Method for Pixelizing the Celestial Sphere. *Appl. J. Lett.*, vol. 470, 2001, p. 81.

2. Wiegert, P.; and Tremaine, S.: The Evolution of Long-Period Comets. *Icarus*, vol. 137, issue 1, Jan. 1999, pp. 84–121.

3. Levison, H. F.; Morbidelli, A.; Dones, L.; Jedicke, R.; Wiegert, P. A.; and Bottke, W. F.: The Mass Disruption of Oort Cloud Comets. *Science*, vol. 296, June 21, 2002, pp. 2212–2215.

Comparison of Detector Technologies for CAPS[9]

JANA L. STOCKUM
San Diego State University
San Diego, California

Introduction

In this paper, several different detectors are examined for use in a Comet/Asteroid Protection System (CAPS), a conceptual study for a possible future space-based system. Each detector will be examined for its future (25 years or more in the future) ability to find and track near-Earth Objects (NEOs) from a space-based detection platform.

Within the CAPS study are several teams of people who each focus on different aspects of the system concept. This study's focus is on detection devices. In particular, evaluations on the following devices have been made: charge-coupled devices (CCDs), charge-injected devices (CIDs), superconducting tunneling junctions (STJs), and transition edge sensors (TESs). These devices can be separated into two main categories; the first category includes detectors that are currently being widely utilized, such as CCDs and CIDs. The second category includes experimental detectors, such as STJs and TESs. After the discussion of the detectors themselves, there will be a section devoted to the explicit use of these detectors with CAPS.

Charge-Coupled Devices Background

Invented in the 1970s, charge-coupled devices (CCDs) and charge-injected devices (CIDs) have made a huge impact on astronomy. Previous observations were primarily made either directly by the human eye or by using photographic plates. Neither the eye nor the photographic plates have a linear response to exposure to light. However, CCDs and CIDs have a highly linear response. The linearity of the device allows scientists to better estimate integration times needed to make observations. As great as the leap of linearity was, these devices have not made such a large impact from this ability alone. CCDs and CIDs have more benefits, such as higher resolution, wider dynamic range, higher sensitivity, and digital capabilities specific to each.

CCDs have become the most widely sold detection devices to date, taking in a projected 79 percent of the market share of solid-state image sensors in the year 2000 (ref. 1). These devices have even made the transition from research tools to commercial use in everyday items, such as digital cameras and pharmaceutical lid detection (ref. 2).

There are several reasons for the large market share taken in by CCDs. These devices have reached a state of maturity allowing for nearly any configuration, and of nearly any size. The working temperature for astronomical uses is relatively easy to maintain. The vast capabilities of CCDs include operating as imagers, photon counters, and with spectrographs. Because of unique specifications, each function requires a different configuration (ref. 3).

One such general configuration used is that of an imager. Operating at nondiscrete wavelengths, the CCD potential wells collect the energy from the incident photons within their electron wells via the photoelectric effect. Once collected, the energy is read out well to well, row by row, and into a computer via charge transfer. The transfer of information from well to well is accomplished by applying a variable

---

[9]Chapter nomenclature available in chapter notes, p. 217.

voltage to neighboring potential wells. This transfer of charge is generally measured and referred to as the charge transfer efficiency (CTE). The CTE can slow the astronomer's ability to observe and therefore a lower CTE is a sign of a higher quality CCD. While the charge is being transferred, the CCD potential well makes no measurement of wavelength or exact photon timing. However, the energy can be used to produce an image of the object(s) being observed (ref. 3).

In order to view colors of the observed image, different pieces of equipment must be introduced to the configuration. A grating or prism is commonly used to separate the differing wavelengths from a source (ref. 3). However, the more commonly used grating (due to the difficulties of creating a perfect prism) works by bending the incoming signal. Once the signal is bent, it is weakened and the signal-to-noise (S/N) ratio drops. To regain a desired S/N ratio, a longer exposure time is required.

Another deficiency of CCDs, called "blooming," refers to the saturation of an electron well. Once an electron well has collected too many photons (the limit is set by CCD composition), incoming information will "spill" onto a neighboring well. The result may be an entire row or column of the array too full of energy to allow for much of an image. Only a streak of bright light across a row or column of pixels can be seen.

Charge-Injected Devices Background

CIDs are similar to CCDs in their general absorption of photons. However, CIDs have a specific configuration that allows for multiple benefits. Instead of reading the charge information via charge transfer, each potential well has the ability to be read individually. This characteristic allows for different modes of controlled operation and minimizes blooming.

The resistance to blooming enables the CID to perform many tasks that a CCD would have trouble completing. When a CID potential well is over-saturated, the energy does not "spill" over to a neighboring potential well, instead just that one, or at most a few, well is affected. For known bright sources, certain wells that would become oversaturated and rendered useless can be turned off. When collecting spectral information, this ability becomes vital for observing the weaker emissions that would otherwise be hidden by a stronger source.

The idea of effectively turning some collection wells off can also be extended into creating different modes of operation. For example, if the entire field of view offered by the device is not required, half the device can be turned off to allow for quicker recovery time. In the commercial world this is used for fast capturing needs, such as tracking. Another scanning method, a "progressive scan," is described by CID Technology on its website: "this readout enables real-time processing by eliminating the delay required to combine odd and even fields. Instead, lines are read sequentially (1, 2, 3, 4, etc.) allowing an image processor to analyze the latest row of video information while readout continues to the next line" (ref. 2). While these abilities are not crucial to near-Earth object (NEO) detection and tracking, they could be helpful in lowering the loss rate of objects.

Superconducting Tunneling Junctions

As valuable as CCDs and CIDs have been in the past, and promise to be in the near future, a new generation of detectors will likely have an even greater impact. Transition edge sensors (TESs) and superconducting tunneling junctions (STJs) have the ability to identify the time each incident photon arrives on the detector with varying time resolution of ≈ 5 µs (ref. 4). The wavelength of each photon, ranging from 1100 Å to 6 µs (ref. 5), and the location of incidence on the detection element are also immediately recorded by TESs and STJs. Armed with these capabilities, these devices have the potential to revolutionize astronomical observation as we know it today.

Currently, there are experimental data being taken by computer simulations (ref. 5) and with small STJ and TES devices on a handful of ground-based telescopes. The William Herschel Telescope is one of the main telescopes being used for such experiments (ref. 6).

The computer simulations of the STJ system performed at the California Institute of Technology demonstrate the enormous potential of these superconducting devices. In simulations described in detail in reference 5, the researchers compare the abilities of a CCD device attached to the Keck Telescope with the abilities of an STJ counterpart. The results include an estimation that, for the CCD arrangement to gain the same information as the STJ system, the observations would take five times as long, with the extra time owing for the need to use several observations with different filters to gain spectral information that was inherently gathered by the STJ. This study further adds that during a simulated galactic search using broadband techniques, the CCD apparatus attached to the simulated Keck telescope found 1018 of 8717 galaxies with redshifts of approximately 3 during a 4-hour observational time, while the same sized STJ device found 2045 galaxies with redshifts of 5 and with higher accuracy in only a 1-hour integration. In reference 5, Mazin and Brunner discuss the future of STJs and other cryogenic imaging spectrophotometers (CISs) as replacing CCDs as "the wide-field detectors of choice." As such, they speculate that if an STJ were used on a next-generation space telescope with "actively cooling detectors and passively cooled mirrors," this configuration would be able to observe as deep as the Hubble Space Telescope Wide-Field Planetary Camera 2 (WFPC2) instrument with higher spectral coverage and resolution, yet it would require only 2 percent of the observation time. With such resounding redictions, it is easy to see why scientists are excited about the possibilities for these cryogenic imaging spectrophotometers.

Several organizations have already made plans to include STJs as part of their future projects. The Boeing Company, for example, has proposed to install an STJ aboard its nonredundant linear array (NRLA) concept (ref. 7). Dr. Peacock, one of the founding scientists of the STJ concept, believes these detectors will soon be used in the microchip industry to identify contaminants in silicon. In fact, the STJ devices have many applications in "commercial and industrial fields, where fast measurements to capture phenomena in wavelength at very low light levels are required" (ref. 8).

Despite nearly limitless uses within the astronomical community and several commercial applications, the STJs and other CISs still need considerable support. The devices themselves need to be cooled to temperatures below 1 K. This refrigeration requirement must be met by some new cooling technology before a long-term space-based implementation can proceed.

Detector size is another hurdle that must be overcome. While research to increase the CIS detector size is currently progressing, much more needs to be done to equal the size of current CCDs. There is considerable optimism that the cooling technology and size limitation problems are solvable in the near future.

Transition Edge Sensors

TESs are among the newly created CISs. These devices appear to have great potential in revolutionizing astronomy and other imaging fields. TES devices are constructed from tungsten cells on a silicon substrate. The operating temperature is well below 100 mK. The sensor is held and monitored at a fixed voltage. Photons are detected when their incidences on the tungsten create a decrease in the current needed to maintain the fixed voltage. This decrease is related to the energy of the incident photon. The exact energy level is recorded as a pulse and sent to a direct current (DC) Superconducting Quantum Interference Detector (SQUID) array, which is one of the most sensitive devices for magnetic field detection (ref. 9). Within the SQUID, the pulse is amplified, digitized, and given a peak height. Then, the information is assigned a time from the Global Positioning System (GPS) receiver and recorded to a computer (ref. 10).

The inherent spectral range of the TES is quite large and may therefore be better suited for a Comet/Asteroid Protection System (CAPS) detection sensor. The sensitivity of the tungsten device begins near 372 nm and ranges to 18600 nm (0.3 eV to 15 eV). However, the quantum efficiency (QE) in the optical, ≈50 percent, is not as high as in other CIS devices, such as STJs (ref. 10).

Detector Functionality: Classification Abilities

Of great importance to an NEO protection system is the determination of an asteroid's size and classification, or composition. These important characteristics cannot be accurately determined by visible observations alone because the amount of reflected sunlight from an object depends on the albedo and the size (ref. 11). Simultaneous measurements of the thermal flux and the visible brightness of an object are necessary to first classify it within the general asteroid compositional groups; then estimates of the size and albedo can be calculated. The hafnium-based STJ device is projected to gain a resolution of wavelength difference on the order of 5 nm over most wavelengths including the near infrared (2 μm on the ground and 6 μm on a space-based platform (ref. 5). According to A. Cellino, this is sufficient to detect the characteristic NEO identifier at 2 μm. However, these values are still insufficient to locate the reradiated peak located between 6 and 10 μm (ref. 12). Other asteroids may only be observable near 12 μm due to their extremely low albedos (ref. 13). A new base material would need to be used for the STJ to reach such wavelengths. TES devices actually have the capability of observing a larger wavelength range than STJs, including this needed range of 12 μm. However, this is at the cost of a 20-percent loss in QE over the optical range (refs. 14 and 15). Yet, even with the decrease in QE, the increased range of wavelength may make the TES detector more favorable. Regardless, the lack of sensitivity in the 8- to 12-μm range is a deficiency of the STJ that would need to be addressed by a CAPS infrared detection capability.

Time Tagging With Automation

Another feature of the STJ and TES that can greatly benefit CAPS would be the ability to time tag. It is possible that time tagging each photon incident to the detector could allow for a reference of movement across the detector's field of view, therefore permitting another (perhaps more accurate) method for a computer to recognize NEOs. To test this hypothesis, more research needs to be done, perhaps a full computer model testing this proposed method versus the streak identification process, or other methods in use today.

Comparison With Streak Method of Detection

Another comparison between the current methods for locating NEOs is presented mathematically. The limiting magnitudes for the systems are calculated using equations developed by Alan Harris. Therefore the larger the calculated number, the better sensitivity of the detection system. The presented mathematical model is taken from a previous comparison by Alan Harris (ref. 16). In his paper, Harris examines the benefits offered by CCDs over the then commonly used photometric plates. In this paper, a comparison of his data with values calculated by assuming an STJ with similar properties of a CCD, but with a more accurate QE, is provided. A description of the symbols used is provided in table 1. The results displayed in table 2 (in the bottom section) show that even when assuming the S/N ratio and limiting magnitudes are the same for the CCD and STJ, the enhanced QE allows an STJ system to find darker targets.

Table 1. Description of Symbols

Description	Symbol
Effective aperture, m	D
Effective focal length	f
Field of view	-
Quantum efficiency	Q
Number of pixels to contain image	N_p
Area of a point image, arcsec^2	A
Exposure time, s	t
Limiting magnitude	m_l

Table 2. Limiting Visual Magnitudes for Detector Technologies

	CCDs	STJs	Assumed for all systems	
S/N	6	6	A	2
Q	0.3	0.8	D	4
m_l	20.5	20.5	t	165
	Photographic limiting magnitudes	CCD limiting magnitudes	STJ limiting magnitudes	Description
Relatively stationary bodies	17.95	23.37	23.83	Object does not move in frame
Relatively short trailed objects	21.10	26.14	26.61	Object moves at rate > $\left[\sqrt{A}/t\right]$
Relatively long trailed objects	22.92	25.19	25.65	Object's trail spans over more than length of detector, here assume they are all the same length (5 cm): rate of movement>length of chip/t

Quantum Efficiency Comparison

It should be noted that the results depicted in table 2 assume conservative input values. To adequately reflect the performance of an STJ, the lack of readout noise needs to be considered when determining the value for the S/N ratio. Due to a lack of available estimates, CCD values were assumed to calculate a lower estimate. When complete STJ data sets become available, better comparisons can be formed. Based on the limited information available, a comparison of developed QEs is shown in figure 1. The higher the QE value, the better the detection device performs at any given wavelength.

Figure 1 is based on information gathered from reference 3 for CCD and eye, and reference 6 for STJs. As shown in the constructed graph, STJs do cover the wavelength range with higher average QE. While some of the wavelength coverage is not currently as efficient, many experts predict improvements with the STJ construction in the near future (next 10 years). Many articles note that hafnium-based STJs have shown improvements in QE (ref. 5). The data in figure 1 have also been limited by the atmospheric interference. One major beneficial change would be a space-based platform. Such improvements would then lead to the next graphical comparison, found in figure 2.

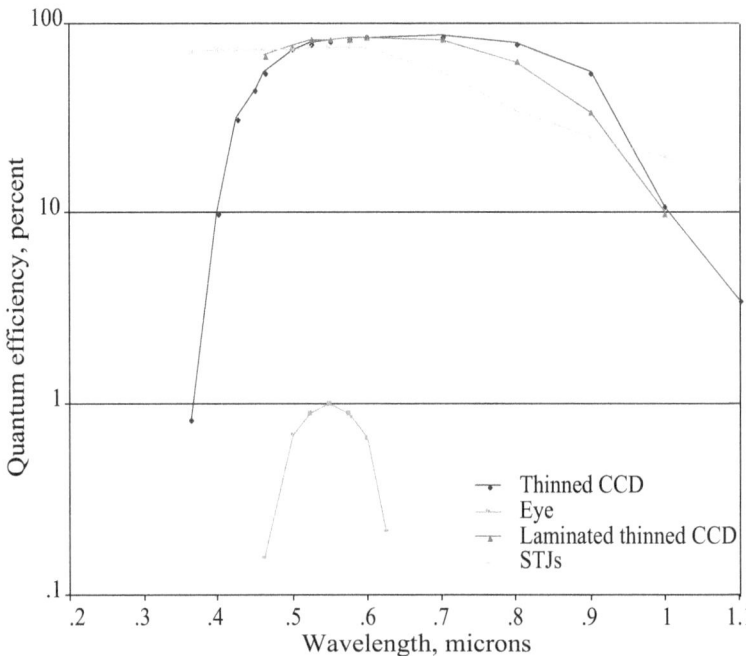

Figure 1. Quantum efficiencies for detector technologies.

Figure 2 is based on information gathered from reference 3 for CCD and eye, reference 6 for STJs, and projected hafnium abilities from personal communication with Mark Cropper (2002) and reference 5. Note that the QE for the projected hafnium-based device is a modest lower estimate. The actual device will have varying QE in relation with wavelength. Figure 2 depicts the projected abilities of the technologies according to Mark Cropper and Dr. Mazin as of the year 2002. When these technologies become available, it is clear to see (the orange line) that the STJ hafnium-based detector placed on a space-based platform will result in the best coverage. According to Dr. Cropper, the STJs will be able to achieve a QE of 90 percent for most of the observable wavelengths.

Near-Earth Object "Tagging"

The idea of assigning an identification description to an NEO apparition based on observable characteristics is theoretically possible. Several distinctions can be made between NEOs. Objects can be "tagged" at two instances: during and after observation.

Currently, "tagging" an object after observation is performed on a regular basis. The objects are separated by the rate at which they cross the frame. This method is performed by using the length of the trail and the time of exposure. In most cases, scientists can use the direction of observation to also infer direction of motion.

However, the trail identification method has some problems. One major concern for NEO tracking missions is the ability to reevaluate the orbit of the target. Many of today's NEO search programs lose objects before complete orbits can be calculated. One method to limit the number of lost targets would be to add more distinguishing characteristics to each description. As a result, any new observations of the previously identified object would be more readily confirmed. With more observations, the object's orbit becomes more accurately determined.

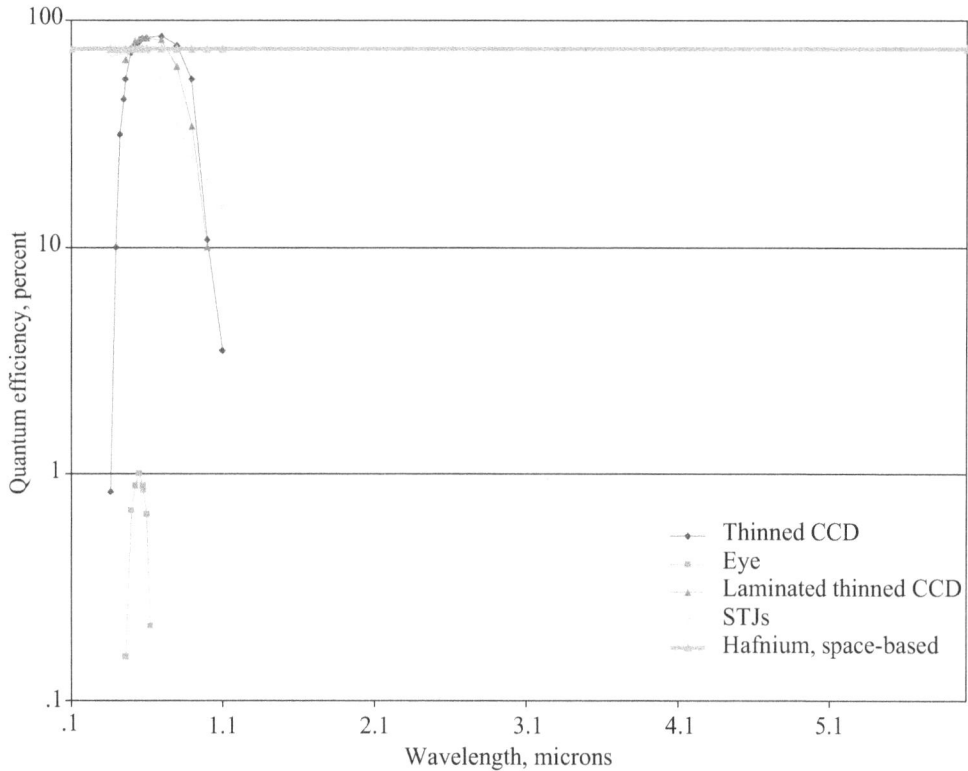

Figure 2. Quantum efficiency percentages of hafnium space-based STJs and other detector technologies.

There are three inherent characteristics of STJs and TESs that could aid in the reacquisition of NEO targets. One such characteristic is the spectral identification of each incident photon. While the objects may streak across the frame, most of the objects do not complete rotations on the scale of scanning observations. Therefore, each incident photon should be from the same side of the target. Ideally, these identified spectral features could allow for quick classification, especially if the aforementioned 2-μm wavelength is observed. Not only is the target less likely to be incorrectly identified as a new object, but also scientists could gain a general idea of the surface composition. Other devices can gain this spectral information as well, but only at the cost of signal detection because other detectors require filters to collect spectral data.

STJs and TESs also offer the same tracking abilities as the previously mentioned CIDs. The key to CID's various modes of operation were in their individualized connections to the computing hardware. STJs and TESs share that individual connecting construction and therefore it seems reasonable that they too could be used in various tracking modes. Being able to split the usage of the actual detection device could also minimize the number of lost targets.

There is one more attribute that the TESs and STJs have that may help tracking abilities: inherent time assignment to each incident photon. Another frame image that may become confusing for an automated computer detection system is that in which two or more objects have crossing trail patterns spanning over the entire detector. Trail patterns could become mixed or simply useless. However, with STJs and TESs, the computer can easily determine the direction of each object by time association. By comparing when each incident photon was detected, the computer can identify the "track" each object created.

The possible benefits of the new cryogenic devices are clear and numerous. While the more developed technologies do offer the same capabilities with the right extra equipment, the newer devices are much more efficient. Once the system has identified many of the NEO targets, research time on telescopes with the cryogenic devices will also be attractive to deep space astronomers. The devices are ideal for such subjects as pulsars or gamma ray bursts.

Array Sizes/Shapes

The possible size and shape of detection devices are also important factors in determining the best system. Currently, CCDs and CIDs are well-developed technologies and can be constructed in nearly any size or shape a reasonable project would require (personal communication with Dr. Bob Leach and Dr. Mark Cropper). Large ($20\,000 \times 20\,000$ pixels) CCDs and CIDs are constructed as mosaics; however, this type of engineering cannot be directly applied to the newer cryogenic devices due to their complicated wiring/magnetic structures. Because STJs and TESs are only in the infancy of development, they are currently no bigger than a postage stamp. Yet, experts such as Dr. Cropper believe with proper funding, a multiplexing configuration or layering process could lead to competitively large arrays within 25 years.

Concluding Remarks

Considering all possible detectors mentioned herein, the benefits of transition edge sensor (TES) and superconducting tunneling junction (STJ) detectors are clear. Not only are the new devices more efficient, but they may offer benefits yet to be fully appreciated. Although they are experimental, assuming the cooling and engineering problems are resolved, usage of the TES/STJ detectors should be recommended. The sensitivity is slightly improved due to lack of readout noise, and the STJs and TESs offer other benefits, such as time tagging, inherent spectral capabilities, and versatility of operation. The largest problem facing the development of STJs and TESs is the cooling technology required to successfully operate the devices. The hafnium-based STJ needs to be kept at 0.01 K.

If a single advanced detector technology is not available or is limited, the solution for the Comet/Asteroid Protection System (CAPS) telescopes is to use multiple detector devices. This approach would use a combined cryotechnology and charge-coupled device (CCD) device, thus allowing for the visible and micrometer wavelength coverage. The refrigeration requirements are more reasonable (≈ 35 K for the system and 10 to 15 K for the devices) than those for STJs or TESs, and a computer model has already been developed. This type of approach has already been proposed for asteroid searches by Edward F. Tedesco and colleagues, and is described in reference 12.

The next step in the comprehensive study of revolutionary detector concepts is to add these described detectors to a working near-Earth object (NEO) detection computer program. After running the software with the various configurations, a general comparison can be made. Once the most effective system is determined, cost needs to be estimated. Finally, a cost/benefits comparison should then be preformed.

References

1. Qi, Hairong: A High-Resolution, Large-Area, Digital Imaging System. Graduate Dissertation in Practical Fulfillment of Requirements for the Degree of Doctor of Philosophy Electrical and Computer Engineering. North Carolina State University, 1999.

2. CID Technology Website: What Are CIDs and What Are Their Advantages? <http://www.cis.rit.edu/research/astro/research/cid1.html> Accessed January 2004.

3. Howell, Steve B.: *Handbook of CCD Astronomy*. Cambridge University Press, 2000.

4. De Bruijne, J. H., et al.: Analysis of Astronomical Data From Optical Superconducting Tunnel Junctions. *Opt. Eng.*, vol. 41, June 2002, pp. 1158–1169.

5. Mazin, B. A., and Brunner, R. J.: Simulated Extragalactic Observations With a Cryogenic Imaging Spectrophotometer. *Astron. J.*, vol. 120, Nov. 2000, pp. 2721–2729.

6. Peacock, T., et al.: Recent Developments in Superconducting Tunnel Junctions for Ultraviolet, Optical and Near-Infrared Astronomy. *Astron. & Astrophys. Supplement Series*, vol. 127, Feb. 1998, pp. 497–504.

7. Non-Redundant Linear Array (NRLA) Technical Appendix Supporting Material, Boeing. <http://planetquest.jpl.nasa.gov/TPF/TPFrevue/FAR2001/Boeing/Appends/A4TechAp.pdf> Accessed January 2004.

8. Press Release ING 0/99 Website, Monday, February 15, 1999: A Totally New Optical Detector Sees First Light at the WHT. <http://www.ing.iac.es/PR/press/Press_Release_ING099.html> Accessed January 2004.

9. European Space Agency (ESA) Technology and Research and Development Website. <http://extids.estec.esa.nl/ATTACHEMENTS/A7884/> Accessed January 2004.

10. Romani, R. W., et al.: First Astronomical Application of a Cryogenic Transition Edge Sensor Spectrophotometer. *Astrophys. J.*, vol. 521, Dec. 1999, pp. L153–L156.

11. Cellino, A.: Physical Properties of Near-Earth Objects: Open Problems. *Adv. Space Res.*, vol. 28, issue 8, 2001, pp. 1103–1106.

12. Tedesco, E. F.; Muinonen, K.; and Price, S. D.: Space-Based Infrared Near-Earth Asteroid Survey Simulation. *Planet. & Space Sci.*, vol. 48, issue 9, Aug. 2000, pp. 801–816.

13. Muller, T. G.: Asteroids in the Infrared—Serendipitous Observations With ISO. *Planet. & Space Sci.*, vol. 49, issue 8, July 2001, pp. 787–791.

14. Cabrera, B., et al.: Detection of Single Infrared, Optical, and Ultraviolet Photons Using Superconducting Transition Edge Sensors. *Appl. Phys. Lett.*, vol. 73, no. 6, 1998, pp. 735–737.

15. Miller, A. J., et al.: Development of Wide-Band, Time and Energy Resolving, Optical Photon Detectors With Application to Imaging Astronomy. *Nuclear Instruments and Methods in Physics Research Section A: Accelerators, Spectrometers, Detectors and Associated Equipment*, vol. 444, issues 1–2, April 2000, pp. 445–448.

16. Harris, A.: CCD Systems for Searching for Near-Earth Asteroids. *ASP Conference Series*, vol. 63, 1994, p. 203.

Survey of Enabling Technologies for CAPS[10]

JEFFREY ANTOL
NASA Langley Research Center

DANIEL D. MAZANEK
NASA Langley Research Center

ROBERT H. KOONS
Swales Aerospace, Inc.

Introduction

The enabling technologies required for the development of a viable Comet/Asteroid Protection System (CAPS) can be divided into two principal areas: detection and deflection/orbit modification. With the proper funding levels, many of the technologies needed to support a CAPS architecture could be achievable within the next 15 to 20 years.

In fact, many advanced detection technologies are currently in development for future in-space telescope systems such as the James Webb Space Telescope (JWST), formerly known as the Next Generation Space Telescope. It is anticipated that many of the JWST technologies would be available for application for CAPS detection concepts.

Deflection/orbit modification technologies are also currently being studied as part of advanced power and propulsion research. However, many of these technologies, such as extremely high-output power systems, advanced propulsion, heat rejection, and directed energy systems, would likely be farther term in availability than many of the detection technologies.

Discussed subsequently is a preliminary examination of the main technologies that have been identified as being essential to providing the element functionality defined during the CAPS conceptual study. The detailed requirements for many of the technology areas are still unknown, and many additional technologies will be identified as future in-depth studies are conducted in this area.

Detection Technologies

Large-aperture, high-resolution advanced telescopes (ultraviolet, optical, and infrared) will be required for detection tasks. An overview of several key technologies needed for the Comet/Asteroid Protection System (CAPS) detection capability is provided. These technologies can be divided into three areas of focus: sensors, optics, and supporting subsystems.

Sensors

The type of sensor utilized for near-Earth object (NEO) detection is critical. To maximize sensitivity, the sensor must exhibit high quantum efficiency (QE) over a bandwidth range consistent with the electromagnetic radiation reflected by comets and asteroids, and the sensor must exhibit low read noise characteristics. Charge-coupled devices (CCDs) have had an enormous impact on astronomy, and have fundamentally enabled the current Earth-based telescopic NEO detection efforts. CCD arrays have been

---

[10]Chapter nomenclature available in chapter notes, p. 217.

rapidly increasing in size and capability, and several advancements in CCD technology could have benefits for CAPS detection concepts. One technology, known as "back-illuminated electron multiplying technology," has demonstrated QEs of up to 95 percent, which is higher than any other detector. A second technology, known as "low light level CCD technology," can effectively eliminate readout noise found in CCDs. Although there are many advances in CCD technology that could assist in NEO detection, susceptibility to radiation and pixel saturation may restrict the effectiveness of CCD arrays for long-duration detection efforts. The remainder of this section focuses on providing an overview of some advanced alternatives to CCD.

Large area mosaic charge-injected device arrays. Charge-injected device (CID) sensor arrays differ in many ways from and have several advantages over CCDs. For example, CIDs exhibit less light bloom from pixel to pixel when subjected to high intensity light, the individual pixels of CIDs are addressable, and CIDs are very radiation tolerant (ref. 1).

Technology requirements. CID sensor arrays can be used for angular measurement of target objects. For the visible band, a 32000 × 32000 silicon detector array with an 8-µm pitch is needed that is sensitive to wavelengths from 0.4 to 1.0 µm and having a QE of 80 percent. For the infrared (IR) band, the array shall be a 4000 × 4000 extrinsic silicon-gallium array with a 64-µm pitch, sensitive to wavelengths from 8 to 18 µm, and having a QE of 50 percent.

Current status of technology. Current CID technology is a 1000 × 1000 silicon detector array with a 27-µm pitch, sensitive to wavelengths from 0.4 to 1.0 µm, and having QEs of 65 to 70 percent in the visible band and ≈50 percent in the IR (ref. 1).

Current technology research activities. CID research is currently being conducted by the Rochester Institute of Technology, Missile Defense Agency (formerly the Ballistic Missile Defense Office), and CID Technologies, Inc., which was recently funded by the National Aeronautics and Space Administration (NASA) through Small Business Innovative Research (SBIR) grants.

Other applications. Other applications include imaging, X-ray imaging, and cameras.

Superconducting camera. The superconducting camera, or S-Cam, is a technology demonstrator developed by the Astrophysics Division of the European Space Agency's Space Science Department (ref. 2). It is an optical camera, using single-photon counting detectors based on superconducting tunnel junctions (STJs), which provide a capability to measure the wavelength, time of arrival, and location of each photon impacting the detector.

Technology requirements. A detector system capable of determining the wavelength and arrival time of each photon would provide the ability to "tag" particular asteroids/comets and aid in identification and tracking efforts.

Current status of technology. System level tests of the first S-Cam were conducted from January to May 1998 followed by integrated systems tests from July through December 1998 at the European Space Research and Technology Centre, Holland. The completed S-Cam was delivered to the William Herschel Telescope (WHT) facility in La Palma in January 1999, with the first astronomical observations being conducted in February (ref. 3). Following successful technical validation, an upgraded system (S-Cam II) was used at WHT for science qualification runs conducted in December 1999 and April 2000. The S-Cam II has completed qualification testing and is the first astronomical cryogenic camera qualified for use (ref. 4). Currently, two fundamental limitations of STJs are small array size (18 × 50 element prototype) and the extreme cooling requirements for the detectors (<1 K).

Current technology research activities. The S-Cam and S-Cam II contain 6 × 6 element STJ array detectors; however, a prototype 18 × 50 element STJ array is currently under development with the goal of creating arrays with thousands and eventually millions of elements. NASA scientists at Goddard Space Flight Center and the Jet Propulsion Laboratory are also developing cryogenic detector technologies using STJ-based detector arrays.

Other applications. Other applications include biotechnology and medical/health.

Optics

The optics used in the detection and tracking telescopes will employ lightweight mirror technologies, requiring active control technologies to form and maintain the mirror shape. Many of these technologies are currently being developed for the JWST program.

Lightweight mirrors. Because launch weight is a major cost driver, lightweight mirror technologies must be developed to reduce systems costs.

Technology requirements. A lightweight primary mirror with at least a 3.2-m diameter is needed for the CAPS survey and tracking telescopes. An areal density of less than 15 kg/m^2 is likely to be required.

Current status of technology. Current technology is represented by the Hubble Space Telescope (HST) primary mirror, which has a 2.4-m glass mirror with an areal density of approximately 180 kg/m^2 (ref. 5). There are no lightweight optical telescope mirrors that are currently mission capable; however, lightweight mirror demonstrators of 1 to 2 m are under development for the James Webb Space Telescope (JWST) program and Goddard Space Flight Center as technology pathfinders. The JWST will have a 6-m class deployable mirror and an areal density of approximately 15 kg/m^2 (ref. 6).

Current technology research activities. The JWST program currently has two lightweight technology development projects for space optics, the Advanced Mirror System Demonstrator (AMSD) and the JWST Mirror System Demonstrator. Through these technology development projects, NASA and Department of Defense (DoD) partners have invested $40 million to demonstrate mirrors with areal densities of less than 15 kg/m^2 that are capable of operating in cryogenic temperatures (ref. 7):

AMSD

- Semirigid low-authority beryllium (Ball Aerospace & Technologies Corporation)

- Semirigid medium-authority ultra-low expansion glass (Eastman Kodak Company)

- Isogrid high-authority fused silica glass (Goodrich Corporation)

JWST Mirror System Demonstrator

- Meniscus very high-authority glass (University of Arizona)

- Rigid hybrid glass composite (Computer Optics Inc., REOSC Optique, France)

    o An isogrid-stiffened glass (Zerodur®) bonded to composite materials

Other projects are developing lightweight mirror technologies with areal densities approaching 1 kg/m^2 (ref. 8):

Nickel metal mirror (Marshall Space Flight Center)

- Electroformed nickel mirror for X-ray detection

Silicon carbide mirror (Ultramet)

- Combination of a precision silicon carbide reflector surface and a high specific strength, low-mass silicon-carbide structural support

Dual Anamorphic Reflector Telescope (Jet Propulsion Laboratory)

- Ultralightweight reflective foil mirrors stretched over a rigid frame having a parabolic contour

Thin film mirrors (Sandia, National Laboratories)

- Smart material mirror that can change shape in response to electron impacts

Nanolamites (Jet Propulsion Laboratory and Lawrence Livermore National Laboratory)

- Thin shell mirrors

Inflatable membrane mirrors

- Polymer membranes—wrinkling properties of material creates difficulties for use in visible and infrared imaging

Other applications. Lightweight mirrors may be used in 4+ m space telescopes or large ground-based telescopes.

Active mirror control. Conventional mirrors use material stiffness to maintain surface quality and mirror shape, whereas ultralightweight mirrors will require active control of the deformable mirror (DM). High quality images will be produced through computer-controlled actuators that adjust the mirror shape to control on-orbit deformations. Mirrors of large diameters will also need to be segmented in order to be packaged into a launch vehicle shroud. These mirrors will require active control for deployment, proper alignment, and reshaping once on-orbit.

Technology requirements. Mirror control systems (including sensors and actuators) must be developed to provide Hubble-type mirror surface/shape accuracy with thin technology mirrors. The systems may have to operate at or near the cryogenic temperatures at which the mirrors and sensor/detector arrays will operate.

Current status of technology. There have been numerous lab demonstrations, but no known on-orbit deployment tests have been conducted at this time. The Wavefront Control Testbed project, also known as the Developmental Comparative Active Telescope Testbed, is located at Goddard Space Flight Center and is being used to validate technologies and demonstrate wavefront sensing and control for segmented optics (ref. 9).

Current technology research activities. The lightweight mirror projects of AMSD and the JWST Mirror System Demonstrator are also developing the associated active control systems consisting of reaction structures and force actuators that are integrated with the mirrors. The Jet Propulsion Laboratory Dual Anamorphic Reflector Telescope project is investigating active control systems for membrane surface control as well as deployment and metrology. Sandia Labs is also exploring active mirror control of its thin film mirrors, which are constructed of a smart material (piezoelectric polyimide) whose shape is altered through the application of electrons from a computer-controlled electron gun.

Active control of a mirror surface at the low temperatures required by some mirrors necessitates the development of cryogenic actuators. Several cryogenic actuator concepts have been developed and evaluated in recent years by NASA, primarily through SBIR contracts. The JWST program is currently evaluating two cryogenic actuator concepts (ref. 10), one by Xinetics, Inc., and one by American Superconductor.

Other applications. Other applications include laser instruments and small satellite actuators.

Supporting Systems

Numerous supporting subsystem technologies will be required for successful operation of the detection and tracking telescopes, including disturbance isolation, active cooling, lightweight sunshade/baffles, precision pointing, accurate position and time knowledge, and advanced data management/communication systems. Development of many of these technologies is currently being funded by NASA for other advanced spacecraft programs.

Disturbance isolation. Spacecraft subsystem components such as reaction control wheels and cryogenic coolers create dynamic disturbances due to imbalances and vibrations. Both passive and active systems are available for isolating disturbances. Passive systems provide damping of disturbances through a fixed stiffness, while active systems use control algorithms and actuators to sense and respond to disturbances. Passive systems offer an advantage over active systems in that they require no power. However, passive systems do not provide adaptability to changing disturbances that active systems provide.

Technology requirements. Dynamic disturbances must be isolated from the telescope during collection of survey and tracking measurements in order to obtain precision pointing requirements. Specific requirements are to be determined.

Current status of technology. A passive isolation system, consisting of a viscous fluid-damped isolator, is currently employed on the HST to isolate disturbances created by the reaction control wheels (ref. 11).

Another passive isolation system designed for space application is the Honeywell D-Strut™, a viscous fluid-damped isolator. The Honeywell Hybrid D-Strut™ integrates an active system with the passive isolator to provide isolation at low frequencies (ref. 12).

Current technology research activities. Current research in disturbance isolation for advanced telescope concepts includes the application of fast steering mirrors for jitter control and low vibration cryocoolers for disturbance mitigation (for examples see the subsequent section on active cooling).

Other applications. Other applications include Earth observation satellites, launch vehicles.

Active cooling. Active cooling is required to achieve optimal performance from the sensors—whether they are CCDs, CIDs, or STJs—with temperature requirements being only a fraction of a degree Kelvin in the case of STJs.

Technology requirements. Visible light detector arrays must be cooled to 230 K with the temperature controlled to 0.1 K. IR detector arrays must be cooled to 10 K with the temperature controlled to 0.001 K.

Current status of technology. The cooling system for the HST Near Infrared Camera and Multi-Object Spectrometer, which was originally a solid nitrogen dewar, was recently upgraded (March 2002) to a Creare, Inc., Reverse-Brayton cooler that maintains detectors between 75 and 85 K with a 0.1 K control capability (ref. 13).

The Jet Propulsion Laboratory has developed a zero-vibration sorption cryocooler that will cool detectors to between 18 and 20 K. Flight units will be delivered in 2004 and 2005 as the NASA contribution to the European Space Agency (ESA) Planck mission, planned for launch in 2007 (ref. 14).

Current technology research activities. JPL's Advanced Cryocooler Technology Development Program is currently developing 6 and 18 K two-stage cooling technologies for the next generation space-based observatories: Terrestrial Planet Finder, JWST, and Constellation-X. Four teams are currently developing preliminary-level designs with an objective of creating engineering model coolers in the 2005 timeframe (ref. 15):

- Mechanical J-T Pre-cooled by Multistage Stirling (Ball Aerospace & Technologies Corporation)
- Two-Stage Turbo-Brayton with 75 K Radiative Pre-cooler (Creare, Inc.)
- Multistage pulse tube (Lockheed Martin)
- Mechanical J-T Pre-cooled by Multistage Pulse Tube (TRW)

Research into advanced cryogenic cooling technologies for low temperature focal planes is also supported by funding through NASA Research Announcements from the NASA enabling concepts and technologies (ECT) program—Advanced Measurement and Detection Element (ref. 16):

- Solid state optical refrigerators (Los Alamos National Laboratory)
- Continuous adiabatic demagnetization refrigerators: 10 K to 50 mK (Goddard Space Flight Center)
- 6 K Vibration-Free Turbo-Brayton Cryocooler (Creare, Inc.)
- Solid state microrefrigerator for 100 mK (Lawrence Livermore National Laboratory)
- Nanocomposite thermomagnetic cryocoolers (Howard University)

Other ECT supported tasks are directed by various NASA centers (ref. 16):

- 4 to 10 K Vibration-Free Coolers Turbo-Brayton
- Ultralow temperature continuous magnetic refrigeration
- Miniloop heat pipe

- Microelectromechanical system microthermal louvers
- Helium-carbon sorption cryocoolers

Other applications. Other applications include space-based IR sensor technology.

Shading/baffle technology. The CAPS telescope must be protected from looking at or near the Sun, or the detector focal plane may be destroyed. In addition, CAPS IR sensors must be maintained at cryogenic temperatures. Finally, making observations near the Sun's observed position is highly desirable to maximize the number of NEOs detected. Shading of the CAPS telescopes from the Sun could be performed using several techniques. For a lunar-based concept, a shelter such as a dome structure could be employed. For free-flying spacecraft, the shading could be in the form of an attached sunshade or a large deployable shade flying in formation with the telescope.

Technology requirements. It is highly desirable to have CAPS detection elements be capable of observing within 20 deg of the detector-sun line to increase the area of sky that can be sampled. In addition, the CAPS IR sensors must be cooled to 10 K.

Current status of technology. Hubble utilizes baffles and a cover door for shading as well as active cooling for IR sensors (see previous discussion on active cooling). A number of subscale deployable shades have been developed for ground testing applications, and a related concept for a flight experiment was developed; however, no in-space sun shield tests or operational deployments have been conducted to date.

Current technology research activities. The proposed JWST design includes a large, five-layer, inflated sunshield attached to the telescope by a thermal isolation astromast. The shade is designed primarily for thermal control as the telescope will not be able to look within 90 deg of the Sun. The back of the shield will maintain a temperature of ≈ 90 K, allowing the mirror, instruments, and associated structure to radiate directly to space and reach cryogenic temperatures. The development of sun shield concepts is being performed by industry partners, International Latex Corporation Dover and L'Garde (ref. 17).

The NASA ECT, Resilient Materials and Structures Element, is also sponsoring research into new space durable polymers, which have potential application to JWST sun shield development (ref. 18).

Other applications. Other applications include any space-based telescope that requires shielding from sunlight and collocation technology for formation flying satellites.

Attitude knowledge and control/precision pointing. Precision spacecraft and detector pointing will be needed to provide star field accuracy for guide stars. Distributed spacecraft technologies, such as formation flying and precise attitude control, will also be needed for an orbital-based interferometry capability.

Technology requirements. Star field accuracy for guide stars must be known to better than 0.001 arcsec, telescopes must be pointed to better than 0.001 arcsec, and attitude control rate management must be to 0.001 arcsec/s. These requirements are for noninterferometric measurements using the tracking telescopes. Significantly higher accuracy will be needed for astrometric interferometry.

Current status of technology. Star field is currently known to 0.002 arcsec. The ESA satellite Hipparcos, which operated from 1989 through 1993, updated the position and distance of more than 100 000 stars (ref. 19).

The HST pointing capability is 0.01 arcsec with a drift of 0.007 arcsec over a 12-hr period (ref. 20).

Current technology research activities. The Full-sky Astrometric Mapping Explorer spacecraft was to continue work begun by the Hipparcos mission by sensing star fields to 50 µas at 15th magnitude. However, the project is currently rescoping the mission due to withdrawal of NASA sponsorship (ref. 21).

The Space Interferometry Mission (SIM) will have a pointing capability of 1 to 4 µas, or 1-µas single measurement accuracy for Narrow Angle Astrometry and 4-µas mission accuracy for Wide Angle Astrometry (ref. 22).

The NASA ECT program is currently funding research for distributed space systems technology within the Distributed and Micro-spacecraft Element, including (ref. 23):

- formation sensing and control
- intersatellite communications
- constellation management and mission operations

Other applications. Other applications include the provision of high accuracy star cataloging for attitude determination and improvement of general celestial knowledge.

Extremely accurate position and time knowledge. CAPS requires accurate knowledge of position and time so that it can precisely acquire targets.

Technology requirements. Linear distance between two interferometry telescopes must be known to within 1 nm.

Current status of technology. Current distances between unmanned spacecraft are measured in meters.

Current technology research activities. Starlight, a technology pathfinder for terrestrial planet finder (TPF), was originally planned for launch in 2006, but has been redirected to focus on ground demonstration of the technologies needed for the formation-flying interferometer concept of TPF (ref. 24):

- Formation interferometer testbed (fringe tracking)
- Metrology technologies
- Prototype autonomous formation flying sensor
- Formation flying algorithm development and simulation

The original goal of the flight program was to maintain position to 1 mm, angular bearing to 3 arcminutes.

SIM also has a number of ground-based test beds (ref. 25):

- Palomar Testbed Interferometer

- Nanometer Testbed - nanometer stability on a flexible structure

- Micro Arcsecond Metrology Testbed—subnanometer metrology

Other applications. Other applications include collocation of microsatellites for phased array radar and stereoscopic imagery.

Advanced data management systems. Advanced data management systems and rapid communications will be needed for processing observation data and cataloging NEOs.

Technology requirements. Significant image data will be generated by multiple large CCD/CID arrays from multiple telescopes potentially at remote locations. These data will have to be processed and downlinked, the resulting image data stored, and an object database created. Although recognized as a very important technology, this area was not investigated in significant depth during the CAPS conceptual study.

Current status of technology. Ultrahigh data rates for downlink may be achievable using optical communications technology. Potential high bandwidth intersatellite communications may also be needed for interferometry or database synchronization.

Current technology research activities. Research activities in this area are ongoing at many NASA centers, academic institutions, and private companies. Data processing and communications speeds are constantly being advanced by ever more demanding space missions, as well as terrestrial based drivers.

Other applications. Other applications include Earth observation missions, next-generation space-based observatories, and advanced solar system based communications systems.

Deflection/Orbit Modification Technologies

To follow is an overview of the two technologies specifically applicable to the enabling of a controlled orbit modification capability for deflecting an impacting NEO or altering an NEO's orbit for resource utilization: high thrust, high specific impulse propulsion systems and high power electrical systems. Advanced thermal management systems to reject large amounts of waste heat are also required to be developed to support the in-space power systems. Specific laser technologies were not investigated in detail during the CAPS study; but reliable, high-power pulsed laser ablation systems would need to be developed. Adaptive laser optics, precision beamwidth focusing, and closed-loop control systems are some of the supporting technologies needed for a laser ablation system. The same, or similar, technologies could also permit the deployment of active laser ranging systems to assist in precision orbit determination. Finally, advanced autonomous or semiautonomous rendezvous and station-keeping capability would be needed to travel to the NEO and engage it at relatively close distances.

High Thrust, High Specific Impulse Propulsion Systems

High thrust, high specific impulse propulsion systems are needed for delivering orbit modification systems to target NEOs.

Systems options. Two options exist, nuclear electric propulsion (NEP), which uses high specific impulse/high voltage thrusters (powered by a nuclear reactor) to accelerate propellants, and nuclear thermal propulsion (NTP), which uses the heat generated by a reactor to directly accelerate the propellants.

Technology requirements. An advanced propulsion system is needed that can provide the capability of rapid rendezvous with an object.

Current status of technology. NASA nuclear technology programs are currently included in the new initiative, Project Prometheus, formerly the Nuclear Systems Initiative (NSI). There are two primary technology areas to be examined as part of Prometheus (ref. 26):

- Radioisotope-based systems

- Nuclear fission-based systems

The main goal of Project Prometheus is to develop technology for increasing spacecraft power capability, thus providing an increased capability for solar system exploration. The primary propulsion approach to be emphasized in this research is NEP.

While NEP is currently NASA's design focus, an extensive amount of research and development of NTP systems has been conducted over the years. The majority of NTP research was conducted between 1955 and 1968, with over $1 billion invested in the Project Rover nuclear rocket program and the nuclear engine for rocket vehicle application (NERVA) research program (ref. 27). Project Rover was primarily a research program while NERVA focused on development and testing. These programs were canceled in 1973 after the development and ground testing of numerous NTP systems, which were never flight tested. NTP research has continued at a reduced scale over the years, with many additional NTP concepts developed (such as NERVA Derivative Nuclear Thermal Rockets and Bimodal NTP).

Current technology research activities. Several thruster technologies are currently being investigated for NEP applications (ref. 28):

- Ion engine

- Magnetoplasmadynamic

- Hall

- Variable specific impulse magnetoplasma rocket (VASIMR)

- Pulsed inductive thruster

The VASIMIR was the primary concept considered for the CAPS analysis because it has the potential for yielding the fastest possible trip time. VASIMR is a high power magnetoplasma rocket that gives continuous and variable thrust at constant power (ref. 29). Hydrogen plasma is heated by radio frequency power to increase exhaust velocity up to 300 km/s. The power output of the engine is kept constant, thus thrust and specific impulse, I_{sp}, are inversely related. Thrust is increased proportional to the power level. The engine can optimize propellant usage and deliver a maximum payload in minimum time by varying thrust and I_{sp} (ref. 30). Therefore, VASIMR can yield the fastest possible trip time with a given amount of propellant by using constant power throttling. A 10-kW space demonstrator experiment has been completed, and a VASIMR engine with 200-MW power could be available around the year 2050. The I_{sp} range of the engine would be 3000 to 30000 s, and the corresponding thrust range would be approximately 5000 N to 500 N (assuming 100 percent power efficiency).

As part of the ongoing research in NEP systems, an end-to-end NEP demonstrator has been developed by Marshall Space Flight Center to demonstrate an integrated NEP system using a simulated fission core (SAFE-30, discussed subsequently), a Stirling power conversion system, and an advanced ion thruster. NEP systems with a specific mass of 1 to 10 kg/kW could be available during the 2020 to 2040 timeframe, while 0.1- to 1-kg/kW class NEP systems are not envisioned until 2030 and beyond. An NEP system utilizing a VASIMR engine is currently estimated to have a maximum overall specific mass of 1.0 kg/kW (ref. 31).

Other applications. Other applications include exploration of the outer planets, human exploration missions to the planets, and human outposts on the Moon and other planets.

High Power Systems (Space Fission Power)

Systems capable of generating significantly high levels of power will be needed for powering the propulsion and laser systems of an intercept spacecraft to quickly reach a target body and then sufficiently modify its orbit to avoid an Earth impact. Nuclear fission power systems are currently the best known method of providing this capability. Extensive research on space fission power systems was conducted in the 1960s and new research into these technologies recently has been initiated under Project Prometheus.

Systems needs. Multimegawatt to gigawattt electrical power systems are needed for propulsion and laser applications.

Technology requirements. Power systems capable of generating 200 MW of electricity or greater are required to provide the energy necessary for the efficient high-thrust propulsion systems as described previously and the laser systems described subsequently.

Current status of technology. As mentioned in the previous section, all NASA nuclear technology programs were recently transitioned from the In-Space Propulsion Program to the NSI (now Project Prometheus), including the Nuclear Fission Power research.

The only flight of a United States fission reactor was the Space Nuclear Auxiliary Power (SNAP)-10A launched into a 1300-km Earth orbit in 1965. It was capable of generating greater than 500 W of power and operated for 43 days until a spacecraft malfunction ended the mission (ref. 32).

Other programs, through the 1990s, have developed nuclear power system technologies but have not flown in space: the Medium-Power Reactor Experiment, Space Power (SP)-100, SNAP-50/SPUR, SPAR/SP-100, Advanced Space Nuclear Power Program, and the Multi-Megawatt Program (ref. 32).

Current technology research activities. The safe affordable fission engine (SAFE)-30 test series is an on-going demonstration of a 30-kW system using nonnuclear testing. Resistance heaters are used to simulate the heat from fission while steel/sodium heat pipes remove the heat from the core, thus providing a close approximation to an actual fission system (ref. 32). The tests also involve a Stirling power conversion system coupled to the SAFE-30 and operated at full power. The primary goal of the tests is to obtain data for validating existing analytical models and for use in the design of future systems. Fifteen restarts were accomplished as of February 2001 and represented the first realistic test of a United States space fission system since 1969. The SAFE-30 tests are being followed by the SAFE-300 test series, a demonstration of a 300-kW system that also uses nonnuclear testing methods. The tests are being conducted by Marshall Space Flight Center in cooperation with Los Alamos National Lab and Sandia National Lab (ref. 32).

Spacecraft and surface power plants with the capability of generating 10 to 100 kW are projected for the 2010 to 2020 timeframe. Multimegawatt (1 to 100 MW) spacecraft and surface power plants could be available during the 2020 to 2040 timeframe, while 100 to 1000 MW class systems are not envisioned until at least 2030 (ref. 32).

Other applications. Other applications include exploration of the outer planets, human exploration missions to the planets, and human outposts on the Moon and planets.

Pulsed Laser Ablation Systems

Multimegawatt pulsed laser ablation systems capable of continuous operation over several months in interplanetary space are needed to impart sufficient momentum to deflect kilometer-class long-period objects that will impact the Earth on their first observed passages through the solar system.

Systems options. Space-based laser systems are necessary because the Earth's atmosphere makes terrestrial-based laser systems ineffective for the application of deflecting kilometer-class long-period objects. Stimulated raman scattering restricts the intensity of the laser beam that would propagate through the atmosphere, atmospheric scintillation requires highly advanced adaptive optics, and the tremendous distances involved require extremely large beam director apertures. A space-based laser could be located in cislunar space, but the power requirements would extend into the multigigawatt range to deflect an NEO approaching the Earth on a collision course. A laser payload carried to the target by a spacecraft with an extremely efficient propulsion system significantly reduces the power required and the complexity of the optics needed to focus a laser with sufficient intensity on the target. Many aspects of pulsed laser systems are classified and the performance of current systems and future technologies development programs is unknown. The goal of the CAPS study was to provide a preliminary estimate of the laser energy levels required to deflect various classes of impactors, not to determine the specific design parameters of a laser ablation system. Given the current state of space-based laser technology, many major breakthroughs would be required for the practical implementation of the high energy levels required for CAPS.

Technology requirements. Approximately 4×10^5 gigajoules (GJs) of laser ablation energy would be required to deflect a 1-km long-period comet (LPC) on a center hit trajectory, with a density of 200 kg/m^3, by only 1 Earth radius if the laser ablation effort could be completed 6 months before a postperihelial impact. This deflection scenario could be accomplished by a 100-MW laser system operating continuously for 45 days. More powerful (gigawatt class) laser systems would be required to deflect an object with greater density or provide additional margin for the final deflection distance. Approximately 6×10^6 GJ of laser ablation energy would be required to deflect a 1-km LPC on a center hit trajectory, with a density of 1000 kg/m^3, by 3 Earth radii if the laser ablation effort could be completed 12 months before a preperihelial impact. This deflection effort would require approximately 70 days of continuous operation by a 1-GW laser system. A less powerful system would require significantly more time on target to provide the necessary incremental velocity change, or multiple laser systems could work cooperatively as a phased laser array to provide the necessary energy.

Current status of technology. Many aspects of high energy laser technology development are classified. Much of the directed energy programs are focused on continuous wave lasers designed to heat and melt the skin of a vehicle or missile. The Mid-Infrared Advanced Chemical Laser is a megawatt-class, continuous wave, deuterium fluoride chemical laser that has accumulated approximately 3500 s of lasing time since 1980 (ref. 33). Laser ablation can be achieved with a continuous wave laser, but a series of high intensity laser pulses is more desirable for producing a vaporized jet of asteroid or comet material and maximizing the laser momentum coupling coefficient. Additionally, a continuous wave laser could heat the material near the target area sufficiently to cause undesirable fragmentation.

In the past 10 years, the peak pulsed laser power levels achieved with terrestrial laser test facilities have increased by a factor of approximately 1000. Most ultrahigh power lasers employ a technique called chirped pulse amplification. This revolutionary pulse-compression technique allows these power levels to be achieved with small table-top lasers (ref. 34). Building-sized lasers were required to produce these power levels just a decade ago. However, these ultrahigh power lasers are only capable of extremely short pulses, low energy, limited pulse repetition, and close range beam focusing.

A CAPS laser system would need much higher energy pulses (10 MJ) repeated many times per second (10 Hz) and applied over long time periods (several million seconds of lasing time applied over a period of a few months). A CAPS high energy laser system would need to interact with the target while separated by many kilometers. Additionally, there are many technology efforts that are necessary to make a high energy laser system applicable for planetary defense, such as beam control, uncooled optics, and precision pointing of the beam on the target.

Current technology research activities. There are currently two major beamed energy weapon programs being conducted by the United States DoD that are widely cited in the unclassified literature. The United States Air Force's Airborne Laser (ABL) is a major weapon system program that has been under development since 1996 (ref. 35). The ABL utilizes a modified 747-400 airplane equipped with several laser systems designed to shoot down a ballistic missile during its boost phase. The ABL has a megawatt-class Chemical Oxygen Iodine Laser that heats and ruptures the missile skin, causing the booster to explode because it is fueled under pressure. The ABL program is expected to be operational by 2007. The Space-Based Laser (SBL) program is currently under development by the Ballistic Missile Defense Organization. It is envisioned that the SBL will consist of a constellation of several orbiting hydrogen fluoride laser platforms (ref. 36). A Space-Based Laser Integrated Flight Experiment is planned to orbit in 2012. This flight experiment could result in an operational SBL system shortly thereafter.

Other applications. Other applications include ablative laser propulsion for spacecraft, Earth-to-orbit vehicles, and orbital debris removal; microwave or optical power beaming; and laser ranging for NEO precision orbit determination.

Concluding Remarks

A preliminary survey of technologies essential to providing a viable Comet/Asteroid Protection System (CAPS) has been conducted. These enabling technologies are divided into two principal areas: detection and deflection/orbit modification. Many of the detection technologies are currently in development for future in-space telescope systems, such as the James Webb Space Telescope (JWST), and should be available in the appropriate timeframe for application to the CAPS. The deflection/orbit modification technologies are also being developed as part of advanced power and propulsion research; however, many of these technologies are farther term in availability than the detection technologies. Future in-depth studies will be necessary to define the detailed requirements for these technology areas and to examine any additional technologies that may exist.

References

1. What is a CID?—CID Research at the Rochester Institute of Technology. <http://www.cis.rit.edu/research/CID/a_cid_is.htm> Accessed February 2004.

2. The STJ Detector Page. <http://astro.estec.esa.nl/SA-general/Research/Stj/STJ_main.html> Accessed February 2004.

3. Rando, N.; Andersson, S.; Collaudin, B.; Favata, F.; Gondoin, P.; Peacock, A.; Perryman, M.; and Verveer, J.: S-Cam: A Technology Demonstrator for the Astronomy of the Future. *European Space Agency Bulletin 98*, June 1999. <http://esapub.esrin.esa.it/bulletin/bullet98/RANDO.PDF> Accessed February 2004.

4. Rando, N.; Verveer, J.; Verhoeve, P.; Peacock, A. J.; Andersson, S.; Reynolds, A.; Favata, F.; and Perryman, M. A.: S-Cam 2: Performance and Initial Astronomical Results. *Proc. Society of Photo-Optical Instrumentation Engineers (SPIE)*, vol. 4008, Optical and IR Telescope Instrumentation and Detectors, August 2000, pp. 646-656. <http://spie.org/scripts/abstract.pl?bibcode=2000SPIE%2e4008%2e%2e646R&page=1&qs=spie> Accessed February 2004.

5. Bilbro, J.: Optics in Orbit. *OE Magazine*, SPIE, August 2001. <http://oemagazine.com/fromTheMagazine/aug01/opticsinorbit.html> Accessed February 2004.

6. James Webb Space Telescope (JWST). <http://ngst.gsfc.nasa.gov/Observatory/> Accessed February 2004.

7. JWST Technologies—Lightweight Optics. <http://ngst.gsfc.nasa.gov/Hardware/text/optics_new.htm> Accessed February 2004.

8. JWST Project—Lightweight Mirror Development. <http://ngst.gsfc.nasa.gov/Hardware/text/optics.html> Accessed February 2004.

9. JWST Wavefront Control Testbed. <http://www.ngst.nasa.gov/Hardware/text/WCT.html> Accessed February 2004.

10. JWST Project—Cryogenic Actuators <http://ngst.gsfc.nasa.gov/Hardware/text/actuator.html> Accessed February 2004.

11. Davis, L. P.; and Wilson, J. F.: Hubble Space Telescope Reaction Wheel Assembly Vibration Isolation System. NASA, Marshall Space Flight Center, AL, March 1986.

12. Foshage, J.; Davis, T.; Sullivan, J.; Hoffman, T.; and Das, A.: Hybrid Active/Passive Actuator for Spacecraft Vibration Isolation and Suppression. SPIE Actuator Technology and Applications Conference, Aug. 1996.

13. Ross, R. G., Jr.: NASA Advanced Cryocooler Technology Development Program. Presentation Slides. Jet Propulsion Laboratory, California Institute of Technology, Pasadena, CA, July 31, 2002. <http://www.submm.caltech.edu/~bradford/SAFIR/safir_meeting_viewgraphs/Ross_presentation.pdf> Accessed February 2004.

14. Planck Surveyor—USA Participation in LFI Consortium. <http://aether.lbl.gov/www/projects/cosa/usa.html#Cryocooling> Accessed February 2004.

15. Ross, R. G., Jr.; Boyle, R. F.; Key, R. W.; and Coulter, D. R.: NASA Advanced Cryocooler Technology Development Program. SPIE Astronomical Telescopes and Instrumentation Conference, Waikoloa, HI, Aug. 22–28, 2002. <http://www.ngst.nasa.gov/public/unconfigured/doc_1008/rev_01/NGST-ARTL-001944.PDF> Accessed February 2004.

16. Krabach, T.: Advanced Spacecraft Systems: Advanced Measurement & Detection Element. Presentation Slides. National Research Council Review of Enabling Concepts and Technologies (ECT) Program, June 11, 2002, pp. 45–46, 75–79.

17. Johnston, J.; Ross, B.; Blandino, J.; Lawrence, J.; and Perrygo, C.: Development of Sunshield Structures for Large Space Telescopes. SPIE Astronomical Telescopes and Instrumentation Conference, Waikoloa, HI, Aug. 22–28, 2002.

<http://www.ngst.nasa.gov/public/unconfigured/doc_1006/rev_01/NGST-ARTL-001923.pdf>
Accessed February 2004.

18. Belvin, W. K.: Resilient Materials and Structures Materials Element. Presentation Slides. National Research Council Review of ECT Program, June 2002, p. 15.

19. The Hipparcos Space Astrometry Mission. <http://astro.estec.esa.nl/Hipparcos/hipparcos.html> Accessed February 2004.

20. The Hubble Space Telescope. <http://www.sciencepresse.qc.ca/clafleur/The-HST.html> Accessed February 2004.

21. Full-Sky Astrometric Mapping Explorer (FAME). <http://www.usno.navy.mil/FAME/> Accessed February 2004.

22. Space Interferometry Mission (SIM). <http://planetquest.jpl.nasa.gov/SIM/sim_index.html> Accessed February 2004.

23. Leitner, J.: Distributed Space Systems, Distributed and Micro-Spacecraft Element. Presentation Slides. National Research Council Review of ECT Program, May 23, 2002.

24. Planet Quest: Missions—StarLight. <http://eis.jpl.nasa.gov/planetquest/technology/formation_flying.html> Accessed February 2004.

25. Space Interferometry Mission: Technology. <http://planetquest.jpl.nasa.gov/SIM/sim_technology.html> Accessed February 2004.

26. NASA - Space Science - Project Prometheus. <http://spacescience.nasa.gov/missions/prometheus.htm> Accessed February 2004.

27. Nuclear Rocket Propulsion—The 1960s. Nuclear Thermal Propulsion. <http://www.fas.org/nuke/space/c04rover.htm> Accessed February 2004.

28. Brophy, J.; Benson, S.; and Hrbud, I: Overview of NEP Thruster Candidates. 2002 Advanced Space Propulsion Workshop, Pasadena, CA, June 4, 2002.

29. Chang-Diaz, F. R., et al.: The Physics and Engineering of the VASIMR Engine. 36th AIAA/ASME/SAE/ASEE Joint Propulsion Conference, Huntsville, AL, July 17–19, 2000, AIAA 2000-3756.

30. Chang-Diaz, F. R.: The VASIMR Rocket. *Sci. American*, Nov. 2000, pp. 90–97.

31. Smith, B.; Knight, T.; and Anghaie, S.: Multimegawatt NEP With Vapor Core Reactor MHD. Space Technology and Applications International Forum (STAIF 2002), Albuquerque, NM, Feb. 3–6, 2002.

32. Van Dyke, M., et. al.: The Safe Affordable Fission Engine (SAFE) Test Series. NASA/JPL/MSFC/UAH 12th Annual Advanced Space Propulsion Workshop, Huntsville, AL, April 3–5, 2001, pp. 7–17.

33. High Energy Laser Systems Test Facility (HELSTF)—The Mid-Infrared Advanced Chemical Laser. <http://helstf-www.wsmr.army.mil/miracl.htm> Accessed February 2004.

34. Chériaux, G.; and Chambaret, J.-P.: Ultra-Short High-Intensity Laser Pulse Generation and Amplification. *Measurement Sci. and Technol.*, vol. 12, 2001, pp. 1769–1776.

35. Dougherty, J. E.: U.S. Developing Airborne Laser Project. *WorldNetDaily Report*. <http://www.worldnetdaily.com/news/article.asp?ARTICLE_ID=15139> Accessed February 2004.

36. Wildt, D.; and Lissit, S.: Space-Based Chemical Lasers for Ballistic Missile Defense. AIAA 93-3150, AIAA 24th Plasmadynamics and Lasers Conference, Orlando, FL, July 6–9, 1993.

Chapter Notes: Nomenclature

1.   Comet/Asteroid Protection System: Concept Study Executive Summary

arcsec	arcsecond
au	astronomical unit ($1.4959787066 \times 10^{11}$ m)
CAPS	Comet/Asteroid Protection System
CCD	charge-coupled device
CID	charge-injection device
ESA	European Space Agency
FOV	field-of-view
LPC	long-period comet
Mt	megaton (1 Mt = 4.184×10^{15} joules of energy)
NASA	National Aeronautics and Space Administration
NEA	near-Earth asteroid
NEO	near-Earth object
RASC	Revolutionary Aerospace Systems Concepts
RMA	reliability, maintainability, and availability
SPC	short-period comet
STJ	superconducting tunneling junction
ΔV	change in velocity

2.   Near-Earth Object (NEO) Hazard Background

ANSMET	Antarctic Search for Meteorites Program
au	astronomical unit ($1.4959787066 \times 10^{11}$ m)
CAPS	Comet/Asteroid Protection System
ECA	Earth-crossing asteroid
ECC	Earth-crossing comet
H	absolute magnitude

IEA	interior-Earth asteroid
K-T	Cretaceous-Tertiary
kt	kiloton (1 kt = 4.184×10^{12} joules of energy)
LINEAR	Lincoln Near-Earth Asteroid Research
LPC	long-period comet
Mt	megaton (1 Mt = 4.184×10^{15} joules of energy)
NEA	near-Earth asteroid
NEAT	Near-Earth Asteroid Tracking
NEO	near-Earth object
SPC	short-period comet
VLT	Very Large Telescope (European Southern Observatory)

3. Accurate Determination of Comet and Asteroid Orbits Leading to Collision With Earth

A	asteroid
A_1, A_2, A_3	comet outgassing coefficients
a	semimajor axis
$\hat{\mathbf{a}}_1, \hat{\mathbf{a}}_2, \hat{\mathbf{a}}_3$	unit vectors used to express comet outgassing force
arcsec	arcseconds
au	astronomical unit ($1.4959787066 \times 10^{11}$ m)
B	position vector from Earth's mass center to comet's mass center, at time of closest approach
B	magnitude of **B** (miss distance)
B_R	dot product of **B** and $\hat{\mathbf{R}}$
B_T	dot product of **B** and $\hat{\mathbf{T}}$
B-plane	body-plane
C	true comet
C'	comet whose orbit is determined from measurements

$[{}^S C^B]$	direction cosine matrix relating $\hat{\mathbf{s}}_1, \hat{\mathbf{s}}_2$, and $\hat{\mathbf{s}}_3$ to $\hat{\mathbf{S}}, \hat{\mathbf{T}}$, and $\hat{\mathbf{R}}$
$[{}^B C^P]$	direction cosine matrix relating $\hat{\mathbf{T}}$ and $\hat{\mathbf{R}}$ to $\hat{\mathbf{p}}_1$ and $\hat{\mathbf{p}}_2$
$\hat{\mathbf{d}}$	unit range vector
$\mathbf{d}$	position vector from observatory to object (range vector)
d	magnitude of $\mathbf{d}$ (range)
$\dot{d}$	range-rate
E	Earth
E	eccentric anomaly
E^*	mass center of Earth
E_k	eccentric anomaly of comet at point of collision
e	eccentricity of object's orbit
$\{e\}$	measurement error matrix
F, F_t, G, G_t	Lagrangian coefficients
$f(x_1, x_2)$	bivariate Gaussian probability density function
$\mathbf{f}_g$	force per unit mass due to comet outgassing
$g(r)$	ratio of vaporization flux for water snow at heliocentric distance r, to flux at heliocentric distance of 1 au
$[H]$	mapping matrix
$[\tilde{H}]$	matrix of partial derivatives of measurements with respect to state
$[I]$	identity matrix
i	inclination of orbital plane to ecliptic plane
K	point where object passes through ecliptic plane at heliocentric distance of 1 au
$[K]$	Kalman gain matrix
LPC	long-period comet
L_0	initial true longitude
m	slope, or number of measurements
N	inertial reference frame
NEA	near-Earth asteroid
n	number of observations, or number of orbital elements
O	observatory

$[P]$	covariance matrix for state deviation estimate error
$[\overline{P}_b]$	partition of $[\overline{P}_B]$ associated with $\hat{\mathbf{T}}$ and $\hat{\mathbf{R}}$
$[\overline{P}_B]$	covariance matrix for directions marked by $\hat{\mathbf{S}}$, $\hat{\mathbf{T}}$, and $\hat{\mathbf{R}}$
$[\overline{P}_k]$	covariance matrix at designed time of collision
$[\overline{P}_p]$	diagonalized covariance matrix containing eigenvalues of $[\overline{P}_b]$
$[\overline{P}_r]$	partition of $[\overline{P}_k]$ associated with position
$[\overline{P}_s]$	covariance matrix including uncertainty due to comet outgassing
$\hat{\mathbf{p}}_1, \hat{\mathbf{p}}_2$	eigenvectors of $[\overline{P}_b]$
Q	sum of squares of weighted residuals
$[Q]$	state noise covariance matrix
$[\overline{Q}]$	average state noise covariance matrix
R	partition of state transition matrix corresponding to derivative of position with respect to initial velocity, evaluated on reference trajectory
$\tilde{R}$	partition of state transition matrix corresponding to derivative of position with respect to initial position, evaluated on reference trajectory
$[R]$	measurement error covariance matrix
R_E	physical radius of Earth
$\mathbf{r}$	position vector from mass center of Sun to mass center of object
r	magnitude of $\mathbf{r}$
$\mathbf{r}'$	position vector resulting from orbit determination
$\mathbf{r}^*$	position vector from mass center of Sun to mass center of object traveling on reference orbit
r_a	radius of aphelion
r_c	effective collision radius of Earth
r_k	heliocentric collision radius, 1 au
r_p	radius of perihelion
r_0	scale heliocentric distance of high comet outgassing activity
S	Sun
$\hat{\mathbf{S}}, \hat{\mathbf{T}}, \hat{\mathbf{R}}$	unit vectors normal to ($\hat{\mathbf{S}}$) and fixed in ($\hat{\mathbf{T}}$, $\hat{\mathbf{R}}$) the B-plane

s_1, s_2	inertial reference frame axes parallel to $\hat{\mathbf{s}}_1$ and $\hat{\mathbf{s}}_2$, respectively
$\hat{\mathbf{s}}_1, \hat{\mathbf{s}}_2, \hat{\mathbf{s}}_3$	unit vectors fixed in an inertial reference frame; $\hat{\mathbf{s}}_1$ is in direction of vernal equinox, $\hat{\mathbf{s}}_3$ is normal to ecliptic plane
t	time
t_b	time of closest approach
t_i	values of time
t_k	designed time of collision
U_i	universal variables
V	partition of state transition matrix corresponding to derivative of velocity with respect to initial velocity, evaluated on reference trajectory
$\tilde{V}$	partition of state transition matrix corresponding to derivative of velocity with respect to initial position, evaluated on reference trajectory
$\mathbf{v}$	velocity in N of object relative to Sun
$\mathbf{v}'$	velocity vector resulting from orbit determination
$\mathbf{v}^*$	inertial time derivative of $\mathbf{r}^*$
$\mathbf{v}^{C/E^*}$	velocity in N of comet relative to mass center of Earth
$\mathbf{v}^{C/S^*}$	velocity in N of comet relative to mass center of Sun
$\mathbf{v}^{E^*/S^*}$	velocity in N of mass center of Earth relative to mass center of Sun
v_∞	magnitude of $\mathbf{v}^{C/E^*}$
$[W]$	weighting matrix
w_j	weighting factor in method of least squares
$\{X\}$	orbital element matrix, also referred to as state vector
$\{\hat{X}\}$	estimate of orbital element matrix
$\{X^*\}$	orbital element matrix associated with reference trajectory
$\{X'\}$	best estimate of orbital elements obtained by sequential filter
$X_1, \ldots, X_6$	orbital parameters
$\hat{X}_1, \ldots, \hat{X}_6$	estimates of orbital parameters
$\hat{X}_1^*, \ldots, \hat{X}_6^*$	best estimate of orbital parameters
$\{x\}$	state deviation matrix

$\{\hat{x}\}$	estimate of state deviation matrix
$\hat{x}_1^*, \ldots, \hat{x}_6^*$	best estimate of state deviations
$\{Y\}$	measurement matrix
$\{Y^*\}$	measurement matrix calculated from reference orbit
$\{y\}$	measurement residual matrix
$\{\hat{y}\}$	estimate of measurement residuals
y_i	measurement residuals
α	comet outgassing normalizing constant
$\boldsymbol{\varepsilon}$	position vector from mass center of Earth to mass center of object at designed time of collision
ε	erroneous predicted miss distance (magnitude of $\boldsymbol{\varepsilon}$)
$\bar{\varepsilon}$	average erroneous predicted miss distance
θ	orientation angle of B-plane error ellipse
λ	latitude
$\tilde{\lambda}$	measured latitude
$\bar{\lambda}$	latitude computed from estimates of orbital parameters
μ	gravitational parameter
ν	true anomaly
ρ	telescope resolution
σ_1	semimajor axis of B-plane error ellipse
σ_2	semiminor axis of B-plane error ellipse
$\sigma_{A_1}, \sigma_{A_2}, \sigma_{A_3}$	uncertainty in comet outgassing coefficients
σ_S	uncertainty in position normal to B-plane
σ_t	uncertainty in time of arrival at B-plane
$[\Phi]$	state transition matrix
ϕ	longitude
$\tilde{\phi}$	measured longitude
$\bar{\phi}$	longitude computed from estimates of orbital parameters
χ	generalized anomaly

Ω	longitude of ascending node
ω	argument of periapsis

4. Detection Element Concepts—Initial Design

Å	angstrom
Al	aluminum
AMSD	Advanced Mirror System Demonstrator
arcsec	arcseconds
au	astronomical unit ($1.4959787066 \times 10^{11}$ m)
Be	beryllium
CAPS	Comet/Asteroid Protection System
CCD	charge-coupled device
CID	charge-injected device
CMG	control moment gyro
CTE	charge transfer efficiency
EFL	effective focal length
FOV	field of view
FPA	focal plane array
GAs	gallium arsenide
HST	Hubble Space Telescope
IR	infrared
JWST	James Webb Space Telescope
LPC	long-period comet
LSST	large-aperture synoptic survey telescope
MOS	metal-oxide semiconductor
NASA	National Aeronautics and Space Administration
NEA	near-Earth asteroid
NEO	near-Earth object

NIR	near infrared
PSF	point spread function
QE	quantum efficiency
RF	radio frequency
RTG	radioisotopic thermoelectric generator
Si	Silicon
TBD	to be decided

5. Near-Earth Object Astrometric Interferometry

au	astronomical unit ($1.4959787066 \times 10^{11}$ m)
B	baseline vector
B	baseline distance
B_{optimal}	optimal baseline distance
C	calibration term
CAPS	Comet/Asteroid Protection System
CCD	charge-coupled device
c	speed of light (2.99792458×10^{8} m/s)
D	delay distance
D_T	telescope diameter
d	diameter of object
I_{MAX}	maximum fringe intensity
I_{MIN}	minimum fringe intensity
J_1	first order Bessel function
M	visual apparent magnitude
mas	milliarcsecond
N	total incoming signal
$\sqrt{N}$	noise estimate
NEO	near-Earth object

R		photon rate
R_0		zero-magnitude photon rate (1.0×10^{-4} photons/s/m^2/Hz)
r		read noise standard deviation
rad		radians
s		position vector
SIM		Space Interferometry Mission
SNR		signal-to-noise ratio
$\hat{\mathbf{s}}$		unit position vector
T		dwell time
V		fringe visibility
x_{det}		detector to object distance
$\delta\nu$		instrument bandwidth
ε		erroneous predicted miss distance
η		interferometer throughput efficiency
θ		observation angle
θ_S		Sun angle
θ_{ud}		target width
λ		average wavelength
μas		microarcseconds
ν		average frequency
$\sigma\dot{D}$		delay rate uncertainty
σ_θ		angular measurement resolution

6.   Mission Functionality for Deflecting Earth-Crossing Asteroids/Comets

a		semimajor axis
au		astronomical unit ($1.4959787066 \times 10^{11}$ m)
b		approach distance of asteroid/comet
b_i		impact parameter of Earth

C_i	equality constraints
C_j	inequality constraints
CTM	chemical transfer module
c_e	exhaust velocity of high-thrust engine
c_m	laser momentum coupling coefficient (assumed in ref. 16 as c_m = 5 dyn-s/J)
c_p	compression wave velocity in asteroid/comet material
D	diameter of asteroid/comet
ECA	Earth-crossing asteroid
ECO	Earth-crossing object
E_{laser}	interceptor energy of laser ablation
e	eccentricity
HPM	hybrid propellant module
i	orbital inclination
J	performance index
kt	kiloton (1 kt = 4.184×10^{12} joules of energy)
LPC	long-period comet
M_{ht}	interceptor mass of high-thrust engine
M_k	interceptor mass of kinetic energy deflection
M_{nso}	interceptor mass of standoff nuclear detonation
M_{nss}	interceptor mass of slightly subsurface nuclear burst
Mt	megaton (1 Mt = 4.184×10^{15} joules of energy)
m	mass of asteroid/comet
NEA	near-Earth asteroids
NLP	nonlinear programming
$NR_\oplus$	N times the radius of the Earth
R	distance between Earth and asteroid/comet
$\dot{R}$	time derivative of R
RASC	Revolutionary Aerospace Systems Concepts

R_m	minimum target miss distance
R_{SOI}	radius of Earth's SOI
$\mathbf{R}_{SOI}$	position vector at Earth's SOI
$R_1(-i)$	rotation matrices about inclination for ECA/ECC
$R_3(-\alpha)$	rotation matrices about angle, α
$R_3(-\Omega)$	rotation matrices about longitude of ascending node for ECA/ECC
$R_3(-\omega)$	rotation matrices about argument of periapsis for ECA/ECC
$R_\oplus$	radius of Earth
$\mathbf{r}_a$	radius vector from Sun to asteroid/comet
$\mathbf{r}_\oplus$	radius vector from Sun to Earth
$\mathbf{r}(t)$	position vector in inertial coordinate frame at time, t
$\mathbf{r}(t_0)$	initial position vector in inertial coordinate frame
SEP	solar electric propulsion
SOI	sphere-of-influence (Earth's)
SPC	short-period comet
t_f	time when $R = R_{SOI}$
t_{impact}	time at collision
$t_{impulse}$	impulse time
$\mathbf{u}$	impulse controls
V_{esc}	escape velocity at R_m
V_∞	magnitude of $\vec{V}_\infty$ vector
$\mathbf{V}_a$	velocity vector of asteroid/comet with respect to Sun
$\mathbf{V}(t)$	velocity vector in inertial coordinate frame at time, t
$\mathbf{V}(t_0)$	initial velocity vector in inertial coordinate frame
$\mathbf{V}_\oplus$	velocity vector of Earth with respect to Sun
$\mathbf{V}_\infty$	vector difference at $\mathbf{R}_{SOI}$ between $\mathbf{V}_a$ and $\mathbf{V}_\oplus$
v	orbital velocity of asteroid/comet
v_e	velocity of ejecta

α	angle between asteroid/comet perifocal coordinate and TNW system
ΔV	impulse change in velocity
ΔV_N	velocity increments normal to ΔV_T in asteroid/comet orbital plane
ΔV_T	velocity increments aligned with asteroid/comet velocity
ΔV_W	velocity increments normal to asteroid/comet orbital plane
$\Delta \mathbf{V}_{XYZ}$	velocity increment vector, inertial coordinate system
ε	fraction that total energy could go into kinetic energy of ejected material
$\mu_\oplus$	gravitational constant of Earth
ρ	assumed density, 200 kg/m³ for comets
ϕ	specific energy that nuclear explosions provide
φ	elevation angle measured from local horizon at SOI
Ω	ascending node
ω	argument of periapsis

7. Orbit Modification of Earth-Crossing Asteroids/Comets Using Rendezvous Spacecraft and Laser Ablation

a	acceleration of spacecraft
a	semimajor axis
au	astronomical unit ($1.4959787066 \times 10^{11}$ m)
CAPS	Comet/Asteroid Protection System
CPT	constant power throttling
e	eccentricity
g	acceleration of gravity
g_o	gravitational parameter at Earth's sea level
GW	gigawatt
H	Hamiltonian function
H_u	partial derivative of H with respect to u
H_x	partial derivative of H with respect to x

H_Λ	partial derivative of H with respect to Λ
$\tilde{H}$	augmented Hamiltonian function
I_{sp}	specific impulse J
i	inclination
J	performance index
J'	augmented performance index
LPC	long-period comet
m	spacecraft mass at any time
m_f	final mass of spacecraft
$\dot{m}$	mass flow rate
MW	megawatt
N	Newton
NEO	near-Earth object
NEP	nuclear electric propulsion
p	power of propulsion system
RF	radio frequency
R^k	k dimension
R^l	l dimension
R^n	n dimension
r	radial distance from Sun to spacecraft
r_a	aphelial distance
r_p	perihelial distance
$\dot{r}$	time derivative of radial distance
T	thrust
TPBVP	two-point boundary-value problem
t	metric ton
t_f	final time
u	radial velocity

$u(t)$	control history
u_{max}	maximum control boundary
u_{min}	minimum control boundary
u^*	extremal controls
$\dot{u}$	time derivative of radial velocity
VASIMR	Variable Specific Impulse Magnetoplasma Rocket
v	tangential velocity
v_e	exhaust velocity
$\dot{v}$	time derivative of tangential velocity
w	normal velocity
w_t	weight for flight time
$\dot{w}$	time derivative of normal velocity
x	state variables
x^*	extremal states
$\dot{x}$	dynamic equations
α	thrust direction angle in-plane
β	thrust direction angle of out-of plane
ΔV	impulsive change in velocity
ε	efficiency of propulsion system
θ	angle measured from x-axis (defined as vernal equinox) in x-y plane
$\dot{\theta}$	time derivative of θ
Λ	Lagrange multipliers
$\dot{\Lambda}$	time derivative of costate variables
Λ^*	extremal costates
λ	costate values
λ_m	costate value of m
λ_r	costate value of r
λ_u	costate value of u

λ_v	costate value of v
λ_w	costate value of w
λ_θ	costate value of θ
λ_ϕ	costate value of ϕ
μ	constant Lagrange multiplier
μ_i	Lagrange multiple
ν	constant multiplier vector
ν_i	constant Lagrange multiplier
ρ	density of NEO
Φ	terminal constraints
ϕ	angle measured from x-y plane
Ω	set of admissible controls

8. CAPS Simulation Environment Development

au	astronomical unit ($1.4959787066 \times 10^{11}$ m)
CAPS	Comet/Asteroid Protection System
CCD	charge-coupled device
EPMD	erroneous predicted miss distance
ERC	external returning comet
FOV	field-of-view
GUI	graphical user interface
LPC	long-period comet
NEA	near-Earth asteroid
NEO	near-Earth object
OD	orbit determination
POD	precision orbit determination
SEE	Synergistic Engineering Environment
S/N	signal-to-noise

STK	Satellite Tool Kit
3-D	three-dimensional

9. Comparison of Detector Technologies for CAPS

A	area of point image (arcsec^2)
CAPS	Comet/Asteroid Protection System
CCD	charge-coupled device
CID	charge-injected device
CIS	cryogenic imaging spectrophotometer
CTE	charge transfer efficiency
D	effective aperture, m
DC	direct current
f	effective focal length
GPS	Global Positioning System
m_l	limiting magnitude
NEO	near-Earth object
N_p	number of pixels to contain image
NRLA	nonredundant linear array
QE	quantum efficiency
S/N	signal-to-noise ratio
SQUID	Superconducting Quantum Interference Detector
STJ	superconducting tunneling junction
TES	transition edge sensor
t	exposure time, s
WFPC2	Wide-Field Planetary Camera 2

10.   Survey of Enabling Technologies for CAPS

ABL	Airborne Laser
AMSD	Advanced Mirror System Demonstrator
arcsec	arcsecond
CAPS	Comet/Asteroid Protection System
CCD	charge-coupled device
CID	charge-injected device
DM	deformable mirror
DoD	Department of Defense
ECT	enabling concepts and technologies
ESA	European Space Agency
GJ	gigajoule
HST	Hubble Space Telescope
IR	infrared
I_{sp}	specific impulse
JWST	James Webb Space Telescope
LPC	long-period comet
NASA	National Aeronautics and Space Administration
NEO	near-Earth object
NEP	nuclear electric propulsion
NERVA	nuclear engine for rocket vehicle application
NSI	Nuclear Systems Initiative
NTP	nuclear thermal propulsion
QE	quantum efficiency
SAFE	safe affordable fission engine
SBIR	Small Business Innovative Research
SBL	Space-Based Laser
S-Cam	superconducting camera

SIM	Space Interferometry Mission
SNAP	Space Nuclear Auxiliary Power
SP	space power
STJs	superconducting tunnel junctions
TPF	terrestrial planet finder
VASIMR	variable specific impulse magnetoplasma rocket
WHT	William Herschel Telescope
µas	microarcsecond
µm	micrometers

www.ingramcontent.com/pod-product-compliance
Lightning Source LLC
Chambersburg PA
CBHW081722170526
45167CB00009B/3663